Lecture Notes in Computer Science 14624

Founding Editors

Gerhard Goos
Juris Hartmanis

The series Lecture Notes in Computer Science (LNCS), including its subseries Lecture Notes in Artificial Intelligence (LNAI) and Lecture Notes in Bioinformatics (LNBI), has established itself as a medium for the publication of new developments in computer science and information technology research, teaching, and education.

LNCS enjoys close cooperation with the computer science R & D community, the series counts many renowned academics among its volume editors and paper authors, and collaborates with prestigious societies. Its mission is to serve this international community by providing an invaluable service, mainly focused on the publication of conference and workshop proceedings and postproceedings. LNCS commenced publication in 1973.

Thomas Neele · Anton Wijs
Editors

Model Checking Software

30th International Symposium, SPIN 2024
Luxembourg City, Luxembourg, April 8–9, 2024
Proceedings

 Springer

Editors
Thomas Neele (ID)
Eindhoven University of Technology
Eindhoven, The Netherlands

Anton Wijs (ID)
Eindhoven University of Technology
Eindhoven, The Netherlands

ISSN 0302-9743   ISSN 1611-3349 (electronic)
Lecture Notes in Computer Science
ISBN 978-3-031-66148-8   ISBN 978-3-031-66149-5 (eBook)
https://doi.org/10.1007/978-3-031-66149-5

Preface

This volume contains the papers presented at SPIN 2024: the 30th International Symposium on Model Checking Software, held on April 10–11, 2024 in Luxembourg City, Luxembourg. SPIN 2024 was co-located with the ETAPS International Joint Conferences on Theory and Practice of Software, hosted by the Interdisciplinary Centre for Security, Reliability and Trust (SnT), at the University of Luxembourg. This edition of SPIN was both the 30th instalment and the 30th anniversary of the symposium: two symposia took place in 1999, but the year 2020 was skipped due to the pandemic.

The symposium sought submissions on the topic of formal verification techniques for automated analysis of (concurrent) software/hardware, including model checking, deductive verification, automated theorem proving (including SAT and SMT solving), abstraction and symbolic execution techniques, static analysis and abstract interpretation, modular and compositional verification techniques, verification of timed and probabilistic systems, automated testing using advanced analysis techniques, program synthesis, derivation of specifications and test cases via formal analysis, formal specification languages, temporal logic, design-by-contract, formal analysis of learned systems or any combination of these. Papers could be submitted to one of three tracks: regular papers, short papers or a special anniversary track. Papers in the anniversary track could be historical accounts, discussion of successful research lines, surveys or position papers.

SPIN 2024 attracted 23 submissions by 63 distinct authors from 13 different countries. Out of these papers, 16 were research papers, 5 were short papers and 2 papers were submitted in the anniversary track. Each submission was reviewed by three program committee members in a single-blind manner. After discussion within the program committee, 13 papers were accepted (9 research papers, 3 short papers and 2 anniversary papers).

To support tool development and reproducibility, SPIN 2024 saw the introduction of an *artifact evaluation committee* (AEC), chaired by Ernst Moritz Hahn and Matthias Volk. The AEC was responsible for reviewing artifacts submitted alongside papers. Submission of an artifact was optional; the authors of 9 papers decided to do so. After each artifact was reviewed by two AEC members, 8 artifacts (7 of which correspond to accepted papers) were granted one or more *badges* from the EAPLS badging scheme.

The program of SPIN 2024 included three keynote talks. First, there was an online talk by Gerard Holzmann (Nimble Research, USA), the founder of the SPIN model checking tool, about the history and development of SPIN. The abstract is included below. Holger Hermanns (Saarland University, Germany) discussed a framework for monitoring of AI software. Kristin Yvonne Rozier (Iowa State University, USA) discussed MoXI, a new intermediate language for symbolic model checkers. Their contributions are included in these proceedings. We are grateful to the invited speakers for sharing the findings of their research and their views on the field.

Finally, we would like to thank several organizations and people. First and foremost, we extend our heartfelt thanks to all PC members and external reviewers for their critical and timely reviewing, all the authors for their submitted work, and all attendees for participating. In addition, we thank the artifact evaluation chairs and committee, and we are very grateful to the local organizers of ETAPS 2024, including Peter Y.A. Ryan, Peter Roenne and Magali Martin, for hosting SPIN 2024 alongside the main ETAPS program. We are also grateful to EAPLS for use of their artifact badging scheme and Springer for publishing these proceedings.

April 2024

Thomas Neele
Anton Wijs

Organization

Program Committee Chairs

Thomas Neele	Eindhoven University of Technology, The Netherlands
Anton Wijs	Eindhoven University of Technology, The Netherlands

Artifact Evaluation Committee Chairs

Ernst Moritz Hahn	University of Twente, The Netherlands
Matthias Volk	Eindhoven University of Technology, The Netherlands

Steering Committee

Georgiana Caltais	University of Twente, The Netherlands
Susanne Graf	Université Grenoble Alpes, CNRS, France
Gerard Holzmann	Nimble Research, USA
Stefan Leue	University of Konstanz, Germany
Jaco van der Pol	Aarhus University, Denmark
Neha Rungta	Amazon, USA
Christian Schilling	Aalborg University, Denmark
Willem Visser	Stellenbosch University, South Africa
Anton Wijs	Eindhoven University of Technology, The Netherlands

Program Committee

Georgiana Caltais	University of Twente, The Netherlands
Deepak D'Souza	Indian Institute of Science, Bangalore, India
Sofie Haesaert	Eindhoven University of Technology, The Netherlands
Matthias Heizmann	University of Freiburg, Germany
Paula Herber	University of Münster, Germany

Gerard Holzmann	Nimble Research, USA
Mitja Kulczynski	Kiel University, Germany
Ondrej Lengal	Brno University of Technology, Czech Republic
Radu Mateescu	Inria Grenoble Rhône-Alpes, France
Rosemary Monahan	Maynooth University, Ireland
Thomas Neele	Eindhoven University of Technology, The Netherlands
Violet Ka I Pun	Western Norway University of Applied Sciences, Norway
Alceste Scalas	Technical University of Denmark, Denmark
Christian Schilling	Aalborg University, Denmark
Stephen F. Siegel	University of Delaware, USA
Marjan Sirjani	Mälardalen University, Sweden
Scott Smolka	Stony Brook University, USA
Yann Thierry-Mieg	LIP6-Sorbonne Université, France
Jaco van de Pol	Aarhus University, Denmark
Anton Wijs	Eindhoven University of Technology, The Netherlands

Artifact Evaluation Committee

Peter Backeman	Mälardalen University, Sweden
Rong Gu	Mälardalen University, Sweden
Sudeep Kanav	LMU Munich, Germany
Martin Kristjansen	Aalborg University, Denmark
Stefano Nicoletti	University of Twente, The Netherlands
Quentin Nivon	Inria Grenoble Rhône-Alpes, France
Reza Soltani	University of Twente, The Netherlands
Ahang Zuo	Inria Grenoble Rhône-Alpes, France

Additional Reviewers

Syed Ali Asadullah Bukhari
Daniel Dietsch
Hubert Garavel
Vojtěch Havlena
Stefan Marksteiner

Fereidoun Moradi
Sumanth Prabhu
Wendelin Serwe
Steffan Sølvsten

Sponsors

European Association for Programming Languages and Systems (EAPLS)

Springer

The Spin on Spin (Abstract of Invited Talk)

Gerard J. Holzmann

Nimble Research, Monrovia, CA, USA

Abstract. The first Spin (then) Workshop was a small event that was organized by INRS-Telecommunications in Montreal, Quebec in 1995. Since then, the event was held 16 times in Europe, 10 times in the USA, twice in Canada, and once in each of China and South Africa. Today, the focus of the Symposium is much broader than a single verification tool or technology, but it is interesting to see how much progress has been made over the years. The development of Spin itself started in 1984 at Bell Labs, a respectable 40 years ago. That first version was 3,508 lines of code (which included a good amount of optional debugging code). The code has grown quite a bit since then, as all code inevitably does until it stops being useful. In this talk I'll reflect a little on what specific problems Spin was originally trying to solve, and how our collective focus has changed over the years.

Contents

Verification Tools

Software Verification

Invited Contributions

Taming the AI Monster: Monitoring of Individual Fairness for Effective Human Oversight

Kevin Baum[1], Sebastian Biewer[2], Holger Hermanns[2(✉)], Sven Hetmank[3],
Markus Langer[4], Anne Lauber-Rönsberg[3], and Sarah Sterz[2]

[1] Neuro-Mechanistic Modeling, DFKI Saarbrücken, Saarbrücken, Germany
[2] Department of Computer Science, Universität des Saarlandes, Saarbrücken,
Germany
hermanns@cs.uni-saarand.de
[3] IRGET, Technische Universität Dresden, Dresden, Germany
[4] Department of Psychology, Universität Freiburg, Freiburg im Breisgau, Germany

Abstract. This invited paper reviews a framework to assist in mitigating societal risks that software can pose. This is to promote effective human oversight, which is a central requirement enforced by the European Union's upcoming AI Act [29]. *The paper advertises fragments of an upcoming journal publication [12], and as such is itself low in genuine originality.* Yet it offers a specific perspective on that original work. Extrapolating earlier work on software doping, we report on the combination of established techniques for runtime monitoring and for probabilistic falsification to arrive at a black-box analysis technique for identifying undesired effects of software. We describe its application to high-risk systems that evaluate humans in a possibly unfair or discriminating way. The approach can assist humans-in-the-loop to make better informed and more responsible decisions. Our technical contribution is complemented by juridically, philosophically, and psychologically informed perspectives on the potential problems caused by AI systems.

Keywords: artificial intelligence · algorithmic fairness · probabilistic falsification · adequate trust · human oversight

1 Introduction

The lack of transparency of many AI-supported systems raises significant societal risks, including the potential for unfair or biased decision-making. This can lead to morally and instrumentally problematic outcomes, to breaches of legal obligations, to unfavourable societal effects, and to the undermining of public trust and acceptance of AI technologies. This is especially true for high-risk applications, which include credit approval [63], decisions on visa applications [54], admissions to higher education [19,85], screening of individuals in predictive policing [35], selection in HR [60–62], judicial decisions (as with COMPAS

S. Sterz—Authors are listed alphabetically.

© The Author(s), under exclusive license to Springer Nature Switzerland AG 2025
T. Neele and A. Wijs (Eds.): SPIN 2024, LNCS 14624, pp. 3–25, 2025.
https://doi.org/10.1007/978-3-031-66149-5_1

[3, 21, 23, 47]), tenant screening [76], and more. In many of these areas, there are legitimate interests and valid reasons for using AI technology, although the risks associated with their use to date are manifold.

One frequently proposed remedy to the problems posed by the high-risk uses of opaque AI is *human oversight* [36, 53, 82] where a human expert is to make sure that the system operates in accordance with the desiderata set out by other human stakeholders. By now, the requirement for human oversight is even reflected in law, such as the *AI Act* of the European Union [29] that is about to be adopted or certain US state laws [84]. However, human oversight is not an unconditional remedy for any and all problems, and the effectiveness of a human overseer can be greatly reduced when certain conditions fail to be met. Notably, if the human overseer cannot gain enough knowledge about the system, their oversight will not achieve the desired aims. For example, if the overseer lacks any means to decide if a system made an unfair decision, they will have no reliable way of intervening when the system is, in fact, unfair. This is the problem that this paper has set out to tackle.

2   Setting the Stage

The challenge to overcome can best be introduced by an exemplary, albeit hypothetical admission system for higher education (inspired by [19, 85]).

Example 1. A large university assigns scores to applicants aiming to enter their computer science PhD program. The scores are computed using an automated, model-based procedure P, which is based on three data points: the position of the applicant's last graduate institution in an official, subject-specific ranking, the applicant's most recent grade point average (GPA), and their score in a subject-specific standardised test taken as part of the application procedure. The system then automatically computes a score for the candidate based on an estimation of how successful it expects them to be as students. A dedicated university employee, Unica is in charge of overseeing the individual outcomes of P and desk-reject candidates whose scores are below a certain, predefined threshold – unless she finds problems with P's scoring. The university pays especial attention to fairness in the scoring procedure, so Unica has to watch out for any signs of potential unfairness. If she suspects unfairness, Unica must decide on the case manually. Without any additional support, Unica, as human overseer in the loop, must manually check *all* cases for signs of unfairness as they are processed. This can be a tedious, complicated, and error-prone task and, as such, constitutes an impediment to the assumed scalability of the automated scoring process for high numbers of applicants. Therefore, she requires tools that assist her in detecting when something is off about the scoring of individual applicants.

Sometimes, we cannot mitigate all risks of high-risk AI in advance by technical measures, and some risk mitigation requires trade-off decisions involving features that are either impossible or difficult to operationalise and formalise. This is why it is arguably essential that a human effectively oversees the system (which is also emphasised by several institutions such as UNESCO [82] and the *European High Level Expert Group* [36]), as well as in applicable law (such as the European AI Act [29] or the Washington State facial recognition law [84]). *Effective* human oversight, however, is only possible with the appropriate technical measures that allow human overseers to better understand the system at runtime [45, 46]. From a technical point of view, this raises the pressing question of what such technical measures can and ought to look like to actually enable humans to live up to their responsibilities. Our contribution is intended to bridge the gap between the normative expectations of law and society and the current reality of technological design. Developing such a technical measure, a software tool supporting Unica, is thus the prime problem we focus on.

Our solution is based on the work developed in [12], in terms of a runtime monitor that provides automated assistance (based on [14]) to the human oversight and itself is based on a probabilistic falsification technique (introduced in [15]). All this is rooted in a suitable formal basis for rolling out runtime monitors for such high-risk systems that can detect and flag discrimination or unfair treatment of humans. We live up to the societal complexity of this example and provide an interdisciplinary situation analysis and an interdisciplinary assessment of the solution we shall propose.

The contributions echoed here from the original article [12] are twofold.

Promoting effective human oversight. We discuss and demonstrate a contribution to effective human oversight of high-risk systems, as required by the AI Act. The hypothetical university admission scenario introduced above will serve as a demonstrator for shedding light on the applicability of our approach and on the principles behind it. On a conceptual level, we consider it important to clarify which duties come with the usage of such a system; from a *legal* perspective, particularly considering the AI Act, substantiated by considering the *ethical* dimension from a philosophical perspective, and from a *psychological* perspective, particularly deliberating on how the overseeing can become *effective*.

Falsification-based test input generation. On a technical level, we describe how recent work [13] on a formal framework for robust cleannesscan be combined with a probabilistic falsification technique to identify problems of fairness and discrimination in AI usages akin to the admission scenario described above. We describe a search procedure that aims at generating synthetic data of (hypothetical) applicants whose parameters are very similar to the individuals currently looked at but who are classified differently by the AI. The approach uses a fairness test procedure, and the problem then is to effectuate test input selection in a meaningful manner. In this, probabilistic falsification supports the testing procedure by guiding it towards test inputs that make the fairness tests likely to fail. Altogether, we arrive at a runtime monitor for individual fairness based on probabilistic falsification. This we consider as a core component for assisting humans who need to oversee scenarios as the one described above.

While the contents of this paper are subsumed by the original contents of [12], *the former has been rearranged in order to directly put in focus the use of the above contributions for the benefit of human oversight. In this respect, this paper offers a distinct value to the interested reader.*

3   Fairness, Discrimination, Explainability

Our contribution draws on and adds to three vibrant topics of current research, namely *Explainable AI (XAI)*, *AI Fairness*, and *Discrimination*.

Explainable AI. Many of the most successful AI systems today are black boxes of some kind [8]. Accordingly, the field of "Explainable AI" [32] focuses on the question of how to provide users (and possibly other stakeholders) with more information via several key perspicuity properties [78] of these systems and their outputs to make them understand these systems and their outputs in ways necessary to meet various desiderata [4,20,44,49,55,59]. The concrete expectations and promises associated with various XAI methods are manifold. Among them are enabling warranted trust in systems [9,39,42,69,73], increasing human-system decision-making performance [43], for instance through increasing human situation awareness when operating systems [71], enabling responsible decision-making and effective human oversight [10,51,75], as well as identifying and reducing discrimination [49]. It often remains unclear what kind of explanations are generated by the various explainability methods and how they are meant to contribute to the fulfilment of the desiderata, even though these questions have become the subject of systematic and interdisciplinary research [44,46].

Our approach can be taxonomised along at least two different distinctions [46, 56,68,69,77]: First, it is *model-agnostic* (not *model-specific*), i.e., it is not tailored to a particular class of models but operates on observable behaviour – the inputs and outputs of the model. Second, our method is a *local method* (not *global*), i.e., it is meant to shed light on certain outputs rather than the system as a whole.

Fairness. Fairness, discrimination, justice, equal opportunity, bias, prejudice, and many more such concepts are part of a meaningfully interrelated cluster that has been analysed and dissected for millennia [5,6]. Many fields are traditionally concerned with the concepts of fairness and discrimination, ranging from philosophy [5,6,25,31,65–67] to legal sciences [18,34,81,83], to psychology [37,88], to sociology [2,40], to political theory [66], to economics [33]. Nowadays, it has also become a technological topic that calls for cross-disciplinary perspectives [30]. It is widely recognised that discrimination by unfair classification and regression models is one particularly important risk of AI-supported decision making. As a result, a colourful zoo of different operationalisations of unfairness has emerged [64,83], which should be seen less as a set of competing approaches and more as mutually complementary [31].

With regard to fairness, two distinctions are especially relevant to our work. First, one distinction is made between *individual fairness*, i.e., that similar individuals are treated similarly [24], and *group fairness*, i.e., that there is adequate group parity [16]. Measures of individual fairness are often close to the Aristotelian dictum to treat like cases alike [5,6]. In a sense, operationalisations of individual fairness are robustness measures [17,79], but instead of requiring robustness with respect to noise or adversarial attacks, measures of individual fairness, such as the one by Dwork et al. [24], call for robustness with respect to highly context-dependent differences between representations of human individuals. Second, recent work from the field of law [83] suggests to differentiate between *bias preserving* and *bias transforming* fairness metrics. Bias preserving fairness metrics seek to avoid adding new bias. For such metrics, historic performances are the benchmarks for models, with equivalent error rates for each group being a constraint. In contrast, bias transforming metrics do not accept existing bias as a given or neutral starting point but aim at adjustment. Therefore, they require to make a "positive normative choice" [83], i.e., to actively decide which biases the system is allowed to exhibit and which it must not exhibit.

Over the years, many concrete approaches have been suggested to foster different kinds of fairness in artificial systems, especially in AI-based ones [49, 52,64,83,86]. Yet, to the best of our knowledge, an approach like ours is still missing. One of the approaches that are closest to ours, namely that by John et al. [41], is not local and, therefore, not suitable for runtime monitoring. Also, it is not model-agnostic. So, to the best of our knowledge, our approach provides a new contribution to the debate on unfairness detection.

It is important to note/recognise that our approach can only be understood as part of a more holistic approach to preventing or reducing unfairness. After all, there are many sources of unfairness [7] (also see Figure 1). Therefore, not every technical measure can detect every kind of unfairness, and eliminating one source of unfairness might not be sufficient to eliminate all unfairness. Our approach tackles only unfairness introduced by the system, but not other kinds of unfairness.

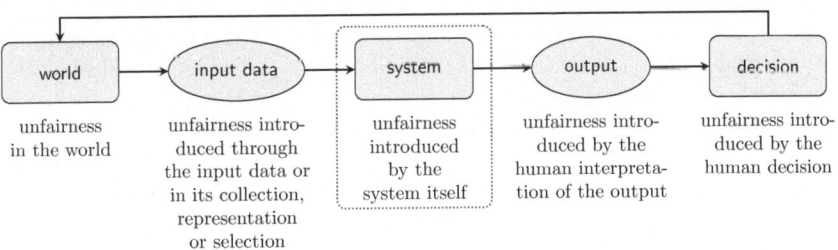

Fig. 1. Sketch of different origins of unfairness in a decision process supported by a system; the dotted box indicates which unfairness our monitoring targets.

Discrimination. We understand discrimination as dissimilar treatment of similar cases or similar treatment of dissimilar cases without justifying reason. This definition can also be found in the law [28, §43]. Our work exclusively focuses on discrimination *qua* dissimilar treatment of similar cases. Discrimination requires a thoughtful and largely not formalisable consideration of "justifying reason." However, we will exploit the relation between discrimination and fairness: Unfairness in a system can arguably be a good proxy of discrimination – even though not every unfair treatment by a system necessarily constitutes discrimination (especially not in the legal sense). Thus, a tool that highlights cases of unfairness in a system can be highly instrumental in detecting discriminatory features of a system. It is not viable, though, to let such a tool rule out unfair treatment fully automatically without human oversight since there could be justifying reasons to treat two similar inputs in a dissimilar way.

4   Individual Fairness of Systems Evaluating Humans

Against the contextual backdrop as described above, we now return to the characteristics of Example 1, where Unicauses an AI system that is supposed to assist her with the selection of applicants for a hypothetical university. A usable fairness analysis can happen no later than at runtime since Unicaneeds to make a timely decision on whether to include the applicant in further considerations. We describe technical measures that help mitigate this challenge by providing her with information from an individual fairness analysis in a suitable, purposeful, and expedient way. To this end, we propose a formal definition for individual fairness extending the one by Dwork et al. [24], extrapolating earlier work on robust cleanness [11]. We develop a runtime monitor that analyses every output of P immediately after P's decision, which strategically searches for unfair treatment of a particular individual by comparing them to relevant hypothetical alternative individuals so as to provide a fairness assessment in a timely manner.

4.1   Individual Fairness

Unicafrom Example 1 should be able to detect individual unfairness. An operationalisation thereof by Dwork et al. [24] is based on the Lipschitz condition to enforce that similar individuals are treated similarly. To measure similarity, they assume the existence of distance functions measuring the distances between system inputs as well as between system outputs. A function $d : X \times X \to \overline{\mathbb{R}}_{\geq 0}$ is a *distance function* if and only if it satisfies $d(x, x) = 0$ and $d(x, y) = d(y, x)$. In this, $\overline{\mathbb{R}}_{\geq 0} := \{x \in \mathbb{R} \mid x \geq 0\} \cup \{\infty\}$ is the set of the non-negative extended real numbers. Two such functions are assumed to exist, namely d_{In} operating on the set In of system inputs, and similarly d_{Out} for system output (from set Out). In the example, the inputs are the data vectors representing human individuals, and outputs correspond to the scores produced.

Dwork et al. [24] assume that both distance functions perfectly measure distances between individuals[1] and between outputs of the system, respectively, but admit that in practice these distance functions are only approximations of a ground truth at best. They suggest that distance measures might be learned, but there is no one-size-fits-all approach to selecting distance measures. Indeed, obtaining such distance metrics is a topic of active research [38,57,87]. Additionally, the Lipschitz condition assumes a Lipschitz constant L to establish a linear constraint between input and output distances.

Definition 1. *A program* $P : In \rightarrow Out$ *is* Lipschitz-fair *w.r.t.* $d_{In} : In \times In \rightarrow \mathbb{R}$, $d_{Out} : Out \times Out \rightarrow \mathbb{R}$, *and a Lipschitz constant* L, *if and only if for all* $i_1, i_2 \in In$, $d_{Out}(P(i_1), P(i_2)) \leq L \cdot d_{In}(i_1, i_2)$.

Lipschitz-fairness comes with some restrictions that limit its suitability for practical application:

d_{In}-d_{Out}-**relation:** High-risk systems are typically complex systems and ask for more complex fairness constraints than the linearly bounded output distances provided by the Lipschitz condition. For example, using the Lipschitz condition prevents us from allowing small local jumps in the output and, at the same time, from forbidding jumps of the same rate of increase over larger ranges of the input space.

Input relevance: The condition quantifies over the entire input domain of a program. This overlooks two things: first, it is questionable whether each input in such a domain is plausible as a representation of a real-world individual. But whether a system is unfair for two implausible and purely hypothetical inputs is largely irrelevant in practice. Secondly, it also ignores that mere potential unfair treatment is at most a threat, not necessarily already a harm [70]. Therefore, even with a restriction to only plausible applicants, the analysis might take into account more inputs than needed for many real-world applications. What is important in practice is the ability to determine whether *actual* applicants are treated unfairly – and for this, it is often not needed to look at the entire input domain.

Monitorability: In a monitoring scenario with the Lipschitz condition in place, a fixed input i_1 must be compared to potentially all other inputs i_2. Since the input domain of the system can be arbitrarily large, the Lipschitz condition is not yet suitable for monitoring in practice (for a related point see John et al. [41]).

We propose a notion of individual fairness that, instead of the constant L, uses a function f to relate input distances and output distances in a more general

---

[1] For easier readability, we will not distinguish between *individuals* and their *representations* unless this distinction is relevant in the specific context. It is nevertheless important to note that inputs are not individuals, but only representations of individuals, since an input could inadequately represent an individual and therefore be unfair.

way. Further, we make it explicit that d_{In}, d_{Out}, and f are parameters of the fairness notion by encapsulating them in triples $\mathcal{F} = \langle d_{\text{In}}, d_{\text{Out}}, f \rangle$ that we call *fairness contracts*. Our fairness definition evaluates fairness for a finite set of individuals $\mathcal{I} \subseteq \text{In}$ (e.g., a set of applicants). A fairness contract specifies certain fairness parameters for a concrete context or situation. Such parameters should generally not already include $\mathcal{I}$ to avoid introducing new unfairness through the monitor by tailoring it to specific inputs individually or by treating certain inputs differently from others. We can operationalise[2] individual fairness as follows:

Definition 2. *A program* $\text{P} : \text{In} \rightarrow \text{Out}$ *is* individually fair *for a set* $\mathcal{I} \subseteq \text{In}$ *of actual inputs w.r.t. a fairness contract* $\mathcal{F} = \langle d_{\text{In}}, d_{\text{Out}}, f \rangle$, *if and only if for all* $\text{i} \in \mathcal{I}$ *and all* $\text{i}' \in \text{In}$, $d_{\text{Out}}(\text{P}(\text{i}), \text{P}(\text{i}')) \leq f(d_{\text{In}}(\text{i}, \text{i}'))$.

The idea behind *individual fairness* is that every individual in set $\mathcal{I}$ is compared to potential other inputs in the domain of P. These other inputs do not necessarily need to be in $\mathcal{I}$, nor do these inputs need to have "physical counterparts" in the real world. Driven by the insights of the *Input relevance* restriction of Lipschitz-fairness, we explicitly distinguish inputs in the following and will call inputs that are given to P by a user *actual inputs*, denoted i_a, and call inputs to which such i_a are compared *synthetic inputs*, denoted i_s. Actual inputs typically[3] are inputs that have a real-world counterpart, while this might or might not be true for synthetic inputs.

At first glance, it might seem sufficient to use only actual inputs. This way, for example, Unica would find out whether one applicant was treated unfairly relative to another applicant. This, however, is not enough: Unica is after *any* unfairness of the system towards a certain applicant, and not just one relative to some other, actual applicant. At the same time, Unica cannot and should not expect that, coincidentally, a candidate has applied who has the very specific properties needed to unveil the system's unfairness towards another candidate. Hence, synthetic inputs are worthwhile to be considered.

Notice that *individual fairness* is a conservative extension of Lipschitz-fairness. With $\mathcal{I} = \text{In}$ and $f(x) = L \cdot x$, *individual fairness* mimics Lipschitz-fairness. Wachter et al. [83] classify the Lipschitz-fairness of Dwork et al. [24] as bias-transforming. As we generalise this and introduce no element that has to be regarded as bias-preserving, our approach arguably is bias-transforming, too.

Individual fairness, with its function f, provides a powerful tool to model complex fairness constraints. How such an f is defined has a profound impact

---

[2] Definition 2 is not a *definition* of individual fairness in the strict sense, since individual fairness already has a meaning, namely that similar individuals are treated similarly, as described above in Section 3. It rather is an *operationalisation* that has to be employed appropriately in order to yield a proper measure of individual fairness. This, for example, includes a parameterisation with a fitting fairness contract that is meaningful in the context of individual fairness, and fixes what similarity is to mean in this context. Nevertheless, Definition 2 is a suitable operationalisation for our purposes. In this paper, it will be clear from the context whether we will talk about individual fairness in its original sense or in terms of the operationalisation.

[3] A case where actual inputs might not have real-world counterparts is testing.

Algorithm 1. FairnessMonitor, with ξ-min $S = (\xi, i_1, i_2)$ only if $(\xi, i_1, i_2) \in S$ and for all $(\xi', i_1', i_2') \in S$, $\xi' \geq \xi$

Falsification Parameters: PS: Proposal scheme, β: Temperature parameter
Input: System P : In $\rightarrow$ Out, Fairness contract $\mathcal{F} = \langle d_{\mathsf{In}}, d_{\mathsf{Out}}, f \rangle$, and set of actual inputs $\mathcal{I}$
Output: A minimal fairness score triple from $\mathbb{R} \times \mathcal{I} \times \mathsf{In}$.
1: $i_s \leftarrow$ any input $i_a \in \mathcal{I}$
2: $(\xi, i_{\min}, i_s) \leftarrow \xi\text{-min}\{(F(i_a, i_s), i_a, i_s) \mid i_a \in \mathcal{I}\}$
3: $(\xi_{\min}, i_1, i_2) \leftarrow (\xi, i_{\min}, i_s)$
4: **while not** timeout **do**
5:    $i_s' \leftarrow \mathsf{PS}(i_s, \mathsf{P}(i_s))$
6:    $(\xi', i_{\min}', i_s') \leftarrow \xi\text{-min}\{(F(i_a, i_s'), i_a, i_s') \mid i_a \in \mathcal{I}\}$
7:    $(\xi_{\min}, i_1, i_2) \leftarrow \xi\text{-min}\{(\xi_{\min}, i_1, i_2), (\xi', i_{\min}', i_s')\}$
8:    $\alpha \leftarrow \exp(-\beta(\xi' - \xi))$
9:    $r \leftarrow \mathsf{UniformRandomReal}(0, 1)$
10:   **if** $r \leq \alpha$ **then**
11:      $i_s \leftarrow i_s'$
12:      $\xi \leftarrow \xi'$
13:   **end if**
14: **end while**
15: **return** $(\xi_{\min}, i_1, i_2)$

---

on the quality of the fairness analysis. A full discussion about which types of functions make a good f goes beyond the scope of this paper. What are suitable choices for f and the distance functions d_{In} and d_{Out} heavily depends on the context in which fairness is analysed – there is no one-fits-it-all solution. *Individual fairness* makes this explicit with the formal fairness contract $\mathcal{F} = \langle d_{\mathsf{In}}, d_{\mathsf{Out}}, f \rangle$.

4.2 Fairness Monitoring

We now develop a fairness monitor based on probabilistic falsification [1]. Given a set of actual inputs, the monitor searches for a synthetic counterexample to falsify a system P w.r.t. a fairness contract $\mathcal{F}$. To this end, we define a *fairness score* as function $F(i_a, i_s) := f(d_{\mathsf{In}}(i_a, i_s)) - d_{\mathsf{Out}}(\mathsf{P}(i_a), \mathsf{P}(i_s))$. With regard to probabilistic falsification F is a quantitative description of *individual fairness* that serves as a robustness estimate. That is, if $F(i_a, i_s)$ is non-negative, then $d_{\mathsf{Out}}(\mathsf{P}(i_a), \mathsf{P}(i_s)) \leq f(d_{\mathsf{In}}(i_a, i_s))$, and if it is negative, then $d_{\mathsf{Out}}(\mathsf{P}(i_a), \mathsf{P}(i_s)) \not\leq f(d_{\mathsf{In}}(i_a, i_s))$. For a set of actual inputs $\mathcal{I}$, the definition generalises to $F(\mathcal{I}, i_s) := \min\{F(i_a, i_s) \mid i_a \in \mathcal{I}\}$, i.e., the overall fairness score is the minimum of the concrete fairness scores of the inputs in $\mathcal{I}$.

Algorithm 1 shows FairnessMonitor, which searches for the minimal fairness score in a system P for fairness contract $\mathcal{F}$. The algorithm stores fairness scores in triples that also contain the two inputs for which the fairness score was computed. The minimum in a set of such triples is defined by the function ξ-min that returns the triple with the smallest fairness score of all triples in the set. The first line of

FairnessMonitor initialises the variable i_s with an arbitrary actual input from $\mathcal{I}$. For this value of i_s, the algorithm checks the corresponding fairness scores for all actual inputs $i_a \in \mathcal{I}$ and stores the smallest one. In line 3, the globally smallest fairness score triple is initialised. In line 5, the algorithm uses a parameterisable proposal scheme to get the next synthetic input i'_s. Line 6 is similar to line 2: for the newly proposed i'_s it finds the smallest fairness score, stores it, and updates the global minimum if it found a smaller fairness score (line 7). Lines 8-13 are the heart of the probabilistic search for an example that violates fairness; it comes from the original algorithm proposed by Abbas et al. [1]. Our variant of the algorithm does not (exclusively) aim to falsify the fairness property, but aims at minimising the fairness score; even if the fair treatment of the inputs in $\mathcal{I}$ cannot be falsified in a reasonable amount of time, we still learn how robustly they are treated fairly, i.e., how far the least fairly treated individual in $\mathcal{I}$ is away from being treated unfairly. After the timeout occurs, the algorithm returns the triple with the overall smallest seen fairness score ξ_{min}, together with the actual input i_1 and the synthetic input i_2 for which ξ_{min} was found. In case ξ_{min} is negative, i_2 is a counterexample for P being individually fair.

FairnessMonitor implements a sound $\mathcal{F}$-unfairness detection as stated in Proposition 1. However, it is not complete, i.e., it is not generally the case that P is individually fairfor $\mathcal{I}$ if ξ is positive. It may happen that there is a counterexample, but FairnessMonitor did not succeed in finding it before the timeout.

Proposition 1. *Let* P $:$ In $\rightarrow$ Out *be a program,* $\mathcal{F} = \langle d_{\text{In}}, d_{\text{Out}}, f \rangle$ *a fairness contract , and* $\mathcal{I}$ *a set of actual inputs. Further, let* (ξ_{min}, i_1, i_2) *be the result of* FairnessMonitor(P, $\mathcal{F}, \mathcal{I}$). *If* ξ_{min} *is negative, then* P *is not individually fairfor* $\mathcal{I}$ *w.r.t.* $\mathcal{F}$.

Moreover, FairnessMonitor circumvents major restrictions of Lipschitz-fairness:

d_{In}-d_{Out}-**relation:** *Individual fairness*defines constraints between input and output distances by means of a function f, which allows the expression of complex fairness constraints. For a more elaborate discussion, see [12, Appendix A].

Input relevance: *Individual fairness*explicitly distinguishes between actual and synthetic inputs. This way, *individual fairness*acknowledges a possible obstacle of the fairness theory when it comes to real-world usage of the analysis, namely that only some elements of the system's input domain might be plausible, and usually, only a few of them become actual inputs that have to be monitored for unfairness.

Monitorability: FairnessMonitor demonstrates that *individual fairness*is monitorable. It resolves the quantification over In using the above concepts from probabilistic falsification using the robustness estimate function F as defined above.

Towards individual fairness in the loop. If a high-risk systemis in operation, a human in the loop must oversee the correct and fair functioning of the outputs of the system. To do this, the human needs real-time fairness information. Figure 2 shows how this can be achieved by coupling the system P and

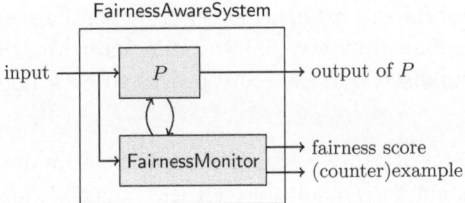

Fig. 2. Schematic visualisation of FairnessAwareSystem

the FairnessMonitor in Algorithm 1 in a new system called FairnessAwareSystem. FairnessAwareSystem is sketched in Algorithm 2.

Algorithm 2. FairnessAwareSystem

Parameters: System $P : In \to Out$, Fairness contract $\mathcal{F} = \langle d_{In}, d_{Out}, f \rangle$
Input: Input $i_a \in In$
Output: Tuple of the system output, normalised fairness score, and synthetic values witnessing the fairness score

1: $(\xi_{min}, i_a, i_s) \leftarrow$ FairnessMonitor$(P, \mathcal{F}, \{i_a\})$

2: **return** $\left(P(i_a), \frac{\xi_{min}}{f(d_{In}(i_a, i_s))}, (i_s, P(i_s)) \right)$

Intuitively, the FairnessAwareSystem is a higher-order program that is parameterised with the original program P and the fairness contract $\mathcal{F}$. When instantiated with these parameters, the program takes arbitrary (actual) inputs i_a from In. In the first step, it does a fairness analysis using FairnessMonitor with arguments P, $\mathcal{F}$, and $\{i_a\}$. To make fairness scores comparable, FairnessAwareSystem normalises the fairness score ξ received from FairnessMonitor by dividing[4] it by the output distance limit $f(d_{In}(i_a, i_s))$. For fair outputs, the score will be between 0 (almost unfair) and 1 (as fair as possible).[5] Outputs that are not individually fair are accompanied by a negative score representing how much the limit $f(d_{In}(i_a, i_s))$ is exceeded. A fairness score of $-n$ means that the output distance of $P(i_a)$ and $P(i_s)$ is $n + 1$ times as high as that limit. Finally, FairnessAwareSystem returns the triple with P's output for i_a, the normalised fairness score, and the synthetic input with its output witnessing the fairness score.

---

[4] For f that can return 0, a division of zero by zero may occurr. The result of this division should be defined depending on the concrete context; reasonable values range from the extreme scores 0 (to indicate that the score is on edge of becoming 'unfair') to 1 (to indicate that more fairness is impossible).

[5] Fairness may be a vague concept that cannot be dichotomised. By its choice of the fairness contract parameters, our approach nevertheless specifies a (non-arbitrary) cut-off point at 0; but it does so for purely instrumental and non-ontological reasons.

Interpretation of monitoring results. Especially when FairnessAwareSystem finds a violation of *individual fairness*, the suitable interpretation and appropriate response to the normalised fairness score proves to be a non-trivial matter that requires expertise.

Example 2. Instead of using P from Example 1 on its own, Unicanow uses FairnessAwareSystem with a suitable fairness contract. and thereby receive a fairness score along with P's verdict on each applicant. (Which fairness contracts are suitable is an open research problem, see *Limitations & Challenges* in Section 6.) If the fairness score is negative, she can also take into account the information on the synthetic counterpart returned by FairnessAwareSystem. Among the 4096applicants for the PhD program, the monitoring assigns a negative fairness score to three candidates: Alexa, who received a low score, Eugene, who was scored very highly, and John, who got an average score. According to their scoring, Alexawould be desk-rejected, while Eugeneand Johnwould be considered further.

Alexa's synthetic counterpart, let's call him Syntbad, is ranked much higher than Alexa. In fact, he is ranked so high that Syntbadwould not be desk-rejected. Unicacompares Alexaand Syntbadand finds that they only differ in one respect: Syntbad's graduate university is the one in the official ranking that is immediately *below* the one that Alexaattended. Unicadoes some research and finds that Alexa's institution is predominantly attended by People of Colour, while this is not the case for Syntbad's institution. Therefore, FairnessAwareSystem helped Unicanot only to find an unfair treatment of Alexa, but also to uncover a case of potential racial discrimination.

John's counterpart, Synclair, is ranked much lower than him. Unicamanually inspects John's previous institution (an infamous online university), his GPA of 1.8, and his test result with only 13%. She finds that this very much suggests that Johnwill not be a successful PhD candidate and desk-rejects him. Therefore, Unicahas successfully used FairnessAwareSystem to detect a fault in the scoring system P whereby Johnwould have been treated unfairly in a way that would have been to his advantage.

Eugenereceived a top score, but his synthetic counterpart, Syna, received only an average one. Unicasuspects that Eugenewas ranked too highly given his graduate institution, GPA, and test score. However, as he would not have been desk-rejected either way, nothing changes for Eugene, and the unfairness he was subject to, is not of effect to him.

The cases of Johnand Eugeneshare similarities with the configuration in (b) in Figure 3, the one of Alexawith (a), and the ones of all other 4093candidates with (c).

If our monitor finds only a few problematic cases in a (sufficiently large and diverse) set of inputs, our monitoring helps Unicafrom our running example by drawing her attention to cases that require special attention. Thereby, individuals who are judged by the system have a better chance of being treated fairly,

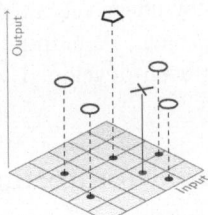
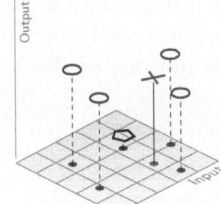
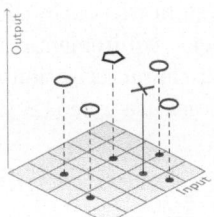

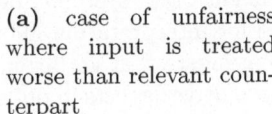

(a) case of unfairness where input is treated worse than relevant counterpart

(b) case of unfairness where input is treated better than relevant counterpart

(c) case of no detected unfairness

Fig. 3. Exemplary illustration of configurations of an input (red cross) and its synthetic counterparts (grey circles) and the synthetic counterpart with the minimal fairness score (blue polygon); with a two-dimensional input space (grid) and a one-dimensional output.

since even rare instances of unfair treatment can be detected. If, on the other hand, the number of problematic cases found is large or Unicafinds especially concerning cases or patterns, this can point to larger issues within the system. In these cases, Unicashould take appropriate steps and make sure that the system is no longer used until clarity is established as to why so many violations or concerning patterns are found. If the system is found to be systematically unfair, it should arguably be removed from the decision process. A possible conclusion could also be that the system is unsuitable for certain use cases, e.g., for the use of individuals from a particular group. Accordingly, it might not have to be removed altogether but only needs to be restricted such that problematic use cases are avoided. In any case, significant findings should also be reported to the developers or deployers of the potentially problematic system. A fairness monitoring such as in FairnessAwareSystem or a fairness analysis as in FairnessMonitor could also be useful to developers, regulating authorities, watchdog organisations, or forensic analysts as it helps them to check the individual fairness of a system in a controlled environment.

Remark 1. Individual fairness is called func-fairness in [12] and is an adaptation of func-cleanness, which has been studied in earlier work [11,22] on *software doping*. In this context, a cleanness property – like func-cleanness – characterises the absence of doped software. Intuitively, software doping relates to the existence of a hidden feature in a software that was added intentionally by the software manufacturer, but which is not in the interest of the user or society. The diesel emissions scandal is by now the archetypal example of software doping: various car manufacturers added defeat devices into their emission cleaning systems to distinguish whether the car is undergoing an emissions test from whether it is used in normal operation on the road. In the former case, the emission cleaning worked as required, while in the latter case, the engine control system was optimising for other objectives instead, thereby effectively infringing legal requirements. A falsification-based monitoring approach, including a logical characterisation of various notions of cleanness, has been developed for the diesel use-case [12,15].

5 Interdisciplinary Assessment of Fairness Monitoring

The upcoming AI act stresses the need for human oversight of AI systems, but its stipulations are not free of ambiguities and the need for interpretation. This raises the question of whether our approach meets requirements that go beyond pre-theoretical deliberations. We here assess some key normative aspects in philosophical and legal terms, and also briefly turn to the related empirical aspects, especially from psychology.

5.1 Psychological assessment

Fairness monitoring promises various advantages in terms of human-system interaction in application contexts – provided it is extended by an adequate user interface – which calls for empirical tests and studies. We will only discuss a possible benefit that closely aligns with the upcoming AI Act: our approach may support effective human oversight. Two central aspects of effective oversight are situation awareness and warranted trust. Our method highlights unfairness in outputs which can be expected to increase users' situation awareness (i.e., "the perception of the elements in the environment within a volume of time and space, the comprehension of their meaning and the projection of their status in the near future" [26, p. 36]), which is a variable central for effective oversight [27]. In the minimal case, this allows users to realise that something requires their attention and that they should check the outputs for plausibility and adequacy. In the optimal case and after some experience with the monitor, it may even allow users to predict instances where a system will produce potentially unfair outputs. In any case, the monitoring should enable them to understand the limitations of the system and to feed back their findings to developers who can improve the system. This leads us to warranted trust, which includes that users are able to adequately judge when to rely on system outputs and when to reject them [39,48]. Building warranted trust strongly depends on users being able to assess system trustworthiness in the given context of use [48,72]. According to their theoretical model on trust in automation, Lee and See [48] propose that trustworthiness relates to different facets of which performance (e.g., whether the system performs reliably with high accuracy) and process (e.g., knowing how the system operates and whether the system's decision-processes help to fulfil the trustor's goals) are especially relevant in our case. Specifically, fairness monitoring should enable users to judge system performance more accurately (e.g., by revealing possible issues with system outputs) and system processes (e.g., whether the system's decision logic was appropriate). In line with Lee and See's propositions, this should provide a foundation for users to judge system trustworthiness better and should thus be a promising means to promote warranted trust. In consequence, our monitoring provides a needed addition to high-risk use contexts of AI because it offers information enabling humans to more adequately use AI-based systems in the sense of possibly better human-system decision performance and with respect to user duties as described in the AI Act.

5.2   Philosophical assessment

More effective oversight promises more informed decision-making. This, in turn, enables morally better decisions and outcomes since humans can morally ameliorate outcomes in terms of fairness and can see to it that moral values are promoted. Fairness monitoring also helps safeguard fundamental democratic values if it is applied to potentially unfair systems used in certain societal institutions of a high-risk character, such as courts or parliaments. It could, for example, make AI-aided court decisions more transparent and promote equality before the law. However, since our approach requires finding context-appropriate and morally permissible parameters for $\mathcal{F}$, moral requirements arise to enable the finding of such parameters. This affects not only developers of such systems but also those who are in a position to enforce that adequate parameters are chosen, such as governmental authorities, supervising institutions, or certifiers.

Apart from that, various parties have arguably a legitimate interest in adequately ascribing moral responsibility for the outcomes of certain decisions to human deciders [10] – regardless of whether the decision-making process is supported by a system. Adequately ascribing moral responsibility is not always possible, though. One precondition for moral responsibility is that the agent had sufficient epistemic access to the consequences of their doing [58,80], i.e., that they have enough and sufficiently well-justified beliefs about the results of their decision. Someone overseeing a university selection process (like Unica) should, for example, have sufficiently well-justified beliefs that, at the very least, their decisions do not result in more unfairness in the world. If the admission process is supported by a black-box system, though, Unicacannot be expected to have any such beliefs since she lacks insight in the fairness of the system. Therefore, adequate responsibility ascription is usually not possible in this scenario. Our monitoring alleviates this problem by providing the decider with better epistemic access to the fairness of the system.

FairnessAwareSystem helps in making Unica's role in the decision process significant and not only that of a mere button-pusher. FairnessAwareSystem makes it possible for her to fulfil some of the responsibilities and duties plausibly associated with her role. For example, she can now be realistically expected to not only detect, but resolve at least some cases of apparent unfairness competently (although she may need additional information to do so). In this respect, she should not be 'automated away' (cf. [50]).

5.3   Legal assessment

A central legislative debate of our time is how to counter the risks AI systems can pose to the health and safety or fundamental rights of natural persons. Protective measures must be taken at various levels: First, before being permitted on the market, it must be ensured *ex-ante* that such high-risk AI-systems are in conformity with mandatory requirements regarding safety and human rights [29, Art. 16, Art 27]. This means in particular that the selection of the properties that a system should exhibit requires a positive normative choice and should not simply replicate biases present in the status quo [83]. In addition, AI-systems must be designed and developed in such a way that natural persons can oversee their functioning. For this purpose, it is necessary for the provider to design and develop the AI system in such a way that it includes appropriate features enabling human oversight before it is placed on the market or put into service [29, Art. 14].

Second, during runtime, the proper functioning of high-risk AI systems that have been legally placed on the market must be ensured. To achieve this goal, a bundle of different measures is needed, ranging from legal obligations to implement and perform meaningful oversight mechanisms to user training and awareness in order to counteract "automation bias'. In particular, such measures should guarantee that the natural persons to whom human oversight has been assigned have the necessary competence, training, and authority to carry out that role [29, Art. 26 (2)]. Furthermore, the AI Act proposal requires deployers to inform the provider or distributor and suspend the use of the system when they have identified any serious incidents or any malfunctioning [29, Art. 26(5)].

Third, and *ex-post*, providers must act and take the necessary corrective actions as soon as they become aware, e.g., through information provided by the deployer, that the high-risk system does not (or no longer) meet the legal requirements [29, Art. 20]. To this end, they must establish and document a system of monitoring that is proportionate to the type of AI technology and the risks of the high-risk AI system [29, Art. 72].

Fairness monitoring can be helpful in all three of the above respects. Therefore, we argue that there is even a legal obligation to use technical measures such as the method presented in this paper if this is the only way to ensure effective human oversight.

6   Conclusion

This invited paper has echoed elements of a forthcoming journal publication [12] that applies runtime monitoring and probabilistic falsification techniques to high-risk (AI) systems.

We have looked at a runtime fairness monitor to promote effective human oversight of high-risk systems. An interdisciplinary evaluation from a psychological, philosophical, and legal perspective complements the development of this monitor. As seen in Figure 1, our fairness monitoring aims to uncover a particular kind of unfairness, namely individual unfairness, that originates from within the system. This does not include group unfairness as well as unfairness from sources other than the system. Another limitation is the need to account for the human's competence to interpret the system outputs. Even though this is not a limitation that is inherent to our approach, it nevertheless will arguably be relevant in some practical cases, and an implementation of the monitoring always has to happen with the human in mind. For example, the design of the tool should avoid creating the false impression that the system is proven to be fair for an individual if no counterexample has been found. Interpretations like this could lead to inflated judgements of system trustworthiness and eventually to overtrusting system outputs [72,74]. Also, it might be reasonable to limit access to the monitoring results: if individuals who are processed by the system have full access to their fairness analysis, they could use this to 'game' the system, i.e., they could use the synthetic inputs to slightly modify their own input such that they receive a better outcome. While more transparency for the user is generally desirable, this has to be kept in mind to avoid introducing new unfairness on a meta-level.

Acknowledgements. This work is partially funded by DFG grant 389792660 as part of TRR 248 – CPEC, by VolkswagenStiftung as part of grants AZ 98514, 98513 and 98512 – EIS, by the European Regional Development Fund and the Saarland within the scope of (To)CERTAIN, and as part of STORM_SAFE, an Interreg project supported by the North Sea Programme of the European Regional Development Fund.

References

1. Abbas, H., Fainekos, G.E., Sankaranarayanan, S., Ivancic, F., Gupta, A.: Probabilistic temporal logic falsification of cyber-physical systems. ACM Trans. Embed. Comput. Syst. **12**(2s), 95:1–95:30 (2013). https://doi.org/10.1145/2465787.2465797
2. Alves, W.M., Rossi, P.H.: Who should get what? fairness judgments of the distribution of earnings. American journal of Sociology **84**(3), 541–564 (1978)
3. Angwin, J., Larson, J., Mattu, S., Kirchner, L.: Machine Bias (2016), https://www.propublica.org/article/machine-bias-risk-assessments-in-criminal-sentencing
4. Arrieta, A.B., Díaz-Rodríguez, N., Del Ser, J., Bennetot, A., Tabik, S., Barbado, A., García, S., Gil-López, S., Molina, D., Benjamins, R., et al.: Explainable artificial intelligence (XAI): Concepts, taxonomies, opportunities and challenges toward responsible AI. Information Fusion **58**, 82–115 (2020)
5. Artistotle: The Nicomachean Ethics. Oxford worlds classics, Oxford University Press, Oxford (1998), translation by W.D. Ross. Edition by John L. Ackrill, and James O. Urmson
6. Artistotle: Politics. Oxford worlds classics, Oxford University Press, Oxford (1998), translation by Ernest Barker. Edition by R. F. Stalley
7. Barocas, S., Selbst, A.D.: Big data's disparate impact. Calif. L. Rev. **104**, 671 (2016)
8. Bathaee, Y.: The artificial intelligence black box and the failure of intent and causation. Harv. JL & Tech. **31**, 889 (2017)

9. Baum, D., Baum, K., Gros, T.P., Wolf, V.: XAI Requirements in Smart Production Processes: A Case Study. In: World Conference on Explainable Artificial Intelligence. pp. 3–24. Springer (2023)

10. Baum, K., Mantel, S., Schmidt, E., Speith, T.: From responsibility to reason-giving explainable artificial intelligence. Philosophy & Technology **35**(1), 12 (2022). https://doi.org/10.1007/s13347-022-00510-w, https://doi.org/10.1007/s13347-022-00510-w

11. Biewer, S.: Software Doping – Theory and Detection. Phd thesis, Universität des Saarlandes (2023). https://doi.org/10.22028/D291-40364, http://dx.doi.org/10.22028/D291-40364

12. Biewer, S., Baum, K., Sterz, S., Hermanns, H., Hetmank, S., Langer, M., Lauber-Rönsberg, A., Lehr, F.: Software doping analysis for human oversight. Formal Methods in System Design (2024). https://doi.org/10.1007/s10703-024-00445-2, to appear; preprint available at https://arxiv.org/abs/2308.06186

13. Biewer, S., D'Argenio, P.R., Hermanns, H.: Doping tests for cyber-physical systems. ACM Trans. Model. Comput. Simul. **31**(3), 16:1–16:27 (2021). https://doi.org/10.1145/3449354, https://doi.org/10.1145/3449354

14. Biewer, S., Finkbeiner, B., Hermanns, H., Köhl, M.A., Schnitzer, Y., Schwenger, M.: On the road with RTLola. Int. J. Softw. Tools Technol. Transf. **25**(2), 205–218 (2023). https://doi.org/10.1007/s10009-022-00689-5, https://doi.org/10.1007/s10009-022-00689-5

15. Biewer, S., Hermanns, H.: On the detection of doped software by falsification. In: Johnsen, E.B., Wimmer, M. (eds.) Fundamental Approaches to Software Engineering - 25th International Conference, FASE 2022, Held as Part of the European Joint Conferences on Theory and Practice of Software, ETAPS 2022, Munich, Germany, April 2-7, 2022, Proceedings. Lecture Notes in Computer Science, vol. 13241, pp. 71–91. Springer (2022). https://doi.org/10.1007/978-3-030-99429-7_4, https://doi.org/10.1007/978-3-030-99429-7_4

16. Binns, R.: On the apparent conflict between individual and group fairness. In: Proceedings of the 2020 Conference on Fairness, Accountability, and Transparency. p. 514-524. FAT* '20, Association for Computing Machinery, New York, NY, USA (2020). https://doi.org/10.1145/3351095.3372864, https://doi.org/10.1145/3351095.3372864

17. Bloem, R., Chatterjee, K., Greimel, K., Henzinger, T.A., Hofferek, G., Jobstmann, B., Könighofer, B., Könighofer, R.: Synthesizing robust systems. Acta Informatica **51**(3-4), 193–220 (2014). https://doi.org/10.1007/s00236-013-0191-5, https://doi.org/10.1007/s00236-013-0191-5

18. Borgesius, F.J.Z.: Strengthening legal protection against discrimination by algorithms and artificial intelligence. The International Journal of Human Rights **24**(10), 1572–1593 (2020). https://doi.org/10.1080/13642987.2020.1743976, https://doi.org/10.1080/13642987.2020.1743976

19. Burke, L.: The Death and Life of an Admissions Algorithm (2020), https://www.insidehighered.com/admissions/article/2020/12/14/u-texas-will-stop-using-controversial-algorithm-evaluate-phd

20. Chazette, L., Brunotte, W., Speith, T.: Exploring explainability: A definition, a model, and a knowledge catalogue. In: 2021 IEEE 29th International Requirements Engineering Conference (RE). pp. 197–208 (2021). https://doi.org/10.1109/RE51729.2021.00025

21. Chouldechova, A.: Fair prediction with disparate impact: A study of bias in recidivism prediction instruments. Big Data **5**(2), 153–163 (2017). https://doi.org/10.1089/big.2016.0047, https://doi.org/10.1089/big.2016.0047

22. D'Argenio, P.R., Barthe, G., Biewer, S., Finkbeiner, B., Hermanns, H.: Is your software on dope? - formal analysis of surreptitiously "enhanced" programs. In: Yang, H. (ed.) Programming Languages and Systems - 26th European Symposium on Programming, ESOP 2017, Held as Part of the European Joint Conferences on Theory and Practice of Software, ETAPS 2017, Uppsala, Sweden, April 22-29, 2017, Proceedings. Lecture Notes in Computer Science, vol. 10201, pp. 83–110. Springer (2017). https://doi.org/10.1007/978-3-662-54434-1_4, https://doi.org/10.1007/978-3-662-54434-1_4

23. Dressel, J., Farid, H.: The accuracy, fairness, and limits of predicting recidivism. Science advances 4(1), eaao5580 (2018)

24. Dwork, C., Hardt, M., Pitassi, T., Reingold, O., Zemel, R.: Fairness through awareness. In: Proceedings of the 3rd innovations in theoretical computer science conference. pp. 214–226 (2012)

25. Dworkin, R.: What is equality? part 2: Equality of resources. Philosophy & Public Affairs 10(4), 283–345 (1981), http://www.jstor.org/stable/2265047

26. Endsley, M.R.: Toward a theory of situation awareness in dynamic systems. Human Factors 37(1), 32–64 (1995). https://doi.org/10.1518/001872095779049543

27. Endsley, M.R.: From here to autonomy: Lessons learned from human-automation research. Human Factors 59(1), 5–27 (2017). https://doi.org/10.1177/0018720816681350, https://doi.org/10.1177/0018720816681350, pMID: 28146676

28. European Court of Justice: C-356/12 - glatzel ecli:eu:c:2014:350 (2014), https://curia.europa.eu/juris/liste.jsf?language=en&num=C-356/12

29. European Union: Regulation laying down harmonised rules on Artificial Intelligence (Artificial Intelligence Act), provisional version that has been adopted by the European Parliament on 13 March 2024 (2024), https://www.europarl.europa.eu/doceo/document/TA-9-2024-0138_EN.pdf

30. Ferrer, X., Nuenen, T.v., Such, J.M., Coté, M., Criado, N.: Bias and discrimination in AI: A cross-disciplinary perspective. IEEE Technology and Society Magazine 40(2), 72–80 (2021). https://doi.org/10.1109/MTS.2021.3056293

31. Friedler, S.A., Scheidegger, C., Venkatasubramanian, S.: The (im)possibility of fairness: Different value systems require different mechanisms for fair decision making. Commun. ACM 64(4), 136-143 (mar 2021). https://doi.org/10.1145/3433949, https://doi.org/10.1145/3433949

32. Gunning, D.: Explainable artificial intelligence (XAI) (darpa-baa-16-53). Tech. rep., Arlington, VA, USA (2016)

33. Guryan, J., Charles, K.K.: Taste-based or statistical discrimination: The economics of discrimination returns to its roots. The Economic Journal 123(572), F417–F432 (2013), http://www.jstor.org/stable/42919257

34. Hartmann, F.: Diskriminierung durch Antidiskriminierungsrecht? Möglichkeiten und Grenzen eines postkategorialen Diskriminierungsschutzes in der Europäischen Union. EuZA - Europäische Zeitschrift für Arbeitsrecht p. 24 (2006)

35. Heaven, W.D.: Predictive policing algorithms are racist. They need to be dismantled. (2020), https://www.technologyreview.com/2020/07/17/1005396/predictive-policing-algorithms-racist-dismantled-machine-learning-bias-criminal-justice/

36. High-Level Expert Group on Artificial Intelligence: Ethics Guidelines for Trustworthy AI (2019), https://digital-strategy.ec.europa.eu/en/library/ethics-guidelines-trustworthy-ai

37. Hough, L.M., Oswald, F.L., Ployhart, R.E.: Determinants, detection and amelioration of adverse impact in personnel selection procedures: Issues, evidence and lessons learned. International Journal of Selection and Assessment 9(1-2), 152–194 (2001)

38. Ilvento, C.: Metric learning for individual fairness. arXiv preprint arXiv:1906.00250 (2019)
39. Jacovi, A., Marasović, A., Miller, T., Goldberg, Y.: Formalizing trust in artificial intelligence: Prerequisites, causes and goals of human trust in AI. In: Proceedings of the 2021 ACM Conference on Fairness, Accountability, and Transparency. pp. 624–635 (2021)
40. Jewson, N., Mason, D.: Modes of discrimination in the recruitment process: formalisation, fairness and efficiency. Sociology **20**(1), 43–63 (1986)
41. John, P.G., Vijaykeerthy, D., Saha, D.: Verifying individual fairness in machine learning models. In: Adams, R.P., Gogate, V. (eds.) Proceedings of the Thirty-Sixth Conference on Uncertainty in Artificial Intelligence, UAI 2020, virtual online, August 3-6, 2020. Proceedings of Machine Learning Research, vol. 124, pp. 749–758. AUAI Press (2020), http://proceedings.mlr.press/v124/george-john20a.html
42. Kästner, L., Langer, M., Lazar, V., Schomäcker, A., Speith, T., Sterz, S.: On the relation of trust and explainability: Why to engineer for trustworthiness. In: Yue, T., Mirakhorli, M. (eds.) 29th IEEE International Requirements Engineering Conference Workshops, RE 2021 Workshops, Notre Dame, IN, USA, September 20-24, 2021. pp. 169–175. IEEE (2021). https://doi.org/10.1109/REW53955.2021.00031, https://doi.org/10.1109/REW53955.2021.00031
43. Lai, V., Tan, C.: On human predictions with explanations and predictions of machine learning models: A case study on deception detection. In: Proceedings of the conference on fairness, accountability, and transparency. pp. 29–38 (2019)
44. Langer, M., Baum, K., Hartmann, K., Hessel, S., Speith, T., Wahl, J.: Explainability auditing for intelligent systems: A rationale for multi-disciplinary perspectives. In: Yue, T., Mirakhorli, M. (eds.) 29th IEEE International Requirements Engineering Conference Workshops, RE 2021 Workshops, Notre Dame, IN, USA, September 20-24, 2021. pp. 164–168. IEEE (2021). https://doi.org/10.1109/REW53955.2021.00030, https://doi.org/10.1109/REW53955.2021.00030
45. Langer, M., Baum, K., Schlicker, N.: Effective human oversight of ai-based systems: A signal detection perspective on the detection of inaccurate and unfair outputs (2023). https://doi.org/10.31234/osf.io/ke256
46. Langer, M., Oster, D., Speith, T., Hermanns, H., Kästner, L., Schmidt, E., Sesing, A., Baum, K.: What do we want from explainable artificial intelligence (XAI)? - A stakeholder perspective on XAI and a conceptual model guiding interdisciplinary XAI research. Artif. Intell. **296**, 103473 (2021). https://doi.org/10.1016/j.artint.2021.103473, https://doi.org/10.1016/j.artint.2021.103473
47. Larson, J., Mattu, S., Kirchner, L., Angwin, J.: How We Analyzed the COMPAS Recidivism Algorithm (2016), https://www.propublica.org/article/how-we-analyzed-the-compas-recidivism-algorithm
48. Lee, J.D., See, K.A.: Trust in automation: Designing for appropriate reliance. Human factors **46**(1), 50–80 (2004)
49. Linardatos, P., Papastefanopoulos, V., Kotsiantis, S.: Explainable AI: A review of machine learning interpretability methods. Entropy **23**(1) (2021). https://doi.org/10.3390/e23010018, https://www.mdpi.com/1099-4300/23/1/18
50. Matthias, A.: The responsibility gap: Ascribing responsibility for the actions of learning automata. Ethics and Information Technology **6**(3), 175–183 (2004). https://doi.org/10.1007/s10676-004-3422-1
51. Mecacci, G., de Sio, F.S.: Meaningful human control as reason-responsiveness: The case of dual-mode vehicles. Ethics and Information Technology **22**(2), 103–115 (2020). https://doi.org/10.1007/s10676-019-09519-w

52. Mehrabi, N., Morstatter, F., Saxena, N., Lerman, K., Galstyan, A.: A survey on bias and fairness in machine learning. ACM Computing Surveys (CSUR) **54**(6), 1–35 (2021)
53. Methnani, L., Aler Tubella, A., Dignum, V., Theodorou, A.: Let me take over: Variable autonomy for meaningful human control. Frontiers in Artificial Intelligence **4** (2021). https://doi.org/10.3389/frai.2021.737072, https://www.frontiersin.org/article/10.3389/frai.2021.737072
54. Meurrens, S.: The Increasing Role of AI in Visa Processing (2021), https://canadianimmigrant.ca/immigrate/immigration-law/the-increasing-role-of-ai-in-visa-processing
55. Mittelstadt, B.D., Allo, P., Taddeo, M., Wachter, S., Floridi, L.: The ethics of algorithms: Mapping the debate. Big Data & Society **3**(2), 2053951716679679 (2016). https://doi.org/10.1177/2053951716679679, https://doi.org/10.1177/2053951716679679
56. Molnar, C., Casalicchio, G., Bischl, B.: Interpretable machine learning - A brief history, state-of-the-art and challenges. In: Koprinska, I., Kamp, M., Appice, A., Loglisci, C., Antonie, L., Zimmermann, A., Guidotti, R., Özgöbek, Ö., Ribeiro, R.P., Gavaldà, R., Gama, J., Adilova, L., Krishnamurthy, Y., Ferreira, P.M., Malerba, D., Medeiros, I., Ceci, M., Manco, G., Masciari, E., Ras, Z.W., Christen, P., Ntoutsi, E., Schubert, E., Zimek, A., Monreale, A., Biecek, P., Rinzivillo, S., Kille, B., Lommatzsch, A., Gulla, J.A. (eds.) ECML PKDD 2020 Workshops - Workshops of the European Conference on Machine Learning and Knowledge Discovery in Databases (ECML PKDD 2020): SoGood 2020, PDFL 2020, MLCS 2020, NFMCP 2020, DINA 2020, EDML 2020, XKDD 2020 and INRA 2020, Ghent, Belgium, September 14-18, 2020, Proceedings. Communications in Computer and Information Science, vol. 1323, pp. 417–431. Springer (2020). https://doi.org/10.1007/978-3-030-65965-3_28, https://doi.org/10.1007/978-3-030-65965-3_28
57. Mukherjee, D., Yurochkin, M., Banerjee, M., Sun, Y.: Two simple ways to learn individual fairness metrics from data. In: III, H.D., Singh, A. (eds.) Proceedings of the 37th International Conference on Machine Learning. Proceedings of Machine Learning Research, vol. 119, pp. 7097–7107. PMLR (13–18 Jul 2020), https://proceedings.mlr.press/v119/mukherjee20a.html
58. Noorman, M.: Computing and Moral Responsibility. In: Zalta, E.N. (ed.) The Stanford Encyclopedia of Philosophy. Metaphysics Research Lab, Stanford University, Spring 2020 edn. (2020)
59. Nunes, I., Jannach, D.: A systematic review and taxonomy of explanations in decision support and recommender systems. User Modeling and User-Adapted Interaction **27**(3), 393–444 (2017)
60. O'Neil, C.: How algorithms rule our working lives (2016), https://www.theguardian.com/science/2016/sep/01/how-algorithms-rule-our-working-lives, Online; accessed: 2023-06-23
61. O'Neil, C.: Weapons of Math Destruction: How Big Data Increases Inequality and Threatens Democracy. Crown Publishing Group, USA (2016)
62. Orcale: AI in human resources: The time is now (2019), https://www.oracle.com/a/ocom/docs/applications/hcm/oracle-ai-in-hr-wp.pdf
63. Organisation for Economic Co-operation and Development (OECD): Artificial intelligence, machine learning and big data in finance: Opportunities, challenges and implications for policy makers. Tech. rep., [París] : (2021), https://www.oecd.org/finance/financial-markets/Artificial-intelligence-machine-learning-big-data-in-finance.pdf

64. Pessach, D., Shmueli, E.: A review on fairness in machine learning. ACM Comput. Surv. **55**(3) (feb 2022). https://doi.org/10.1145/3494672, https://doi.org/10.1145/3494672

65. Rawls, J.: Justice as fairness: Political not metaphysical. Philosophy & Public Affairs **14**(3), 223–251 (1985), http://www.jstor.org/stable/2265349

66. Rawls, J.: A theory of justice: Revised edition. Harvard university press (1999)

67. Rawls, J.: Justice as fairness: A restatement. Harvard University Press (2001)

68. Ribeiro, M.T., Singh, S., Guestrin, C.: Model-agnostic interpretability of machine learning. CoRR **abs/1606.05386** (2016), http://arxiv.org/abs/1606.05386

69. Ribeiro, M.T., Singh, S., Guestrin, C.: "Why should I trust you?": Explaining the predictions of any classifier. In: Proceedings of the 22nd ACM SIGKDD International Conference on Knowledge Discovery and Data Mining. p. 1135-1144. KDD '16, Association for Computing Machinery, New York, NY, USA (2016). https://doi.org/10.1145/2939672.2939778, https://doi.org/10.1145/2939672.2939778

70. Rowe, T.: Can a risk of harm itself be a harm? Analysis **81**(4), 694–701 (2022). https://doi.org/10.1093/analys/anab033

71. Sanneman, L., Shah, J.A.: A situation awareness-based framework for design and evaluation of explainable AI. In: International Workshop on Explainable, Transparent Autonomous Agents and Multi-Agent Systems. pp. 94–110. Springer (2020)

72. Schlicker, N., Langer, M.: Towards warranted trust: A model on the relation between actual and perceived system trustworthiness. In: Mensch und Computer 2021, pp. 325–329 (2021)

73. Schlicker, N., Langer, M., Ötting, S.K., Baum, K., König, C.J., Wallach, D.: What to expect from opening up 'black boxes'? comparing perceptions of justice between human and automated agents. Comput. Hum. Behav. **122**, 106837 (2021). https://doi.org/10.1016/j.chb.2021.106837, https://doi.org/10.1016/j.chb.2021.106837

74. Schlicker, N., Uhde, A., Baum, K., Hirsch, M., Langer, M.: Calibrated trust as a result of accurate trustworthiness assessment – introducing the trustworthiness assessment model. PsyArXiv Preprints (2022). https://doi.org/10.31234/osf.io/qhwvx

75. Santoni de Sio, F., van den Hoven, J.: Meaningful human control over autonomous systems: A philosophical account. Frontiers in Robotics and AI **5** (2018). https://doi.org/10.3389/frobt.2018.00015, https://www.frontiersin.org/article/10.3389/frobt.2018.00015

76. Smith, E., Vogell, H.: How Your Shadow Credit Score Could Decide Whether You Get an Apartment (2021), https://www.propublica.org/article/how-your-shadow-credit-score-could-decide-whether-you-get-an-apartment, Online; accessed: 2023-06-23

77. Speith, T.: A review of taxonomies of explainable artificial intelligence (XAI) methods. In: 2022 ACM Conference on Fairness, Accountability, and Transparency. p. 2239-2250. FAccT '22, Association for Computing Machinery, New York, NY, USA (2022). https://doi.org/10.1145/3531146.3534639, https://doi.org/10.1145/3531146.3534639

78. Sterz, S., Baum, K., Lauber-Rönsberg, A., Hermanns, H.: Towards perspicuity requirements. In: Yue, T., Mirakhorli, M. (eds.) 29th IEEE International Requirements Engineering Conference Workshops, RE 2021 Workshops, Notre Dame, IN, USA, September 20-24, 2021. pp. 159–163. IEEE (2021). https://doi.org/10.1109/REW53955.2021.00029, https://doi.org/10.1109/REW53955.2021.00029

79. Tabuada, P., Balkan, A., Caliskan, S.Y., Shoukry, Y., Majumdar, R.: Input-output robustness for discrete systems. In: Proceedings of the 12th International Conference on Embedded Software, EMSOFT 2012, part of the Eighth Embedded Systems Week, ESWeek 2012, Tampere, Finland, October 7-12, 2012. pp. 217–226. ACM (2012), http://doi.acm.org/10.1145/2380356.2380396

80. Talbert, M.: Moral Responsibility. In: Zalta, E.N. (ed.) The Stanford Encyclopedia of Philosophy. Metaphysics Research Lab, Stanford University, Winter 2019 edn. (2019)

81. Thüsing, G.: European Labour Law, § 3 Protection against discrimination. C.H. Beck (2013)

82. United Nations Educational, Scientific and Cultural Organization (UNESCO): Recommendation on the ethics of artificial intelligence (2021), https://unesdoc. unesco.org/ark:/48223/pf0000380455

83. Wachter, S., Mittelstadt, B., Russell, C.: Bias preservation in machine learning: the legality of fairness metrics under eu non-discrimination law. W. Va. L. Rev. **123**, 735 (2020). https://doi.org/10.2139/ssrn.3792772, http://dx.doi.org/10.2139/ssrn. 3792772

84. Washington State: Certification of Enrollment: Engrossed Substitute Senate Bill 6280 ('Washington State Facial Recognition Law') (2020), https://lawfilesext. leg.wa.gov/biennium/2019-20/Pdf/Bills/Senate%20Passed%20Legislature/6280-S.PL.pdf?q=20210513071229

85. Waters, A., Miikkulainen, R.: Grade: Machine learning support for graduate admissions. AI Magazine **35**(1), 64 (2014). https://doi.org/10.1609/aimag.v35i1.2504, https://ojs.aaai.org/index.php/aimagazine/article/view/2504

86. Zehlike, M., Yang, K., Stoyanovich, J.: Fairness in ranking: A survey. CoRR **abs/2103.14000** (2021), https://arxiv.org/abs/2103.14000

87. Zemel, R., Wu, Y., Swersky, K., Pitassi, T., Dwork, C.: Learning fair representations. In: International conference on machine learning. pp. 325–333. PMLR (2013)

88. Ziegert, J.C., Hanges, P.J.: Employment discrimination: the role of implicit attitudes, motivation, and a climate for racial bias. Journal of applied psychology **90**(3), 553 (2005)

MoXI: An Intermediate Language
for Symbolic Model Checking

Kristin Yvonne Rozier[1]([✉]), Rohit Dureja[2], Ahmed Irfan[3], Chris Johannsen[1],
Karthik Nukala[3], Natarajan Shankar[3], Cesare Tinelli[4], and Moshe Y. Vardi[5]

[1] Iowa State University, Ames, IA, USA
{kyrozier,cgjohann}@iastate.edu
[2] Advanced Micro Devices, Inc., Santa Clara, USA
rohit.dureja@amd.com
[3] SRI International, Menlo Park, USA
{ahmed.irfan,karthik.nukala,shankar}@sri.com
[4] The University of Iowa, Iowa City, USA
cesare-tinelli@uiowa.edu
[5] Rice University, Houston, USA
vardi@cs.rice.edu

Abstract. Three progressive challenges stand in between the popular,
"push-button," industrially valuable technique of symbolic model checking and the level of widespread adoption achieved by other verification
techniques: (1) the specification bottleneck; (2) the state-space explosion
problem; and (3) the lack of standardization and open-source implementations limiting the impact of advances in (1) and (2). We address this
third challenge. Learning from past definitions of intermediate languages
and common interfaces, as well as input from the international research
community, we define a new, extensible intermediate language for hardware symbolic model checking. Our contributions include: (a) defining
the syntax and semantics of **MoXI**, the **Mo**del e**X**change **I**nterlingua
designed to become a standard for the international research community; (b) demonstrating that an initial implementation of symbolic model
checking through MoXI performs competitively with current state-of-
the-art symbolic model checkers; (c) reframing the next symbolic model
checking research challenges considering this new community standard.

Keywords: Model Checking · Intermediate Language · SMT

1   Introduction

Symbolic model checking has made foundational changes to impactful, real-world
system designs, yet its ascent to a common-place verification technique is currently most limited by a few barriers to adoption, centering on its lack of standardization. For just one example, in our own work, symbolic model checking

This work was funded by NSF:CCRI Award #2016592, #2016597, #2016656.
The GitHub organization provides full artifacts: https://github.com/ModelChecker.
Thanks to our international Technical Advisory Board for invaluable feedback; see the
project website: https://modelchecker.temporallogic.org.

T. Neele and A. Wijs (Eds.): SPIN 2024, LNCS 14624, pp. 26–46, 2025.
https://doi.org/10.1007/978-3-031-66149-5_2

with NUXMV pinpointed hard-to-find, requirements-violating control sequences, and thus changed the design of NASA's Automated Airspace Concept [56,57]. This led to NASA using NUXMV for the next, much larger, stages of the project, including model checking with fault-tree analysis of a large set of possible safe configurations of the next-generation system [41], and design-space exploration of over 20,000 possible air traffic control designs [28].

The collection of impactful success stories of symbolic model checking in real-world system development are far too many and too diverse to cite in a single paper; there is no question that model checking provides value beyond its cost in verifying a wide range of systems uphold requirements for safety, security, and other desirable properties (such as consistency or financial soundness). However, model checking is still not as commonly-used as informal verification techniques such as simulation and testing. One primary reason for this is the specification bottleneck [51]: creating and validating system models and temporal logic specifications remains a challenging undertaking. The second barrier to adoption is the famous state-space explosion problem: the combination of the modeling technique used to represent the relevant system characteristics and the back-end model-checking algorithm can result in an untenable search space. However, the third barrier is perhaps both the broadest impediment and the easiest to overcome: a lack of standardization in symbolic model checking prevents the propagation of techniques aimed at lowering to the first two barriers.

The SMV modeling language represents an advancement in ameliorating barrier (1), the specification bottleneck: it is an expressive modeling language that, due to its appealing syntax that intuitively represents many common systems, continues to be successfully used in a wide range of industrial verification efforts [7,8,13,21,22, 27,28,31,38,41,44,45,49,54–57]. Two freely-available model checkers previously provided viable research platforms for

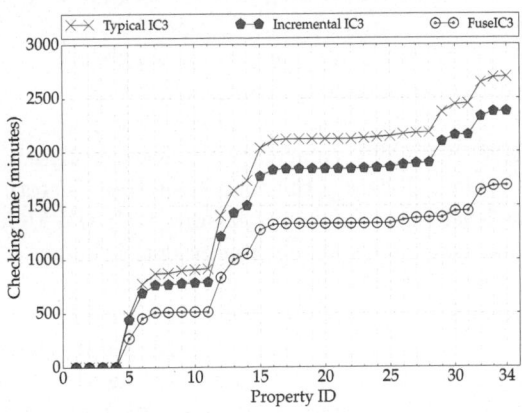

Fig. 1. FuseIC3 [22]: model checking **34 formulas** over **1,620 models** is **5.48x faster**.

checking SMV language models: CadenceSMV [43] and NUSMV [14] (which is integrated into today's NUXMV [47]). Yet, today CadenceSMV's 32-bit pre-compiled binary and NUXMV's increasingly restricted, closed-source releases are no longer suitable for research, e.g., into improved model-checking algorithms. How can we continue the progression of high-level language model checking in SMV with no open-source research platforms that allow new algorithms under the hood, e.g., to continue mitigating the state-space explosion problem?

Continuing with our example of symbolic model checking in NASA's next-generation transportation system, the next step of the process was to narrow down the design space (system designers reduced the possible configurations from 20,250 to 1,620), add details, and continue design-space model checking to refine the (safe) design set. To improve performance (per barrier (2), the state space explosion problem), we designed a new algorithm for model checking large design spaces, FuseIC3 [22]. However, to compare FuseIC3 with the state-of-the-art symbolic model-checking algorithms (namely, those claimed to be used in NUXMV [24] and in industry [4]), we had to **re-implement both of these existing model checkers from scratch**—our results appear in Fig. 1. In other words, there was no easy way to compare state-of-the-art model checking back-end algorithms because the existing implementations were closed source, and did not check models in the same modeling languages. No tool accepted our SMV-language models and implemented the best back-end model-checking algorithm for the industrial verification task at hand, and there was no way to add additional algorithms to NUXMV because it is closed source. (It would also have been useful to check how closely NUXMV implemented the state-of-the-art algorithms described in the literature, and to be able to modify the existing implementation.) In addition to being a time-consuming engineering task, this hurdle also curtailed the use of model checking in this NASA project. The lack of standardization in symbolic model checking meant that it was no longer the case that the benefits of model checking outweighed the cost. While we successfully ameliorated barriers (1) and (2), at least in the context of the task at hand, we could not apply our advances efficiently, or in the context of existing, and trusted, model-checking infrastructures.

This is not a unique story. The Hardware Model Checking Competition (HWMCC) [6] regularly spurs back-end algorithmic advances in symbolic model checking. However, the competing tools historically only accepted models in the bit-level input language AIGER [1,2]; HWMCC only recently advanced to the word-level, yet still machine-oriented, language BTOR2 [46]. Such languages do not support direct modeling of modern complex systems—the way SMV does—and hinder adoption; it is very hard to convince industrial system designers that AIGER or BTOR2 models correctly capture their higher-level systems. Most existing pathways for translating from high-level models to AIGER or BTOR2 focus on hardware designs, and do not provide a natural means to describe realistic systems. Even outside of HWMCC, many new, advanced, model-checking algorithms get developed for open-source, award-winning, model-checking engines, like ABC [10], which do not accept models in popular languages that would benefit from model checking using those algorithms. Then, we also face the open problem of translating counterexamples produced, e.g., by ABC, back into meaningful counterexamples for a non-hardware-centric higher-level language model, such as a model in SMV.

For another example, the PANDA algorithm and tool achieved up to exponential performance improvement using the same symbolic model-checking back-end, just by encoding LTL input specifications differently [50]. Yet PANDA's impact is limited since the implementation is only compatible with tools that accept models written in SMV. If, instead, PANDA translated LTL specifications into a common intermediate language, we could use it to improve model-checking performance broadly.

To alleviate the language issues above we introduce **Model Exchange Interlingua (MoXI)**, an intermediate language meant to be a common input and output standard for model checkers for finite- and infinite-state systems. MoXI was designed to be general enough to be an intermediate target language for a variety of user-facing specification languages for model checking. At the same time, MoXI is simple enough to be easily comparable to lower-level languages or be directly supported by model-checking tools based on SAT/SMT technology.

Models expressed in MoXI are expected to be produced and processed by tools. Thus MoXI provides: a simple, easily-machine-parsable syntax; a rich set of data types; minimal syntactic sugar (at least initially); a well-understood formal semantics; a small but comprehensive set of commands; and simple translations to lower-level modeling languages, such as BTOR2 [46]. Based on these principles, MoXI does not provide direct support for many of the expressive features offered by current hardware modeling languages such as VHDL [33,35] and Verilog [34], or more general-purpose system modeling languages such as SMV [42,48], TLA+ [39], PROMELA [32], Simulink [19], SCADE [17,20], and Lustre [11,36]. However, it is sufficiently expressive that problems defined in reasonably large fragments of those languages can be reduced to problems in MoXI. MoXI closely resembles the SMT-LIB language [3,53], but with new commands to define and verify systems. It allows the definition of multi-component synchronous or asynchronous reactive systems. It also allows the specification and checking of reachability conditions (or, indirectly, state and transition invariants) and deadlocks, possibly under fairness conditions on input values.

With MoXI, it is now possible to introduce or refine expressive modeling languages, model check them using existing and future state-of-the-art back-end algorithms, and return counterexamples in the original, high-level modeling language, simply by creating a translation between the modeling language and MoXI. Advances in back-end model checking algorithms can now be leveraged for models in any number of high-level modeling languages, simply by creating a translation between their low-level representations and MoXI. Most importantly, innovations mitigating the specification bottleneck and the state-space explosion problem will now push forward symbolic model checking as a whole, rather than advancing a single tool.

We introduce the notation and trace semantics underlying MoXI in Sect. 2. Section 3 defines MoXI, including the semantics for system definition and system checking. We demonstrate a prototype implementation translating the SMV modeling language through MoXI for model checking with the top tools from the last HWMCC in Sect. 4. Section 5 concludes with a concrete list of future research directions for the international symbolic model checking community, given the standardization provided by MoXI.

2   Preliminaries

The base logic of MoXI is the same as that of SMT-LIB: many-sorted first-order logic with equality and `let` binders. We refer to this logic simply as FOL. When we say *formula*, with no further qualifications, we refer to an arbitrary formula of FOL (possibly with quantifiers and `let` binders).

We say that a formula is *quantifier-free* if it contains no occurrences of the quantifiers $\forall$ and $\exists$. We say that it is *binder-free* if it is quantifier-free and also contains no occurrences of the `let` binder. The *scope* of binders and the notion of *free* and *bound* (occurrences of) variables in a formula are defined as usual.

2.1   Notation

If F is a formula and $\boldsymbol{x} = (x_1, \ldots, x_n)$ a tuple of distinct variables, we write $F[\boldsymbol{x}]$ or $F[x_1, \ldots, x_n]$ to express the fact that every variable in $\boldsymbol{x}$ occurs free in F (although F may have additional free variables). Let $\sigma(x_i)$ denote the type or *sort* of variable x_i in $\boldsymbol{x}$. We denote by $\boldsymbol{x}'$ the tuple $(x_1', \ldots, x_n')$ such that $\sigma(x_i) = \sigma(x_i')$. We write $\boldsymbol{x}, \boldsymbol{y}$ to denote the concatenation of tuple $\boldsymbol{x}$ with tuple $\boldsymbol{y}$. When it is clear from the context, given a formula $F[\boldsymbol{x}]$ and a tuple $\boldsymbol{t} = (t_1, \ldots, t_n)$ of terms of the same type as $\boldsymbol{x} = (x_1, \ldots, x_n)$, we write $F[\boldsymbol{t}]$ or $F[t_1, \ldots, t_n]$ to denote the formula obtained from F by simultaneously replacing each occurrence of x_i by t_i for all $i = 1, \ldots, n$. A formula may contain *uninterpreted* constant and function symbols, that is, symbols with no constraints on their interpretation. For most purposes, we treat uninterpreted constant and function symbols as free (rigid) variables respectively of first and second order.

Definition 1 (Transition system). *A transition system S is a pair of predicates of the form $S = (I_S[\boldsymbol{i}, \boldsymbol{o}, \boldsymbol{l}],\ T_S[\boldsymbol{i}, \boldsymbol{o}, \boldsymbol{l}, \boldsymbol{i}', \boldsymbol{o}', \boldsymbol{l}'])$ where*

1. *$\boldsymbol{i}$ and $\boldsymbol{i}'$ are tuples of* input *variables;*
2. *$\boldsymbol{o}$ and $\boldsymbol{o}'$ are tuples of* output *variables;*
3. *$\boldsymbol{l}$ and $\boldsymbol{l}'$ are tuples of* local *variables;*
4. *I_S is a formula representing the* initial state *condition; and*
5. *T_S is a formula representing the* transition *condition.*

2.2   Trace Semantics

A transition system implicitly defines a model (i.e., a Kripke structure) of First-Order Linear Temporal Logic (FO-LTL). The language of FO-LTL extends that of FOL with the same modal operators of time as in standard (propositional) LTL: **always, eventually, next, until, release.** For our purposes of defining the semantics of transition systems, it is enough to consider just the always and **eventually** operators.

The set of non-temporal operators depends on the particular theory, in the sense of SMT, considered (e.g., linear integer/real arithmetic, bit vectors, strings, and so on, and their combinations). The meaning of theory symbols (such as arithmetic operators) and theory sorts (such as Int, Real, Array(Int, Real), BitVec(3), ...) is fixed by the theory $\mathcal{T}$. With a fixed theory $\mathcal{T}$, the meaning of a FO-LTL formula F is provided by an interpretation of the uninterpreted (constant and function) symbols of F, if any, as well as an infinite sequence of valuations for the free variables of F.

Let tuple $x = (x_1, \ldots, x_n)$ denote distinct *state* variables, meant to represent the state of a computation system. We write formulas of the form $F[f, x, x']$ where f is a tuple of uninterpreted (constant and function) symbols. If F has free occurrences of variables from x but not from x' we call it a *one-state* formula; otherwise, we call it a *two-state* formula.

A *valuation* of x, or a *state over* x, is a function mapping each variable x in x to a value of x's sort. Let κ be a positive ordinal up to ω, the cardinality of the natural numbers. A *trace* (of length κ over x) is any state sequence $\pi = (s_j \mid 0 \le j < \kappa)$. Note that π is the finite sequence $s_0, \ldots, s_{\kappa-1}$ when $\kappa < \omega$, and is the infinite sequence $s_0, s_1, \ldots$ otherwise. For all i such that $0 \le i < \kappa$, we denote by $\pi[i]$ the state s_i and by π^i the subsequence $(s_j \mid i \le j < \kappa)$.

Infinite Trace Semantics. Let $F[f, x, x']$ be a formula as above. If $\mathcal{I}$ is an interpretation of f in the theory $\mathcal{T}$ and π is an infinite trace, then $(\mathcal{I}, \pi)$ *satisfies* F, written $(\mathcal{I}, \pi) \models F$, *iff* one of the following holds:

1. F is atomic and $\mathcal{I}[x \mapsto \pi[0](x), x' \mapsto \pi[1](x)]$ satisfies F as in FOL;
2. $F = \neg G$ and $(\mathcal{I}, \pi) \not\models G$;
3. $F = G_1 \wedge G_2$ and $(\mathcal{I}, \pi) \models G_i$ for $i = 1, 2$;
4. $F = \exists z\, G$ and $(\mathcal{I}[z \mapsto v], \pi) \models G$ for some value v for x;
5. $F = \textbf{eventually } G$ and $(\mathcal{I}, \pi^i) \models G$ for some $i \ge 0$;
6. $F = \textbf{always } G$ and $(\mathcal{I}, \pi^i) \models G$ for all $i \ge 0$.

The semantics of the propositional connectives $\vee, \rightarrow, \leftrightarrow$, and the quantifier $\forall$ can be defined by reduction to the connectives above (e.g., by defining $G_1 \vee G_2$ as $\neg(\neg G_1 \wedge \neg G_2)$, and so on). Note that $\exists z$ is a *static*, or *rigid*, quantifier: the meaning of the variable it quantifies does not change over time, i.e., from state to state in π. Uninterpreted symbols are rigid in the same sense: their meaning does not change over time.[1] Given a transition system $S = (I_S, T_S)$, the infinite trace semantics of S is the set of all pairs $(\mathcal{I}, \pi)$ of interpretations $\mathcal{I}$ in $\mathcal{T}$ and infinite traces π such that $(\mathcal{I}, \pi) \models I_S \wedge \textbf{always } T_S$. We call any such pair an *execution of S*.

Finite Trace Semantics. Given formula $F[f, x, x']$, interpretation $\mathcal{I}$ in theory $\mathcal{T}$, infinite trace π, and $n \ge 0$, $(\mathcal{I}, \pi)$ *n-satisfies* F, written $(\mathcal{I}, \pi) \models_n F$, *iff*

1. F is atomic and $\mathcal{I}[x \mapsto \pi[0](x), x' \mapsto \pi[1](x)]$ satisfies F as in FOL;

---

[1] Another way to understand the difference between rigid and non-rigid symbols is that state variables are mutable over time, whereas quantified variables, theory symbols, and uninterpreted symbols are all immutable.

2. $F = \neg G$ and $(\mathcal{I}, \pi) \not\models_n G$;
3. $F = G_1 \wedge G_2$ and $(\mathcal{I}, \pi) \models_n G_i$ for $i = 1, 2$;
4. $F = \exists z\, G$ and $(\mathcal{I}[z \mapsto v], \pi) \models_n G$ for some value v for z;
5. $F = \textbf{eventually } G$ and $(\mathcal{I}, \pi^i) \models_{n-i} G$ for some $i = 0, \ldots, n$; or
6. $F = \textbf{always } G$ and $(\mathcal{I}, \pi^i) \models_{n-i} G$ for all $i = 0, \ldots, n$.

The semantics of the propositional connectives $\vee, \rightarrow, \leftrightarrow$, and the quantifier $\forall$ is defined by reduction to the connectives above. Intuitively, n-satisfiability specifies when a formula is true over a trace's first n states. Note that this notion is well defined even when $n = 0$ regardless of whether F has free occurrences of variables from $\boldsymbol{x}'$ or not. This is true in the atomic case because the infinite trace π contains the state $\pi[1]$. The claim can be shown in the general case by a simple inductive argument.

The notion of n-satisfiability is useful in state reachability. A state satisfying a (non-temporal) state property R is reachable in a system S only if the temporal formula $\textbf{eventually } R$ is n-satisfied by an execution of S for some n. Note that the converse does not hold; R can be reachable in a system S without being n-satisfied by an execution of S.

3   The MoXI Intermediate Language

MoXI assumes a discrete and linear notion of time and adopts the trace-based semantics defined in the previous section. It builds on the SMT-LIB language, extending it with commands to represent transition systems and to specify properties or *queries*. It also standardizes a format for witnesses generated by back-end algorithms. Table 1 shows the supported SMT-LIB commands in MoXI. Since enumerated sorts are useful in modeling real-world systems, MoXI introduces a `declare-enum-sort` command. For example, (`declare-enum-sort` s ($c_1 \cdots c_n$)) declares s to be an enumerative type with (distinct) values $c_1, \ldots, c_n$.

Table 1. Supported SMT-LIB commands in MoXI

(`declare-sort` s n)
Declares s to be a sort symbol (i.e., type constructor) of arity n.
(`define-sort` s ($u_1 \cdots u_n$) τ)
Defines S as a synonym of a parametric type τ with parameters $u_1 \cdots u_n$.
(`declare-const` c σ)
Declares a constant c of sort σ.
(`define-fun` f ((x_1 σ_1) $\cdots$ (x_n σ_n)) σ t)
Defines a function f with inputs $x_1, \ldots, x_n$ (of respective sort $\sigma_1, \ldots, \sigma_n$), output sort σ, and body t.
(`set-logic` L)
Defines the model's *data logic*, i.e., the background theories of relevant data types (e.g., integers, reals, bit vectors, and so on) as well as the language of allowed logical constraints (e.g., quantifier-free, linear, etc.).

3.1   System Definition

MoXI allows the definition of a model as the composition of one or more *systems*. MoXI's system definition commands follow the SMT-LIB syntax for attribute-value pairs. Each system definition:

1. defines a *transition system* via the use of SMT formulas, imposing minimal syntactic restrictions on those formulas;
2. is parameterized by a *state signature*, a sequence of typed variables;
3. partitions *state* variables into *input*, *output*, and *local* variables;
4. can be expressed as the (a)synchronous composition of other systems.

Atomic Systems. MoXI defines an *atomic transition system* that has m inputs, n outputs, and p local variables via the command:

    (define-system S
     :input ($(i_1\ \delta_1)$ $\cdots$ $(i_m\ \delta_m)$) :output ($(o_1\ \tau_1)$ $\cdots$ $(o_n\ \tau_n)$)
     :local ($(l_1\ \sigma_1)$ $\cdots$ $(l_p\ \sigma_p)$) :init I :trans T :inv P), where

- S is the system's identifier;
- each i_j is an *input* variable of sort δ_j;
- each o_j is an *output* variable of sort τ_j;
- each l_j is a *local* variable of sort σ_j;
- each i_j, o_j, l_j denote *current-state* values;
- I, *the initial condition*, is a one-state formula over the unprimed system's variables (input, output, and local state variables) that expresses a constraint on the initial states of S;
- T, *the transition condition*, is a two-state formula over all of the system's variables (primed and unprimed) that expresses a constraint on the state transitions of S;
- P, *the invariance condition*, is a one-state formula over all of the *unprimed* system's variables that expresses a constraint on all reachable states of S.

Next-state variables are not provided explicitly but are denoted by convention by appending $'$ to the names of the current-state variables i_j, o_j, and l_j. Note that all attributes are optional but can occur at most once, and the order of the attributes is immaterial except that :input, :output, and :local must occur before :init, :trans, and :inv. The default value for a missing attribute is the empty list () for :input, :output, and :local; and true for :init, :trans, and :inv. Syntactically, the system identifier, the input, output, and local variables are SMT-LIB symbols. In contrast, the sorts δ_j, τ_j, σ_j are SMT-LIB sorts, while the formulas I, T, and P are SMT-LIB terms of type Bool.

Semantics. Let $\boldsymbol{i} = (i_1, \ldots, i_m)$, $\boldsymbol{o} = (o_1, \ldots, o_n)$, $\boldsymbol{l} = (l_1, \ldots, l_p)$, and $\boldsymbol{x} = \boldsymbol{i}, \boldsymbol{o}, \boldsymbol{l}$. A system S introduced by the **define-system** command above is a transition system whose behavior consists of all the (infinite) executions $(\mathcal{I}, \pi)$ over $\boldsymbol{x}$ such that $(\mathcal{I}, \pi) \models I[\boldsymbol{x}] \land \textbf{always}\ (P[\boldsymbol{x}] \land T[\boldsymbol{x}, \boldsymbol{x}'])$. We call $I_S = I[\boldsymbol{x}]$ *the initial state predicate* of S and $T_S = P[\boldsymbol{x}] \land T[\boldsymbol{x}, \boldsymbol{x}']$ *the transition predicate* of S.

Composite Systems. Transition systems can be defined as the synchronous[2] composition of other systems using the command:

> (define-system S :input ($(i_1 \ \delta_1) \ \cdots \ (i_m \ \delta_m)$)
> :output ($(o_1 \ \tau_1) \ \cdots \ (o_n \ \tau_n)$) :local ($(l_1 \ \sigma_1) \ \cdots \ (l_p \ \sigma_p)$)
> :subsys ($N_1 \ (S_1 \ \boldsymbol{x}_1 \ \boldsymbol{y}_1)$) $\cdots$:subsys ($N_q \ (S_q \ \boldsymbol{x}_q \ \boldsymbol{y}_q)$)
> :init I :trans T :inv P), where

- :input, :output, :local, :init, :trans, and :inv are as in atomic system definitions;
- $q > 0$ and each S_i is the name of a system other than S;
- the names $S_1 \ldots, S_q$ need not be all distinct;
- each N_i is a local synonym for S_i, with $N_1, \ldots, N_q$ distinct;
- each $\boldsymbol{x}_i$ consists of S's variables of the same sort as S_i's input;
- each $\boldsymbol{y}_i$ consists of S's local/output variables of the same sort as S_i's output;
- the directed subsystem graph rooted at S is acyclic.

Semantics. For $k = 1, \ldots, q$, let $S_k = (I_k[i_k, o_k, l_k], T_k[i_k, o_k, l_k, i'_k, o'_k, l'_k])$, with the elements of $l_1, \ldots, l_q$ all mutually distinct. Let $i = (i_1, \ldots, i_m)$, $o = (o_1, \ldots, o_n)$, $l = (l_1, \ldots, l_p), l_1, \ldots, l_q$, and $\boldsymbol{x} = i, o, l$. A composite system S introduced by the define-system command above is a transition system whose behavior consists of all the (infinite) executions $(\mathcal{I}, \pi)$ over $\boldsymbol{x}$ such that $(\mathcal{I}, \pi) \models I_S[\boldsymbol{x}] \wedge \textbf{always } T_S[\boldsymbol{x}, \boldsymbol{x}']$, where

- $I_S[\boldsymbol{x}] = I[\boldsymbol{x}] \wedge \bigwedge_{k=1,\ldots,q} I_k[\boldsymbol{x}_k, \boldsymbol{y}_k, l_k]$ and
- $T_S[\boldsymbol{x}, \boldsymbol{x}'] = P[\boldsymbol{x}] \wedge T[\boldsymbol{x}, \boldsymbol{x}'] \wedge \bigwedge_{k=1,\ldots,q} T_k[\boldsymbol{x}_k, \boldsymbol{y}_k, l_k, \boldsymbol{x}'_k, \boldsymbol{y}'_k, l'_k]$.

Sanity Requirements on I_S and T_S. Every system defined in MoXI is expected to execute forever. This is not a limitation in practice because systems meant to reach a final state can be modeled with states that cycle back to themselves and produce stuttering outputs. In such semantics, the reachability of a *deadlocked state* (i.e., a state with no successors in the transition relation) indicates the presence of an error in the system's definition. For a system definition to define a deadlock-free system, the following must hold for the variables $\boldsymbol{x} = i, o, l$ and their primed versions.

(1) Every assignment of values to the input variables i can be extended to an assignment to $\boldsymbol{x}$ that satisfies $I_S[\boldsymbol{x}]$.
(2) For every reachable state s (i.e., assignment to the variables $\boldsymbol{x}$), every assignment to the primed input variables i' can be extended to an assignment s' to $\boldsymbol{x}'$ so that s, s' satisfies $T_S[\boldsymbol{x}, \boldsymbol{x}']$.

The first restriction guarantees that the system can start. The second ensures that from any reachable state and for any new input, the system can move to another state (*and* also produce output). Given a specified background theory,

---

[2] The asynchronous composition of systems is planned for a later version of MoXI.

– A sufficient condition for (1) is the validity of the formula

$$\forall i\, \exists o\, \exists l\, I_S[i, o, l]$$

– A sufficient condition for (2) is the validity of the formula

$$\forall i\, \forall o\, \forall l\, \forall i'\, \exists o'\, \exists l'\, T_S[i, o, l, i', o', l']$$

Note that this is not a necessary condition as it needs not apply to unreachable states. Let Reachable$[i, o, l]$ denote the (possibly higher-order) formula satisfied exactly by the reachable states of S. Then, a more accurate sufficient condition for (2) above would be the validity of the formula

$$\forall i\, \forall o\, \forall l\, \forall i'\, \exists o'\, \exists l'\, \text{Reachable}[i, o, l] \Rightarrow T_S[i, o, l, i', o', l']$$

3.2   System Checking

The properties to check for a (possibly composite) defined system are specified using the following command for defining *queries* on the system's behavior.

```
(check-system S
  :input ( (i₁ δ₁) ⋯ (iₘ δₘ) ) :output ( (o₁ τ₁) ⋯ (oₙ τₙ) )
  :local ( (l₁ σ₁) ⋯ (lₚ σₚ) ) :assumption (a A ) :fairness (f F)
  :reachable (r R) :current (c C)  :query (q (g₁ ⋯ g_q) )
  :queries ( ( (q₁ (g₁,₁ ⋯ g₁,ₙ₁) ) ⋯ (q_t (g_t,₁ ⋯ g_t,ₙ_t) ) ) )), where
```

– S is the identifier of a system with m inputs, n outputs, and p local variables;
– $i = (i_1, \ldots, i_m)$ is a renaming of input variables in S of sort $\delta = (\delta_1, \ldots, \delta_m)$;
– $o = (o_1, \ldots, o_n)$ is a renaming of output variables in S of sort $\tau = (\tau_1, \ldots, \tau_n)$;
– $l = (l_1, \ldots, l_p)$ is a renaming of local variables in S of sort $\sigma = (\sigma_1, \ldots, \sigma_p)$;
– $a, r, f, c, q, q_1, \ldots, q_k$ are identifiers;
– A is a formula over i, o, l, i' expressing an *assumption* on i';
– F is a formula over i, o, l, i' expressing a *fairness condition* on i';
– R is a formula over i, o, l, i', o', l' expressing a state *reachability condition*;
  C is a formula over i, o, l expressing a state *initiality condition*;
– each g_j and $g_{j,k}$ ranges over the a, r, f, c identifiers;
– $(q\ (g_1 \cdots g_q))$ defines a query q as consisting of the formulas named by $g_1, \ldots, g_q$; the same holds for each $(q_j\ (g_{j,1} \cdots g_{j,n_j}))$.

Note that A, F, R, and C are all non-temporal formulas. Each of the attributes :assumption, :reachable, :query, and :queries can occur zero or more times. Moreover, a query can contain more than one assumption, fairness condition, and reachability condition but at most one initiality condition.

Semantics. Each query (q and each q_j) in the `check-system` command asks for the existence of a trace. The query is to be evaluated with infinite-state semantics if it includes at least one fairness condition, and finite-state semantics otherwise. Specifically, for a system S, let I_S and T_S be the initial state and transition predicates of S *modulo* the variable renamings in the `check-system` command. Let $t, u, v \geq 0$. The semantics of a query are defined as follows:

(1) A query of the form $(a_1 \cdots a_t \ r_1 \cdots r_u)$, where each a_j and r_j identify an assumption A_j and reachability condition R_j, respectively, is *satisfiable* iff

$$I_S \wedge \mathbf{always}\ T_S$$
$$\wedge\ \mathbf{always}\ (A_1 \wedge \cdots \wedge A_t)$$
$$\wedge\ \mathbf{eventually}\ R_1 \wedge \cdots \wedge \mathbf{eventually}\ R_u$$

is *n-satisfiable* in LTL for some $n \geq 0$.

(2) A query of the form $(c \ a_1 \cdots a_t \ r_1 \cdots r_u)$, where c is the initiality condition C, and each a_j and r_j identify an assumption A_j and reachability condition R_j, respectively, is *satisfiable* iff

$$C \wedge \mathbf{always}\ T_S$$
$$\wedge\ \mathbf{always}\ (A_1 \wedge \cdots \wedge A_t)$$
$$\wedge\ \mathbf{eventually}\ R_1 \wedge \cdots \wedge \mathbf{eventually}\ R_u$$

is *n-satisfiable* in LTL for some $n \geq 0$.

(3) A query of the form $(a_1 \cdots a_t \ r_1 \cdots r_u \ f_1 \cdots f_v)$, where each a_j, r_j, and f_j identify an assumption A_j, a reachability condition R_j, and a fairness condition F_j, respectively, is *satisfiable* iff

$$I_S \wedge \mathbf{always}\ T_S$$
$$\wedge\ \mathbf{always}\ (A_1 \wedge \cdots \wedge A_t)$$
$$\wedge\ \mathbf{always\ eventually}\ F_1 \wedge \cdots \wedge \mathbf{always\ eventually}\ F_v$$
$$\wedge\ \mathbf{eventually}\ R_1 \wedge \cdots \wedge \mathbf{eventually}\ R_u$$

is **satisfiable** in LTL.

(4) A query of the form $(c \ a_1 \cdots a_t \ r_1 \cdots r_u \ f_1 \cdots f_v)$, where c is the initiality condition C and each a_j, r_j, and f_j identify an assumption A_j, a reachability condition R_j, and a fairness condition F_j, respectively, is *satisfiable* iff

$$C \wedge \mathbf{always}\ T_S$$
$$\wedge\ \mathbf{always}\ (A_1 \wedge \cdots \wedge A_t)$$
$$\wedge\ \mathbf{always\ eventually}\ F_1 \wedge \cdots \wedge \mathbf{always\ eventually}\ F_v$$
$$\wedge\ \mathbf{eventually}\ R_1 \wedge \cdots \wedge \mathbf{eventually}\ R_u$$

is **satisfiable** in LTL.

Let $\mathcal{T}$ be the background theory specified for a MoXI model. For each satisfiable query in the `check-system` command, the back-end model checking algorithm is expected to produce (1) a $\mathcal{T}$-interpretation $\mathcal{I}$ of the (global) free symbols in the script; (2) a witnessing trace in $\mathcal{I}$. For each unsatisfiable query, the model checker may return a *proof certificate* for that query's unsatisfiability.

The interpretation $\mathcal{I}$ *must be the same* for all queries in the same :queries attribute. In contrast, queries in different attributes may each interpret the free symbols differently. Regardless of where it occurs, each query may have its own witnessing trace.

3.3 System Checking Response

MoXI also defines the content and format of possible responses (from the backend model checker) to a check-system command. Witness traces returned by the model checker are currently limited to *lasso* traces, that is, traces of the form pl^ω, where p and l are finite sequences of state, or *trails*. Each witness is then represented by two trails: (1) a *prefix trail* p, and (2) a *lasso trail* l. In contrast, a proof certificate for a trace represents a proof of the unsatisfiability of the query. Currently, MoXI does not specify the format of proof certificates except for requiring them to be SMT-LIB S-expressions. Figure 2 shows the format for a check-system command with input i, output o, local variable s, and queries q1, q2, q3, where q1 has a reachability condition r and fairness condition f.

```
 1 (check-system-response
 2  :verbosity full
 3  :query (q1 :result sat :model m :trace t)
 4  :query (q2 :result unsat :certificate c)
 5  :query (q3 :result unknown)    ; for timeouts and other cases
 6  :trace (t :prefix p :lasso l) ; t = pl^w
 7  :model (m M)                  ; M is an interpr. in SMT-LIB format
 8  :trail (p ((0 (i i0) (o o0) (s s0) (r r0) (f f0)); first state in p
 9             ...
10              (j (i ij) (o oj) (s sj) (r rj) (f fj)); last state in p
11           )
12         )
13  :trail (l ( ( ... ) ... ( ... ) )) ; similar to p
14  :certificate (c ... )
15 )
```

Fig. 2. Format of a model checker's response to a check-system command.

3.4 Example

As an illustrative example of a MoXI model, consider a timed switch with a single Boolean input and output where the output *switches* from its current value if the next input is true or the output has been true for at least 10 consecutive steps. Figure 3 shows a full definition of such a system in MoXI.

```
1    (set-logic QF_LIA)
2    (declare-enum-sort LightStatus (on off))
3
4    (define-system TimedSwitch :input ( (press Bool) )
5     :output ( (sig Bool) )
6     :local ( (s LightStatus) (n Int) )
7     :inv (= sig (= s on))
8     :init (and (= n 0) (= s (ite press on off)))
9     :trans (let (; transitions
10        (turn-on  (and (= s off) press' (= s' on) (= n' n)))
11        (stay-on  (and (= s on) (< n 10) (not press')
12                       (= s' on) (= n' (+ n 1))))
13        (turn-off (and (= s on) (or (>= n 10) press')
14                       (= s' off) (= n' 0)))
15        (stay-off (and (= s off) (not press') (= s' off) (= n' n)))
16        )
17       (or turn-on stay-on turn-off stay-off)
18     )
19   )
20
21   (check-system TimedSwitch :input ( (press Bool) )
22     :output ( (sig Bool) )
23     :local ( (s LightStatus) (n Int) )
24     :reachable (r1 (and press (not sig) (= s off)))
25     :query (q1 (r1))
26   )
```

Fig. 3. Example MoXI model of a timed switch with a 10-step timeout.

To start, we select the SMT-LIB logic QF-LIA, which restricts the language of formulas to quantifier-free formulas over linear integer arithmetic. Then, we declare an enumeration sort to represent the internal state of the switch. Next, we define the timed switch itself. This begins by declaring its input variable press, output variable sig, and local variables s and n to track, respectively, the switch's current state (on or off) and the number of steps the switch has been continuously set to on. In the same definition, on line 8, we define an invariant stating that sig is true exactly when the value of s is on. The initial condition, on line 9, sets n to 0 and s to the enumeration value corresponding to the initial value of press. We finish up the definition of TimedSwitch with the transition relation on lines 9–19, provided in *named transition* style,[3] where we define each possible transition separately and one of the transitions gets non-deterministically chosen each time. In our case, we have four possibilities based on the pre- and post-state of the transition as follows.

---

[3] MoXI's syntax also supports other styles, such as equational or condition-action.

turn-on: when s is off in the pre-state and **press** is true in the post-state.

stay-on: when s is on and n < 10 in the pre-state, and **press** is false in the post-state.

turn-off: when s is on in the pre-state and either n is at least 10 in the pre-state or **press** is false in the post-state.

stay-off: when s is off in the pre-state and **press** is false in the post-state.

Finally, on lines 21–26, we issue a check request on the system, asking whether it can reach a state where **press** is true, **sig** is false, and s is off. Figure 4 provides a possible response to this request. It shows a finite trace, represented as a lasso trace with an empty lasso, where the transition **turn-off** is taken immediately from an initial state (state 0) where s is on.

```
1 (check-system-response TimedSwitch
2   :query (q1 :result sat :trace w1)
3   :trace (w1 :prefix t1)
4   :trail (t1 (0 (n 0) (s on) (sig true) (press true))
5             (1 (n 0) (s off) (sig false) (press true)))
6 )
```

Fig. 4. A possible response to the **check-system** command from Fig. 3.

4 Translator Toolchain

A core application for MoXI is as an intermediate language in a toolchain integrating a high-level modeling language and a low-level representation for a model-checking back end. As Fig. 5 shows, this allows users to write models in an intuitive, high-level language while also leveraging state-of-the-art algorithms provided by model checkers that accept a low-level language as input.

Prototype Implementation. We provide a prototype Python implementation of such a toolchain, using SMV and BTOR2 as the high-level and low-level languages, respectively [37]. We selected SMV due to its ubiquity in modeling symbolic transition systems [7, 8, 13, 21, 22, 27, 28, 31, 38, 41, 44, 45, 49, 54 57] and BTOR2 due to it being the input language for the most-recent 2020 Hardware Model Checking Competition [6]. The toolchain translates SMV models to behaviorally equivalent MoXI models, and subsequently to BTOR2 models. Then, the tool translates responses from a BTOR2 model checker back to SMV-style responses via MoXI responses.

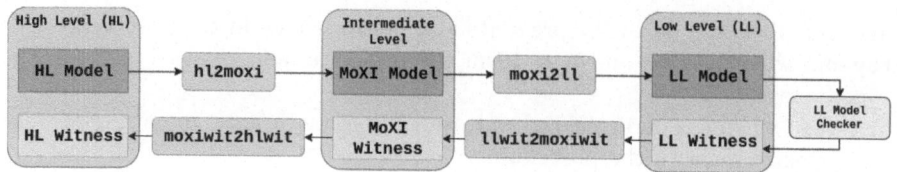

Fig. 5. The abstract translator toolchain provides translators from a high-level model to low-level model via MoXI, and back for generated witnesses. Blue denotes models and their queries, green denotes a translator, and yellow denotes a witness. Examples of high-level languages include SMV and Verilog; low-level languages include BTOR2 and AIGER. All blue boxes are behaviorally equivalent, as are all yellow boxes. (Color figure online)

Experimental Evaluation. To evaluate the effectiveness of our toolchain, we ran it over the set of 960 `QF_BV` and `QF_ABV` benchmarks provided with the NUXMV public release and compared the performance of a variety of back-end model checkers: AVR [30] (`hwmcc20 GitHub branch`), NUXMV [12] (`version 2.0`—latest public release), and PONO [40] (`commit #b243cef`—latest development head).

We ran two experiments: the first ran each solver using IC3-based [9] algorithms and the second ran each solver using a portfolio approach of BMC, K-Induction, and IC3-based algorithms. In the latter case, we collected the best time among the three techniques. For each experiment, we ran NUXMV directly on the SMV benchmarks, i.e., SMV → NUXMV, and ran the other two solvers on BTOR2 generated from the toolchain presented above, i.e., SMV → MoXI → BTOR2 → AVR/PONO and back. Here are some details about each tool.

- AVR (Yices2 [23] as the back-end SMT solver): BMC [5], K-induction based on [52], IC3 based on [29].
- NUXMV (MathSAT5 [16] as the back-end SMT solver): BMC [5], K-induction based on [25], and IC3 based on [15].
- PONO (Boolector [46] and MathSAT5 [16] as back-end SMT solvers): BMC [5], K-induction based on [25], IC3 based on [15].

In all cases, no discrepancies were found, i.e., no two model checkers returned conflicting *safe* and *unsafe* results, and all generated BTOR2 was well-formed according to the reference checker CATBTOR [46].

As the top of Fig. 6 shows, the three model checkers NUXMV, AVR, and PONO complement each other when using their IC3-based algorithms; each solver performs similarly, but the virtual-best outperforms each individual by a noticeable margin. Similarly, the bottom of Fig. 6 shows that while the model checkers perform similarly using a portfolio approach, they again complement each other, as shown by the virtual-best outperforming each model checker individually. Importantly, our initial, non-optimized translation of SMV-language benchmarks through MoXI does not inhibit the model checking performance of either AVR

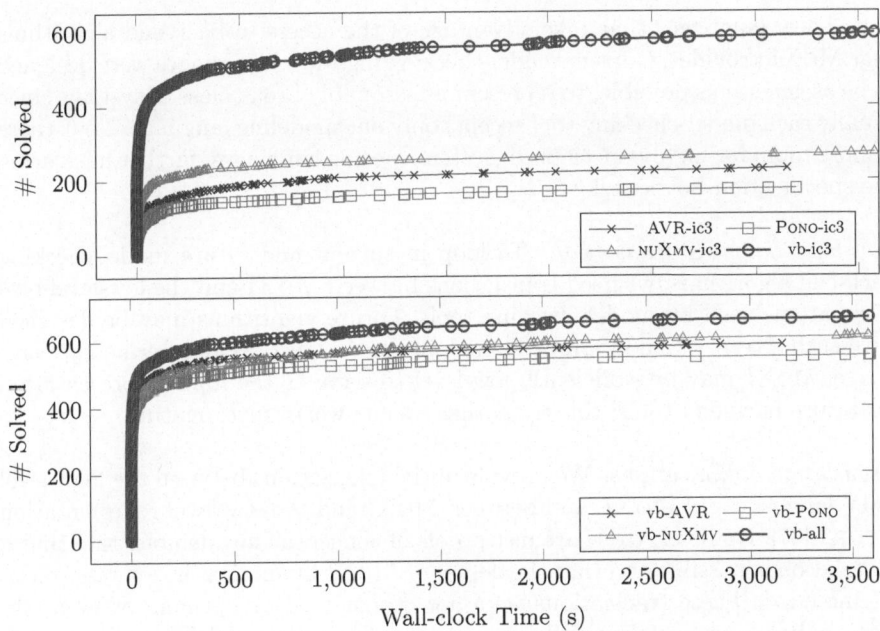

Fig. 6. Performance comparison on SMV-language benchmark queries using IC3 (Top) and portfolio approach (Bottom) across three different model checkers: SMV benchmarks → NUXMV, a translation of SMV benchmarks → MoXI → BTOR2 → AVR (and back), and a translation of SMV benchmarks → MoXI → BTOR2 → PONO (and back). The vb-* represents the virtual-best back-end solver for each model checker. The portfolio approach represents the best time using BMC, k-induction, and IC3 algorithms. Wall-clock time for the non-NUXMV plots include translation time.

or PONO, compared to using NUXMV directly on the same SMV-language benchmarks distributed with that tool.

5 Future Directions for Symbolic Model Checking

Creating an international standard for symbolic model checking changes the landscape of research next steps, from low-hanging fruit to future challenges. There is even more reason to continue and expand on current research directions aimed at breaking down the barriers to the wider adoption of symbolic model checking by addressing (1) the specification bottleneck; and (2) the state-space explosion problem. Having laid a foundation for standardization, we hope that the community can converge on MoXI as a common intermediate language and leverage this standard in pursing research directions that make advances in areas (1) and (2).

High-level language translation. In addition to creating translations for existing high-level modeling languages to/from MoXI, there is now an opportunity to create new languages that take advantage of the access to back-end algorithms that MoXI provides. Current model-checking high-level languages were designed to be as general as possible, to represent a reasonably broad class of systems since usually each model-checking tool accepts only one modeling language. Now, there is more room for highly-specialized, system-specific languages, further mitigating the specification bottleneck.

Low-level language translation. To loop in current and future model-checking back-end algorithms, we need translations between MoXI and the low-level representations used by model-checking tools. Future algorithms may be designed with such a translation in mind. Though it was designed as an intermediate language, MoXI may be sufficiently low-level to serve as the input representation for future back-end tools; this is another avenue worth investigating.

Translation optimizations. We provide initial translations between the high-level SMV language and MoXI, and between MoXI and the low-level representation BTOR2 [37]. However, these are just proofs of concept. They demonstrate that a translation is possible and that the design of MoXI maintains the expressiveness we intended. These translations were not designed to be optimal, or even efficient, only correct and, hopefully, transparent. Our initial toolchain encodes LTL specifications by using PANDA [50] to translate them into SMV, then translating the PANDA-generated SMV models to MoXI; exploring direct LTL-to-MoXI translations could improve model-checking performance. Section 4 demonstrates that model checking through MoXI does not exacerbate the state-space explosion problem, but there is certainly ample room for improvement. We expect a progression of future papers creating increasingly performant translations, improving upon our translations and those contributed by others (see above).

Proofs and benchmarks. While we have extensively investigated the correctness of our initial MoXI translations, we have yet to prove them correct formally, for instance by using an interactive theorem prover. We believe it is possible, though challenging, to state the semantic equivalence of representations in a high- or low-level language and MoXI as theorems, prove them correct using a theorem prover like PVS, and then generate verified translators using a tool like PVS2C [18,26].[4] Lower-hanging fruit involves creating, packaging, and releasing benchmarks for MoXI translations that help others check their new translations and serve as performance checkpoints for translation tools.

Extensions. Though we have initially addressed hardware model checking of finite-state systems, MoXI is extensible by design. Future research directions include further extending MoXI representations to infinite-state model checking, investigating efficient representations for highly-expressive high-level modeling

---

[4] Thanks to Laura Gamboa Guzman, Katherine Kosaian, and Yi Lin for their initial investigations into this possibility.

languages, and even exploring the uses of MoXI for applications in software model checking.[5]

References

1. Biere, A.: The AIGER and-inverter graph (AIG) format version 20071012. http:// fmv.jku.at/aiger/FORMAT. Accessed 25 July 2016
2. Biere, A.: AIGER 1.9 and beyond. http://fmv.jku.at/hwmcc11/beyond1.pdf. Accessed 25 July 2016
3. Barrett, C., Stump, A., Tinelli, C.: The SMT-LIB standard: version 2.0. In: Gupta, A., Kroening, D. (eds.) Proceedings of the 8th International Workshop on Satisfiability Modulo Theories, Edinburgh, UK (2010)
4. Beer, I., Ben-David, S., Eisner, C., Landver, A.: RuleBase: an industry-oriented formal verification tool. In: Design Automation Conference, pp. 655–660. IEEE (1996)
5. Biere, A., Cimatti, A., Clarke, E.M., Zhu, Y: Symbolic Model Checking without BDDs. In: Cleaveland, W.R. (eds) Tools and Algorithms for the Construction and Analysis of Systems, TACAS 1999. Lecture Notes in Computer Science, vol. 1579. Springer, Berlin, Heidelberg (1999). https://doi.org/10.1007/3-540-49059-0_14
6. Biere, A., Froleyks, N., Preiner, M.: Hardware model checking competition (HWMCC) (2020). https://fmv.jku.at/hwmcc20/index.html
7. Bozzano, M., et al.: Formal design and safety analysis of AIR6110 wheel brake system. In: Kroening, D., Păsăreanu, C.S. (eds.) Computer Aided Verification, pp. 518–535. Springer, Cham (2015). https://doi.org/10.1007/978-3-319-21690-4_36
8. Bozzano, M., Cimatti, A., Katoen, J.P., Nguyen, V.Y., Noll, T., Roveri, M.: The COMPASS approach: correctness, modelling and performability of aerospace systems. In: Buth, B., Rabe, G., Seyfarth, T. (eds.) Computer Safety, Reliability, and Security, pp. 173–186. Springer, Berlin, Heidelberg (2009). https://doi.org/10.1007/978-3-642-04468-7_15
9. Bradley, A.R.: SAT-based model checking without unrolling. In: VMCAI, pp. 70–87 (2011)
10. Brayton, R., Mishchenko, A.: ABC: an academic industrial-strength verification tool. In: Touili, T., Cook, B., Jackson, P. (eds.) Computer Aided Verification, pp. 24–40. Springer, Berlin, Heidelberg (2010). https://doi.org/10.1007/978-3-642-14295-6_5
11. Caspi, P., Pilaud, D., Halbwachs, N., Plaice, J.: LUSTRE: a declarative language for programming synchronous systems. In: Proceedings 14th Annual ACM Symposium on Principles of Programming Languages, pp. 178–188 (1987)
12. Cavada, R. et al.: The NUXMV symbolic model checker. In: Biere, A., Bloem, R. (eds.) Proceedings 26th International Conference on Computer Aided Verification, CAV 2014. Lecture Notes in Computer Science, vol. 8559, pp. 334–342. Springer, Cham. (2014). https://doi.org/10.1007/978-3-319-08867-9_22
13. Choi, Y., Heimdahl, M.: Model checking software requirement specifications using domain reduction abstraction. In: IEEE ASE, pp. 314–317 (2003)
14. Cimatti, A. et al.: NuSMV 2: an opensource tool for symbolic model checking. In: CAV 2002, Proceedings 14th International Conference. LNCS, vol. 2404, pp. 359–364. Springer, Berlin, Heidelberg (2002). https://doi.org/10.1007/3-540-45657-0_29

---

[5] Thanks to Dirk Beyer for initial ideas in this direction.

15. Cimatti, A., Griggio, A., Mover, S., Tonetta, S.: IC3 modulo theories via implicit predicate abstraction. In: Tools and Algorithms for the Construction and Analysis of Systems: 20th International Conference, TACAS 2014, Held as Part of the European Joint Conferences on Theory and Practice of Software, ETAPS 2014, Grenoble, France, April 5–13, 2014. Proceedings 20, pp. 46–61. Springer (2014). https://doi.org/10.1007/978-3-642-54862-8_4

16. Cimatti, A., Griggio, A., Schaafsma, B.J., Sebastiani, R.: The MathSAT5 SMT solver. In: TACAS, pp. 93–107 (2013)

17. Colaço, J.L., Pagano, B., Pouzet, M.: Scade 6: a formal language for embedded critical software development. In: 2017 International Symposium on Theoretical Aspects of Software Engineering (TASE), pp. 1–11. IEEE (2017)

18. Beyer, D., Zufferey, D. (eds.): Verification, Model Checking, and Abstract Interpretation: 21st International Conference, VMCAI 2020, New Orleans, LA, USA, January 16–21, 2020, Proceedings. Springer, Cham (2020). https://doi.org/10.1007/978-3-030-39322-9

19. Simulink Documentation: Simulation and model-based design (2020). https://www.mathworks.com/products/simulink.html

20. SCADE Documentation: Ansys SCADE suite (2023). https://www.ansys.com/products/embedded-software/ansys-scade-suite

21. Dureja, R., Rozier, E.W.D., Rozier, K.Y.: A case study in safety, security, and availability of wireless-enabled aircraft communication networks. In: Proceedings of the 17th AIAA Aviation Technology, Integration, and Operations Conference (AVIATION). American Institute of Aeronautics and Astronautics (2017). https://doi.org/10.2514/6.2017-3112

22. Dureja, R., Rozier, K.Y.: FuseIC3: an algorithm for checking large design spaces. In: Proceedings of Formal Methods in Computer-Aided Design (FMCAD). IEEE/ACM, Vienna, Austria (2017)

23. Dutertre, B.: Yices 2.2. In: International Conference on Computer Aided Verification, pp. 737–744. Springer, Cham (2014). https://doi.org/10.1007/978-3-319-08867-9_49

24. Een, N., Mishchenko, A., Brayton, R.: Efficient implementation of property directed reachability. In: FMCAD, pp. 125–134 (2011)

25. Eén, N., Sörensson, N.: Temporal induction by incremental SAT solving. Electr. Notes Theoret. Comput. Sci. **89**(4), 543–560 (2003)

26. Férey, G., Shankar, N.: Code generation using a formal model of reference counting. In: Rayadurgam, S., Tkachuk, O. (eds.) NASA Formal Methods, pp. 150–165. Springer, Cham (2016). https://doi.org/10.1007/978-3-319-40648-0_12

27. Gan, X., Dubrovin, J., Heljanko, K.: A symbolic model checking approach to verifying satellite onboard software. Sci. Comput. Program. **82**, 44–55 (2013). http://dx.doi.org/10.1016/j.scico.2013.03.005

28. Chaudhuri, S., Farzan, A. (eds.): Computer Aided Verification: 28th International Conference, CAV 2016, Toronto, ON, Canada, July 17-23, 2016, Proceedings, Part II, pp. 1–541. Springer, Cham (2016). https://doi.org/10.1007/978-3-319-41528-4

29. Goel, A., Sakallah, K.: Model checking of Verilog RTL using IC3 with syntax-guided abstraction. In: NASA Formal Methods: 11th International Symposium, NFM 2019, Houston, TX, USA, May 7–9, 2019, Proceedings 11, pp. 166–185. Springer, Cham (2019). https://doi.org/10.1007/978-3-030-20652-9

30. Goel, A., Sakallah, K.: AVR: abstractly verifying reachability. In: Biere, A., Parker, D. (eds.) Tools and Algorithms for the Construction and Analysis of Systems, pp. 413–422. Springer, Cham (2020). https://doi.org/10.1007/978-3-030-45190-5_23

31. Gribaudo, M., Horváth, A., Bobbio, A., Tronci, E., Ciancamerla, E., Minichino, M.: Model-checking based on fluid petri nets for the temperature control system of the ICARO co-generative plant. In: Anderson, S., Felici, M., Bologna, S. (eds.) Computer Safety, Reliability and Security, pp. 273–283. Springer, Berlin, Heidelberg (2002). https://doi.org/10.1007/3-540-45732-1_27

32. Holzmann, G.J.: The SPIN Model Checker: Primer and Reference Manual. Addison-Wesley (2003)

33. IEEE: IEEE standard multivalue logic system for VHDL model interoperability (Std_logic_1164) In: IEEE Std 1164-1993, pp. 1–24 (1993). https://doi.org/10.1109/IEEESTD.1993.115571

34. IEEE: IEEE standard for Verilog hardware description language (2005)

35. IEEE: IEEE standard for VHDL language reference manual (2019)

36. Jahier, E., Raymond, P., Halbwachs, N.: The LUSTRE V6 Reference Manual. Verimag, Grenoble (2016)

37. Johannsen, C., et al.: Symbolic model-checking intermediate-language tool suite. In: Proceedings of 36th International Conference on Computer Aided Verification (CAV). LNCS, Springer (2024)

38. Lahtinen, J., Valkonen, J., Björkman, K., Frits, J., Niemelä, I., Heljanko, K.: Model checking of safety-critical software in the nuclear engineering domain. Reliab. Eng. Syst. Saf. **105**, 104–113 (2012)

39. Lamport, L.: Specifying Systems: The TLA+ Language and Tools for Hardware and Software Engineers. Addison-Wesley (2002)

40. Mann, M., et al.: Pono: a flexible and extensible SMT-based model checker. In: Silva, A., Leino, K.R.M. (eds.) Computer Aided Verification, pp. 461–474. Springer, Cham (2021). https://doi.org/10.1007/978-3-030-81688-9_22

41. Mattarei, C., Cimatti, A., Gario, M., Tonetta, S., Rozier, K.Y.: Comparing different functional allocations in automated air traffic control design. In: Proceedings of Formal Methods in Computer-Aided Design (FMCAD 2015). IEEE/ACM, Austin, Texas, U.S.A (2015)

42. McMillan, K.: The SMV language. Technical report, Cadence Berkeley Lab (1999)

43. McMillan, K.: Symbolic Model Checking. Kluwer Academic Publishers (1993)

44. Miller, S.: Will this be formal? In: TPHOLs 5170, pp. 6–11. Springer (2008). http://dx.doi.org/10.1007/978-3-540-71067-7_2

45. Miller, S.P., Tribble, A.C., Whalen, M.W., Per, M., Heimdahl, E.: Proving the shalls. STTT **8**(4–5), 303–319 (2006)

46. Niemetz, A., Preiner, M., Wolf, C., Biere, A.: Btor2, BtorMC, and Boolector 3.0. In: Proceedings 30th International Conference on Computer Aided Verification. Lecture Notes in Computer Science, vol. 10981, pp. 587–595. Springer, Cham (2018). https://doi.org/10.1007/978-3-319-96145-3_32

47. The nuXmv model checker (2015). available at https://nuxmv.fbk.eu/

48. Cavada, R., et al.: NuSMV 2.4 user manual. Technical report, CMU/ITC-IRST (2005)

49. Raimondi, F., Lomuscio, A., Sergot, M.J.: Towards model checking interpreted systems. In: FAABS 02, LNAI 2699, pp. 115–125. Springer, Cham (2002). https://doi.org/10.1145/860575.86079

50. Rozier, K.Y., Vardi, M.Y.: A multi-encoding approach for LTL symbolic satisfiability checking. In: 17th International Symposium on Formal Methods (FM2011). Lecture Notes in Computer Science (LNCS), vol. 6664, pp. 417–431. Springer, Verlag (2011). https://doi.org/10.1007/978-3-642-21437-0_31

51. Rozier, K.Y.: Specification: the biggest bottleneck in formal methods and auton-omy. In: Blazy, S., Chechik, M. (eds.) Verified Software. Theories, Tools, and Experiments, pp. 8–26. Springer, Cham (2016). https://doi.org/10.1007/978-3-319-48869-1_2

52. Sheeran, M., Singh, S., Stålmarck, G.: Checking safety properties using induc-tion and a SAT-solver. In: Hunt, W.A., Johnson, S.D. (eds.) Formal Methods in Computer-Aided Design, pp. 127–144. Springer, Berlin, Heidelberg (2000). https://doi.org/10.1007/3-540-40922-X_8

53. SMTLib. https://smtlib.cs.uiowa.edu/

54. Tribble, A., Miller, S.: Software safety analysis of a flight management system vertical navigation function-a status report. In: DASC, vol. 1, p. 1.B.1-1.1-9 (2003)

55. Yoo, J., Jee, E., Cha, S.: Formal modeling and verification of safety-critical soft-ware. Softw. IEEE **26**(3), 42–49 (2009)

56. Zhao, Y., Rozier, K.Y.: Formal specification and verification of a coordination protocol for an automated air traffic control system. In: Proceedings of the 12th International Workshop on Automated Verification of Critical Systems (AVoCS 2012). Electronic Communications of the EASST, vol. 53, pp. 337–353. European Association of Software Science and Technology (2012)

57. Zhao, Y., Rozier, K.Y.: Formal specification and verification of a coordination protocol for an automated air traffic control system. Sci. Comput. Program. J. **96**(3), 337–353 (2014)

Model Checking

Synchronisation in Language-Level Symmetry Reduction for Probabilistic Model Checking

Ivaylo Valkov[1]([⊠])(iD), Alastair F. Donaldson[2]([⊠])(iD), and Alice Miller[1]([⊠])(iD)

[1] University of Glasgow, Glasgow, UK
2064491v@student.gla.ac.uk, alice.miller@glasgow.ac.uk
[2] Imperial College London, London, UK
alastair.donaldson@imperial.ac.uk

Abstract. The *generic representatives* (or *counter abstraction*) approach has been shown to be an effective symmetry reduction method for model checking. This method was extended to a probabilistic setting via a specialised language, *Symmetric Probabilistic Specification Language* (SPSL) and an associated tool, GRIP, for use with the PRISM model checker. However, SPSL does not support synchronisation-based communication, making this method inapplicable to systems that require synchronisation. We show how synchronisation can be added to SPSL, and develop new counter abstraction translation rules for synchronous statements. We extend GRIP accordingly and demonstrate the feasibility and effectiveness of the new abstraction rules via a range of examples. This extends the applicability of the generic representatives technique to the wide class of probabilistic systems that rely on synchronisation. Experimental results show that our approach works well for systems that are composed of a large number of simple symmetric modules that feature a small amount of synchronisation-based communication.

Keywords: Probabilistic model checking · Symmetry reduction · Generic representatives · Counter abstraction · Synchronisation · PRISM

1 Introduction

Model checking [8,9,26] is an automatic technique for verifying hardware and software systems by checking temporal logic properties against a finite state model of a system. Explicit state model checkers [21] such as SPIN [20] store each state individually, whereas symbolic model checkers such NuSMV [7] use a symbolic representation of states, typically through the use of Binary Decision Diagrams (BDDs) [3,4]. Probabilistic model checkers, such as PRISM [24] and Storm [19] incorporate probabilities and quantitative aspects into the (symbolic) verification process by using Multi-terminal Binary Decision Diagrams (MTB-DDs) [10] to store vectors of probabilities.

© The Author(s), under exclusive license to Springer Nature Switzerland AG 2025
T. Neele and A. Wijs (Eds.): SPIN 2024, LNCS 14624, pp. 49–66, 2025.
https://doi.org/10.1007/978-3-031-66149-5_3

Model checking suffers from the so-called *state-space explosion problem*—the number of states increases exponentially with the number of components in a system. The symbolic approach goes some way to address this issue [5]. However, in many cases replicated components in the system under analysis can lead to large portions of the state-space that are symmetric, and a symbolic representation does nothing to avoid redundancy in analysis arising from this symmetry.

Symmetry reduction [11,16,25,28] is a technique that was originally introduced for explicit state model checking to combat state-space explosion arising from this kind of replication of components. Symmetries of the system are used to partition the state-space into equivalence classes. The model checker then only needs to explore one representative state from each equivalence class. The construction of the equivalence classes is done by identifying a suitable relation known as the *orbit relation*.

Working with the orbit relation can be challenging—especially in the case of symbolic model checking where it has been shown that a BDD encoding of the orbit relation has size exponential in the number of replicated components [11]. An alternative approach using *generic representatives* [16,17], also known as *counter abstraction*, allows symmetry reduction to be applied without the construction of the orbit relation. A system specification is translated into a reduced form known as *generic form*. In the generic form, the full set of individual local variables is replaced by a much smaller set of variables called *counters*, which record the number of components in each local state. Both the reduced specification and resulting model can be significantly smaller than the original.

Symmetry reduction tools have been developed for a variety of model checkers [2,14,18]. In the probabilistic context there have been two notable tools: PRISM-symm [23] and GRIP (Generic Representatives In PRISM) [12,15]. PRISM-symm uses an efficient algorithm for the construction of quotient models from an original, non-reduced model. Property checking on the reduced model can then be performed more efficiently in comparison to property checking on the non-reduced model. However, this approach depends on it being feasible to construct the non-reduced model in the first place. In contrast, the GRIP tool is based on an extension of the generic representatives approach to the probabilistic setting. While PRISM-symm can work well for model specifications that consist of a small number of complex modules, GRIP excels in the context of a large number of relatively simple modules.

The extension of generic representatives to a probabilistic setting on which GRIP is based leverages a specialised language, *Symmetric Probabilistic Specification Language* (SPSL) [13]. This allows for the specification of a probabilistic system comprising multiple communicating modules in such a way that the applicability of the generic representatives technique is guaranteed. An algorithm for direct translation of SPSL specifications of symmetric multi-module probabilistic systems into generic form is also provided [13].

A major limitation of SPSL and the associated GRIP tool is that they only support symmetric systems in which modules communicate with one another through the use of shared variables. An alternative, widely-used communication mechanism involves *synchronisation,* where when multiple modules are ready to take a particular named action they all execute a statement related to this action simultaneously, in a synchronous fashion. This limitation means that there is a large class of systems to which SPSL and GRIP cannot be applied. This problem is more than just a tooling limitation: as we explain in this paper, the problem of how to encode inter-module synchronisation using generic representatives is difficult and—to our knowledge—has not been studied before.

In this paper we show how SPSL and the associated translation algorithm [13] can be extended to include synchronisation. To allow experimenting with the feasibility of the translation in practice, we have extended the GRIP tool to use our new translation method, and present experimental results comparing our updated version of GRIP with PRISM-symm (which already supports synchronisation) on three case studies that rely on inter-module synchronisation.

The experimental results show that our approach enables the symmetry reduction of specifications dependant on synchronisation but comes with an additional overhead for each synchronisation instance. We conclude that the technique works well for systems composed of a large number of simple symmetric modules that feature a small amount of synchronisation-based communication.

2   Background

2.1   Symmetry Reduction via Generic Representatives

The generic representatives approach [16,17], also known as *counter abstraction,* allows symmetry reduction to be applied without the construction of an orbit relation. Specifically, this process involves replacing a specification in which multiple symmetric modules[1] are each individually represented by a single *generic module.* The variables of the generic module represent the number of symmetric modules at a given local state.

We illustrate the generic representatives approach using an example. Consider a mutual exclusion algorithm for six identical modules, each with three local states: neutral (N), trying (T) and critical (C). The global states (N, N, N, T, T, C), (N, T, T, N, N, C) and (C, N, N, T, N, T) are symmetrically equivalent and have generic representative $(3N, 2T, 1C)$. A generic representative indicates how many modules are in each local state, without referring to individual modules. In our example, the generic representative $(3N, 2T, 1C)$ merely records that there are three modules in the N state, two modules in the T state and one module in the C state, without keeping track of which particular module is in each state.

The idea of the generic representatives approach is to rewrite a specification initially expressed as multiple individual symmetric modules into one that is

---

[1] Throughout the paper we use the term "module" to mean what is often called a "process" in the model checking literature. This is because the implementation of our ideas is in the context of the PRISM model checker, which uses the term "module" for this concept.

based on counter variables. The resulting specification represents the original symmetric modules via a *single* module that uses the counter variables to keep track of the number of original symmetric modules that are in each state. This has the effect of exploiting symmetry at the source code level. As a result, there is no need to do symmetry reduction when actually model checking, and so a symbolic approach can be directly applied. This avoids the need to construct a BDD for the orbit relation (which, as discussed above, is prohibitively expensive). It also avoids the need to build an unreduced model and then apply symmetry reduction to it (which is the approach taken by PRISM-symm), allowing the verification of systems for which constructing an unreduced model in the first place is intractable.

2.2   GRIP: Generic Representatives in PRISM

The generic representatives approach has been extended to probabilistic model checking [13,15]. A language, *Symmetric Probabilistic Specification Language* (SPSL), was introduced which allowed for the specification of probabilistic systems in such a way that the applicability of the technique is guaranteed. SPSL specifications can be directly translated into generic form using a defined algorithm.

This translation has been implemented for the probabilistic model checker PRISM and its modelling language [24] via the as the GRIP (Generic Representatives in PRISM) tool. GRIP supports specifications that are defined in *Symmetric PRISM* (henceforth SP). This is a subset of the PRISM modelling language that is analogous to SPSL. Specifications are reduced using a translation corresponding to the SPSL translation rules. Figure 1 shows the structure of the workflow of GRIP (before we applied our modifications). When the source code for GRIP is compiled, an abstract syntax tree representation of the model is created. This is then translated to the reduced specification.

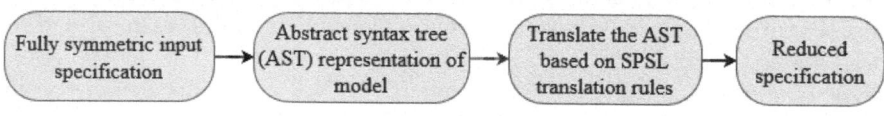

Fig. 1. GRIP workflow.

Symmetric PRISM (SP). Input specifications for GRIP must be written in SP for the translation process to be applicable. An example input specification for GRIP can be seen in Listing 1. Note that, as is required by PRISM, the first line denotes the model type, which in this case is mdp—i.e. a Markov Decision Process (MDP). Other model types include dtmc for a Discrete Time Markov Chain (DTMC) and ctmc for a Continuous Time Markov Chain (CTMC). All model types are described in full in [24].

Consider the following simple model. Two devices call heads or tails for the flip of a coin. They make their decisions at random and with equal probability, at which point they terminate (we do not model any consequence of the coin toss). All devices start at a state 0 (*initial state*), and can move to state 1 (*chosen to call heads*) or state 2 (*chosen to call tails*). Once the devices have both reached state 1 or state 2 they can move to state 3 (*end state*). An SP specification for this system is shown in Listing 1. Note that si=j is a guard that will be true if and only if device i is currently in state j. The model is defined by specifying one concrete module, device1, and then an additional module, device2, by *renaming* device1.

```
1  dtmc
2
3  module device1
4    s1 : [0..3] init 0;
5
6    []  s1=0 -> 0.5 : (s1'=1) + 0.5 : (s1'=2);
7    []  s1=1 -> (s1'=3);
8    []  s1=2 -> (s1'=3);
9    []  s1=3 -> (s1'=3);
10
11 endmodule
12 module device2 = device1[s1=s2,s2=s1] endmodule
```

Listing 1. Example: coin toss model.

The core syntax of SPSL is shown in Table 1. The table is based on the original syntax present in [13] with updates (highlighted in red) made to support the translation of synchronised statements. The updates allow synchronisation labels to be optionally attached to statements. Statements with the same label are executed simultaneously, i.e. all symmetric modules must be in a state that satisfies the guard of at least one such statement, and the updates of those statements get executed simultaneously. The basic structure of an SPSL specification otherwise remains unchanged. A detailed explanation of SP and example specifications can be found in [27].

Translation Process. We describe how a model specified in SP is translated into generic form. First we consider how the local variables for each individual module are replaced by counter variables. We then consider the translation of transitions between states of the original specification to transitions between states defined as values of the counter variables.

An abstract syntax tree is constructed from the original specification based on the SP grammar. As shown in Fig. 2, a walk is then performed on this tree and each element is translated into the reduced specification. The model type (i.e. *DTMC, CTMC, MDP*, etc.) of the two specifications is the same. Similarly, the global variables and the non-symmetric module declarations are directly copied from the original model.

Translating the symmetric modules into a single generic module is more complicated. The local variables are substituted by counter variables, one for each state a symmetric module can be in. Each counter variable keeps track of

Table 1. Syntax of Symmetric Probabilistic Specification Language (SPSL). PCTL-specific syntax is omitted. Updates shown in red.

$$
\begin{aligned}
\text{specification} &::= \text{global-variables}^? \text{ module}^+ \\
\text{global-variables} &::= \texttt{globals } \{ \text{ var-decl}^+ \} \\
\text{module} &::= \texttt{module } M[\text{number}]\{ \text{ var-decl}^* \text{ statement}(M)^+ \} \\
\text{var-decl} &::= \text{name} : \texttt{type init constant} \\
\text{type} &::= [\text{number..number}] \mid \texttt{bool} \\
\text{constant} &::= \texttt{true} \mid \texttt{false} \mid \text{number} \\
\text{statement}(M) &::= [] \text{ expr}(M_i) \rightarrow \text{stoch-update}(M) \\
&\quad\mid\ [\text{name}] \text{ expr}(M_i) \rightarrow \text{update}(M) \\
\text{stoch-update}(M) &::= \text{expr}(M_i):\text{update}(M) + \ldots + \text{expr}(M_i):\text{update}(M) \\
\text{update}(M) &::= \texttt{skip} \mid (\text{name} := \text{expr}(M_i)) \| \ldots \| (\text{name} := \text{expr}(M_i)) \\
\text{symm-expr} &::= \text{constant} \mid \text{global-name} \\
&\quad\mid\ \bigcirc_{1 \leq j \leq \#N} \text{loc-expr}(N)_j \ (\text{for some module type } N) \\
&\quad\mid\ \text{symm-expr} \bowtie \text{symm-expr} \mid \neg\text{symm-expr} \mid (\text{symm-expr}) \\
\text{loc-expr}(M) &::= \text{constant} \mid \text{local-name} \mid \text{loc-expr}(M) \bowtie \text{loc-expr}(M) \\
&\quad\mid\ \neg\text{loc-expr}(M) \mid (\text{loc-expr}(M)) \\
\text{expr}(M_i) &::= \text{loc-expr}(M)_i \mid \text{symm-expr} \mid \bigcirc_{1 \leq j \neq i \leq \#M} \text{loc-expr}(M)_j \\
&\quad\mid\ \text{expr}(M_i) \bowtie \text{expr}(M_i) \mid \neg\text{expr}(M_i) \mid (\text{expr}(M_i))
\end{aligned}
$$

how many symmetric modules are in the state associated with it. Each transition statement of the original symmetric module is translated into one or more reduced statements. These update the counter variables according to the original statement. Listing 2 shows the output produced by GRIP based on the model specification from Listing 1.

```
1  probabilistic
2
3  module generic_process
4    no_0 : [0..2] init 2;   // No modules in state (0)
5    no_1 : [0..2] init 0;   // No modules in state (1)
6    no_2 : [0..2] init 0;   // No modules in state (2)
7    no_3 : [0..2] init 0;   // No modules in state (3)
8
9    [] (no_0>0) -> 0.5:(no_0'=no_0-1)&(no_1'=min(no_1+1,2))
10              + 0.5:(no_0'=no_0-1)&(no_2'=min(no_2+1,2));
11   [] (no_0>1) -> 0.5:(no_0'=no_0-1)&(no_1'=min(no_1+1,2))
12              + 0.5:(no_0'=no_0-1)&(no_2'=min(no_2+1,2));
13   [] (no_1>0) -> (no_1'=no_1-1)&(no_3'=min(no_3+1,2));
14   [] (no_1>1) -> (no_1'=no_1-1)&(no_3'=min(no_3+1,2));
15   [] (no_2>0) -> (no_2'=no_2-1)&(no_3'=min(no_3+1,2));
16   [] (no_2>1) -> (no_2'=no_2-1)&(no_3'=min(no_3+1,2));
17   [] (no_3>0) -> true;
18   [] (no_3>1) -> true;
19 endmodule
```

Listing 2. Example output specification for a model of a coin tossing scenario. Output is generated by GRIP from the input specification shown in Listing 1.

Lines 4 to 7 declare the counter variables replacing the local variables of the eight symmetric modules. Lines 9 to 18 are the translated transition statements. Note that each transition's guard checks the number of symmetric modules in a state associated with a particular counter variable. The update denotes the transfer of one module from one state to another by incrementing/decrementing the associated counter variables.

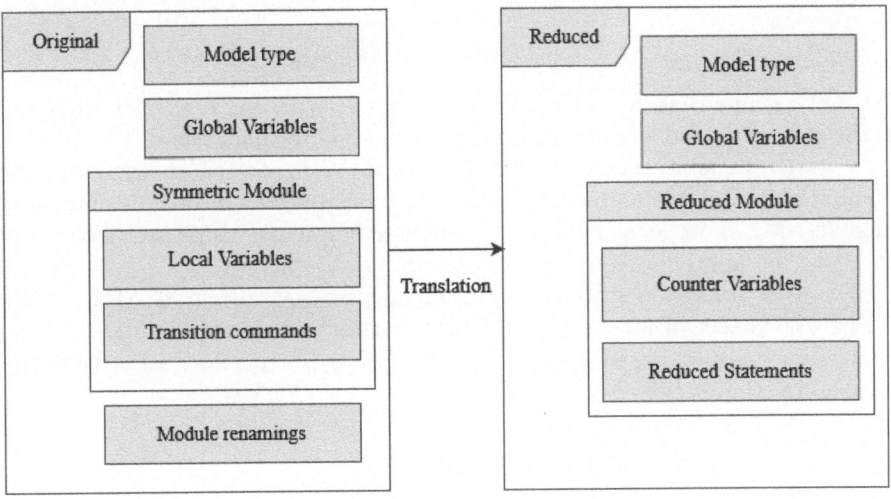

Fig. 2. Visualisation of the translation steps of the SPSL symmetry reduction algorithm. Entities with the same name in the two specifications are direct copies. We include a family of symmetric modules but no asymmetric modules (for simplicity).

For MDPs there is a one-to-one relationship between statements and translated statements. In the case of DTMCs however (as for this example) each statement is translated into a number of reduced statements (one for each symmetric module present in the original specification). This is necessary to correctly model the fact that a counter variable with a higher value is more likely to change in the next transition.

3   Synchronisation and Generic Representatives

Our goal is to add synchronisation to both the grammar and translation rules of SPSL. To do this we first introduce some notation (Sect. 3.1). We then give some initial intuition towards the development of our new translation rules for SPSL via an example in SP (Sect. 3.2). The new translation rules for SPSL are defined in Sect. 3.3.

3.1   Notation

For ease of presentation We assume that our model is for a single family of symmetric modules $M = \{M_1, M_2, \ldots, M_m\}$ that have access to a set of shared global variables.

If the set of states of M is $S(M)$, the set of local states of each M_i is $S(M_i)$, and the set of realisations of the global variables is G, then $S(M) = (\prod_{1 \leq i \leq m} S(M_i)) \times G$. Every state is a tuple of the form

$$(s_1, s_2 \ldots s_m, g)$$

We may assume that, for each i, $S(M_i) = \{s_{i,1}, s_{i,2}, \ldots, s_{i,r}\}$, and for any i and j, the states $s_{i,k}$ and $s_{j,k}$ are identical after module renaming for any $1 \leq k \leq r$.

Counter variables count_$M\_k : k \in \{1, 2, \ldots, r\}$ record at any state the number of modules in a given local state. Specifically they record the number of modules, M_i whose local state is $s_{i,k}$. Counter function f maps each element of $S(M)$ to the appropriate r-tuple of counter variable values.

For each transition statement c in M_i, let $\mathsf{SAT}_{M_i}(c) = \{l \in S(M_i) : l \models c\}$, i.e. the subset of local states of M_i that satisfy its guard. Similarly, for an expression e appearing in a guard, we use $\mathsf{SAT}_{M_i}(e)$ to refer to the subset of local states that satisfy e.

3.2   Basic Synchronisation Example

Consider the basic coin toss example from Listing 1; however, this time with synchronisation labels present. The two devices again make their decisions at random and with equal probability, but now must simultaneously reveal their choices, at which point they terminate (we again do not model any consequence of the coin toss).

```
1  dtmc
2
3  module device1
4     s1 : [0..3] init 0;
5
6     []   s1=0 -> 0.5 : (s1'=1) + 0.5 : (s1'=2);
7     [a]  s1=1 -> (s1'=3);
8  .  [a]  s1=2 -> (s1'=3);
9     []   s1=3 -> (s1'=3);
10
11 endmodule
12 module device2 = device1[s1=s2,s2=s1] endmodule
```

Listing 3. Example: coin toss with synchronisation

Without synchronisation labels, each device could progress to state 3 (c.f. lines 7 and 8 of Listing 3) as soon as it enters state 1 or 2. With the labels, they would need to wait until the other device were ready. Furthermore, synchronisation requires all updates to be executed simultaneously.

Lines 13 to 16 of Listing 2 have been created by GRIP for the reduced specification of the coin toss model without synchronisation. The updates are simple: in each of them the value of the no_3 counter is increased by one, i.e. a module moves to state 3. In two of them that particular module has arrived at that state from state 1, in the other two it has arrived from state 2. The guard of line 13 is equivalent to "there are 1 or 2 modules in state 1" and of line 14 to "there are 2 modules in state 1". Similarly the guards of lines 15 and 16 are equivalent to "there are 1 or 2 modules in state 2", and "there are 2 modules in state 2" respectively. Out of these, lines 14 and 16 would be acceptable even with synchronisation. However, the guards of the other two would need to be tightened. Merely requiring that one module is in state 1 (or in state 2) is insufficient – in this case the guard should also state that the other module is in the corresponding (other) state. That is: when exactly one module is in state 1, the other module would need to be in state 2 and vice versa. Listing 4 shows what a reduced version of the specification could look like if the guards were strengthened to accommodate this observation.

```
1  probabilistic
2  global total : [ 0 .. 2 ] init 0 ;
3
4  module generic_process
5     no_0 : [0..2] init 2;      // No modules in state (0)
6     no_1 : [0..2] init 0;      // No modules in state (1)
7     no_2 : [0..2] init 0;      // No modules in state (2)
8     no_3 : [0..2] init 0;      // No modules in state (3)
9
10    []  (no_0>0)  -> 0.5:(no_0'=no_0-1)&(no_1'=min(no_1+1,2))
11                       + 0.5:(no_0'=no_0-1)&(no_2'=min(no_2+1,2));
12    []  (no_0>1)  -> 0.5:(no_0'=no_0-1)&(no_1'=min(no_1+1,2))
13                       + 0.5:(no_0'=no_0-1)&(no_2'=min(no_2+1,2));
14    [a] (no_1=0) & (no_2=2) -> (no_2'=0)&(no_3'=min(no_3+2,2));
15    [a] (no_1=1) & (no_2=1) -> (no_1'=0)&(no_2'=0)&(no_3'=min(no_3+2,2));
16    [a] (no_1=2) & (no_2=0) -> (no_1'=0)&(no_3'=min(no_3+2,2));
17    []  (no_3>0) -> true ;
18    []  (no_3>1) -> true ;
19 endmodule
```

Listing 4. Example: reduced coin toss specification with synchronisation

Note that in this example, the updates for all synchronised statements in the output specification are the same, suggesting that the three statements could be combined. However this is not true in general, as synchronised statements may have different updates. The reduced updates will therefore consist of a number of distinct assignments each requiring a distinct statement. For each synchronisation label, guard and update combination, we must consider the possible ways of allocating the symmetric modules between the states accepted by the guards. The number of reduced statements arising from a synchronised block of statements is exponential in both the number of symmetric modules and in the number of local states that satisfy the guards of the local statements. This is an important observation (hence we state it again below). It means that adding synchronisation to our translation algorithm comes at a significant cost - although its inclusion is necessary when synchronous protocols are to be modelled.

Observation. Although the method we present allows the generic representatives approach to symmetry reduction to be applied in the presence of synchronised statements, its scalability is limited. This is because our method results in an exponential blow-up in the size of the text of the reduced specification: the blow-up is exponential in both the number of symmetric modules, and the number of statements that use a particular synchronisation label. Although the state space associated with the reduced specification will be smaller than the original state space (thanks to symmetry reduction), the overhead associated with processing the larger specification text in order to build this state space may outweigh the benefits brought by symmetry reduction.

3.3   Translating Synchronisation

We now develop new SPSL translation rules for synchronised statements to be added to the original rules introduced in [13]. For notation see Sect. 3.1. As the translation process is more complex for DTMCs, we assume that our model type is either an MDP or a CTMC for ease of presentation. However, our implementation in the GRIP tool does support DTMCs.

Counter abstraction replaces each family M of m symmetric modules by a single generic module with a set of r counter variables, each of which ranges from 0 to m. All counter variables are initialised to 0, except for that corresponding to the initial state, which is initialised to m.

Without loss of generality we can assume that a synchronised statement has the form:

$$[label] \quad \text{local-expr}(M) \wedge \text{symm-expr}(M) \rightarrow \text{stoch-update}(M) \tag{1}$$

where local-expr(M) is an expression over local variables only and symm-expr(M) may also include some global variables. Either can be set to *true* if no such expression is present. Expression symm-expr(M) must be fully symmetric for translation to be applicable.

Consider a statement in the original specification with local-expr$(M) = e$, and symm-expr$(M) = s$. Without synchronisation, GRIP splits the translation process into cases, one per $l \in \text{SAT}_M(e)$ (see Fig. 3). For each case, a separate reduced generic statement is generated by the following process. Expression e is replaced with a condition count_M_$f_M(l) > 0$. (Although we are assuming that our model is an MDP, it is worth noting that, in the case of DTMCs, each statement gets translated into m statements which are identical except that the ith statement has guard count_M_$f_M(l) > i$, for $0 \leq i \leq m - 1$.) This condition asserts that some member of M has a local state required to satisfy the local part of the guard of this statement. Both s and the stochastic update stoch-update(M) are translated in the context of the state l, so their reduced counterparts do not make use of any variables local to M. Updates affecting variables local to M are replaced with updates to two counter variables: if l is the local state considered in the current case, and l' is the local state reached by performing all local variable updates on l, then the resulting reduced update must decrement count_M_$f_M(l)$ and increment count_M_$f_M(l')$ (representing modules leaving l and arriving at l' respectively).

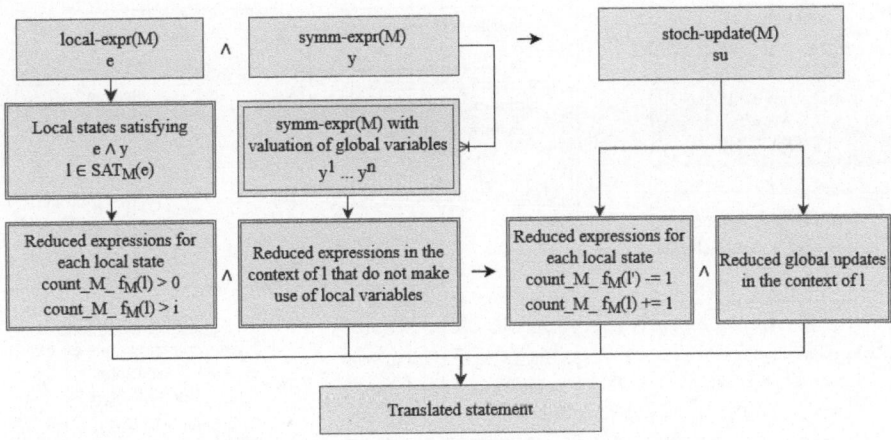

Fig. 3. Translation of a single non-synchronised statement. The top row represents the components a statement in the original specification, while the bottom row represents the corresponding translated components that form a reduced statement.

We approach synchronised statements using a similar methodology (see Fig. 4). First, we assume that all updates of synchronised statements will only change the local state of a module. Note that PRISM does not allow synchronised statements to perform updates to global variables, so this is a reasonable assumption. For brevity, we only consider synchronised statements with a single update (i.e. of the form $p_1 : u_1$ where $p_1 = 1$) rather than a stochastic choice of updates. A discussion of how multiple updates could be included is given in [27].

To translate statements with synchronisation we must consider all statements with the same label at the same time, rather than translating them individually. Again we assume that M consists of a single family of symmetric modules M_i, $1 \leq i \leq m$. For any statement in M_1 (say), the same statement (under module renaming) is present in all modules. For this reason we base our reasoning on statements in M_1.

All statements in M_1 with label α have the form $[\alpha]\ \ e \wedge y \rightarrow p_1 : u_1$ (c.f. Eq. 1). If there are z such statements, with local parts $e_1, e_2, \ldots, e_z$, then define

$$\mathsf{SAT}_{M_1}(\alpha) = \bigcup_{1 \leq j \leq z} \mathsf{SAT}_{M_1}(e_j)$$

Synchronisation over α can only take place at state $s = (s_1, s_2, \ldots, s_m, g)$ if, for each module, at least one statement with label α is enabled at s, i.e. if there exists an $s_{i,k} \in \mathsf{SAT}_{M_i}(\alpha)$ for $1 \leq i \leq m$. For each j and local state $l \in \mathsf{SAT}_{M_1}(e_j)$ we define w_j^l to be the number of modules in their corresponding local state before the synchronised transition is executed, where $0 \leq w_j^l \leq m$. Similarly for any state $l \in \mathsf{SAT}_{M_1}(\alpha)$ define $x^l = \sum_{j:l \in \mathsf{SAT}_{M_1}(e_j)} w_j^l$. Each translated statement then has the condition $\bigwedge_{l \in \mathsf{SAT}_{M_1}(\alpha)} \mathsf{count\_}M\_f_M(l) = x^l$.

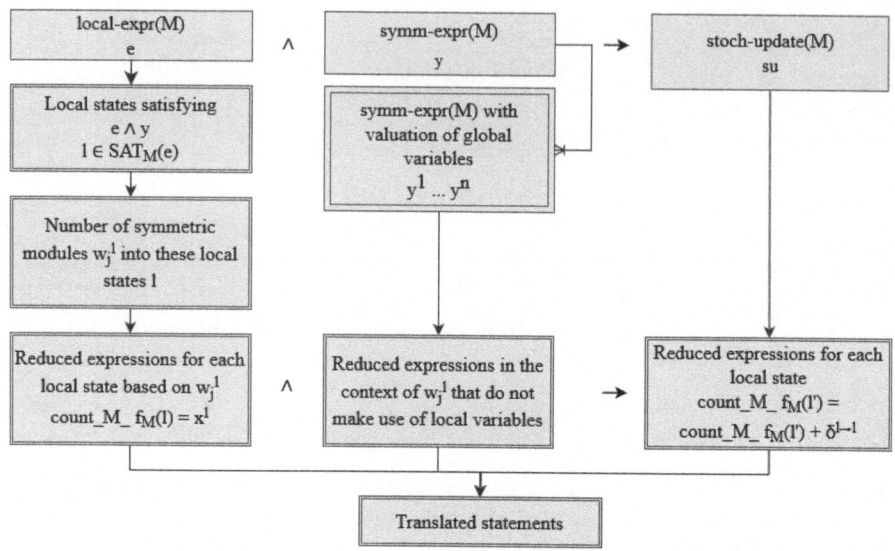

Fig. 4. Translation of a block of synchronised statements. Top and bottom rows represent components in a statement in a synchronised block in the original specification in the reduced model respectively.

The global part of the guards, symm-expr(M), is then translated by considering only those parts which belong to a statement that is being executed, and evaluated in the context of the corresponding local state l. For each local state l with $w_j^l > 0$, we translate the global guard corresponding to e_j. We then combine the translated global guards. The individual translation process here is the same as for the global part of a non-synchronised statement. The translation of updates follows a similar idea: for each $w_j^l > 0$, we find the states resulting from applying the update associated with the statement whose guard is e_j. For any such updated state l' we are interested in two quantities: the number of modules that were in state l' before the updates were applied, $w_j^{l'}$, and the number of modules in states which change to state l' after the updates are applied. The translated updates have the following form $\bigwedge_l \texttt{count\_M\_} f_M(l) \texttt{:=} \texttt{count\_M\_} f_M(l) + \delta^{l \to l'}$, where $\delta^{l \to l'}$ is the difference between the modules changing into state l' and the number of modules changing out of it.

4   Experimental Results

We have implemented our new translation techniques in an updated version of GRIP: GRIP 3.0. The input PRISM specification (in SP) may now contain any number of synchronised statements across multiple modules. There may be multiple blocks of synchronised statements (each with a different label).

We now investigate how well GRIP 3.0 performs compared to PRISM and PRISM-symm. Our first example (Rock-Paper-Scissors) is a purpose-built small example to show a complete SP specification with synchronisation. The other two examples (CSMA/CD, and a synchronous version of a randomised Byzantine agreement protocol) are based on examples from [23] and demonstrate where GRIP does badly compared to PRISM-symm, and where it does well. The experiments were performed on a 2.60 GHz PC with 16 GB RAM, running PRISM version 4.8 under Windows. The maximum memory of the CUDD library was set to 1 GB (PRISM default) and the Java maximum memory was set to 6 GB.

We note that all of the PRISM case studies [1] suitable for symmetry reduction using GRIP have already been considered in the previous version of the tool (for more information see [27]). Many of the other PRISM case studies are not suitable for the generic representatives approach as they possess *ring symmetry*, rather than the full symmetry we require [11].

Rock-Paper-Scissors. We first consider a model of a Rock-Paper-Scissors game. Modules represent participants, who choose between three options: rock, paper and scissors. When all choices have been made, they are evaluated. If all choices are different or all are the same, the game continues for another round. Otherwise an outcome is announced, based on the choices made. Synchronised statements are required to ensure that all choices are made before the result is evaluated. The PRISM model is shown in Listing 5.

```
1  dtmc
2  global r : [0..1] init 0;
3  global p : [0..1] init 0;
4  global s : [0..1] init 0;
5
6  module player1
7  // choice: 0-undecided, 1-rock, 2-paper, 3-scissors
8  ch1 : [0..3];
9  // local phase
10 ph1 : [1..2];
11 // winner: 1-rock, 2-paper, 3-scissors
12 res1 : [0..3];
13 // make choice
14 [](( ch1=0)&(ph1=1)&(res1=0))  -> 1/3: (ch1'=1) & (r'=1)
15                                  + 1/3: (ch1'=2) & (p'=1)
16                                  + 1/3: (ch1'=3) & (s'=1);
17 // determine outcome
18 [decided] ((ph1=1) & (res1=0)) -> (ph1'=2) ;
19 []((ph1=2)&(res1=0))&((r=1)&(p=0)&(s=1))->(ch1'=0)&(res1'=1);
20 []((ph1=2)&(res1=0))&((r=1)&(p=1)&(s=0))->(ch1'=0)&(res1'=2);
21 []((ph1=2)&(res1=0))&((r=0)&(p=1)&(s=1))->(ch1'=0)&(res1'=3);
22 []((ph1=2)&(res1=0))&((r=0)&(p=0)&(s=0))->(ch1'=0);
23 []((ph1=2)&(res1=0))&((r=1)&(p=0)&(s=0))->(ch1'=0)&(r'=0)&(p'=0)&(s'=0);
24 []((ph1=2)&(res1=0))&((r=0)&(p=1)&(s=0))->(ch1'=0)&(r'=0)&(p'=0)&(s'=0);
25 []((ph1=2)&(res1=0))&((r=0)&(p=0)&(s=1))->(ch1'=0)&(r'=0)&(p'=0)&(s'=0);
26 []((ph1=2)&(res1=0))&((r=1)&(p=1)&(s=1))->(ch1'=0)&(r'=0)&(p'=0)&(s'=0);
27 // reset for next round if needed
28 [reset] ((ch1=0)&(ph1=2)&(res1=0)) -> (ph1'=1) ;
29 endmodule
30 module player2=player1[ch1=ch2,ch2=ch1,ph1=ph2,ph2=ph1,res1=res2,res2=res1]
       endmodule
```

Listing 5. Rock-Paper-Scissors model. Multiple module renamings are not shown.

Table 2 shows the model sizes and execution times for the Rock-Paper-Scissors model described above for m participants, for $2 \leq m \leq 10$, using

PRISM, PRISM-symm and GRIP respectively. The property verified is: "what is the probability that the winning outcome is *rock?*".

Compared to PRISM, the GRIP specification is more complex in all cases but the resulting model has far fewer states (when $m > 2$). Consequently the build times increase for reduced models and the times taken for model checking decrease. On this example GRIP is significantly out-performed by PRISM-symm. We suspect that this is because, given the size of the example, the synchronisation dominates. (Recall from the observation in Sect. 3.2 that synchronisation results in an exponential increase in translated statements). We expect our approach to be most beneficial for models that involve a majority of non-synchronised statements. However, we do achieve a significant improvement in comparison to standalone PRISM. The example also serves to demonstrate the correctness of GRIP's new support for synchronisation.

Table 2. Model size and build times for the Rock-Paper-Scissors model for m participants, obtained by PRISM, PRISM-symm and GRIP 3.0.

RPS	Model size (MTBDD)			Model build time (sec.)			Model check time (sec.)		
m	PRISM	PRISM-symm	GRIP	PRISM	PRISM-symm	GRIP	PRISM	PRISM-symm	GRIP
2	453	280	605	0.03	0.066	0.01	0.01	0.03	0.03
3	1774	749	1029	0.04	0.109	0.15	0.01	0.05	0.03
4	4311	1513	2156	0.05	0.112	0.35	0.02	0.04	0.04
5	8021	2394	2880	0.07	0.185	0.88	0.05	0.04	0.05
6	12902	3335	3672	0.08	0.207	2.98	0.15	0.06	0.13
7	18951	4360	4593	0.09	0.339	6.66	0.59	0.07	0.13
8	26153	5442	7240	0.15	0.428	20.73	5.16	0.08	0.08
9	34526	6593	8611	0.16	0.545	30.68	148.21	0.09	0.33
10	44067	7816	10198	0.18	0.712	54.09	261.71	0.12	0.55

Carrier Sense, Multiple Access with Collision Detection Protocol (CSMA/CD). PRISM-symm has been applied to a variety of case studies [23]. However, until now, it has not been possible to compare the performance of PRISM-symm to GRIP for one of those case studies due to the lack of support for synchronisation in GRIP. That example is the IEEE 802.3-2002 CSMA/CD (Carrier Sense, Multiple Access with Collision Detection) communication protocol (*csma*) [22]. We now investigate applying GRIP 3.0 for that example.

The PRISM specification for *csma* is of the type that GRIP is not well suited for. GRIP excels at a larger number of symmetric copies of a simpler module, while PRISM-symm is best for a smaller number of more complex modules [13]. Hence we would not expect GRIP to compare favourably to PRISM-symm in this case.

Examining the *csma* specification closely, we note that the symmetric module has $|S(M)| = 118$ local states, an order of magnitude larger than the typical GRIP case studies [13]. Additionally, most of its synchronised statements have loose guards; for example, a single synchronised statement can have its guard satisfied by modules in 60 out of the 118 possible states. As the number of reduced statements increases with the number of local states satisfying the guards of synchronised statements (again, see the observation in Sect. 3.2), the number of reduced statements is very large.

Our attempt to apply GRIP 3.0 to this example resulted in approximately eight million reduced statements being generated for a single synchronisation label of the *csma* specification. A complete model would take over an hour to build and is unlikely to offer any improvement over PRISM. We conclude that while GRIP 3.0 could now be applied to the *csma* case study, it would be ill-advised.

Randomised Byzantine Agreement Protocol. Our final example is an adaptation of a Byzantine agreement protocol [6]. The original protocol was an example for which GRIP performed favourably compared to PRISM-symm [13]. We have added a common synchronisation label to three of the statements that have updates to local states only and compare performance again. Results for a range of numbers of participants m are shown in Table 3. Despite the additional synchronisation, GRIP still performs well for this example compared with PRISM-symm.

Table 3. Model size and build times for the Byzantine model obtained by PRISM, PRISM-symm and GRIP 3.0. OOM signifies models which resulted in an Out-of-Memory error.

Byz	Model size (MTBDD)			Model build time (sec.)			Model check time (sec.)		
	PRISM	PRISM	GRIP	PRISM	PRISM	GRIP	PRISM	PRISM	GRIP
m		-symm			-symm			-symm	
6	130,145	54,512	21,018	1.36	0.207	1.29	2.07	0.30	0.26
8	592,630	214,293	54,887	6.20	0.428	1.92	>10m	1.39	0.66
12	OOM	OOM	218,153	OOM	OOM	2.98	OOM	OOM	4.653
16	OOM	OOM	343,941	OOM	OOM	6.06	OOM	OOM	13.84

5   Conclusion

We have defined new translation rules for synchronised statements in SPSL, and shown that they are sound. We have discussed the limitations of the method and shown that, in worst case, the counter-abstraction approach does not achieve a reduction in the size of the state space. We have updated GRIP (to version

3.0) to include support for specifications including synchronised statements. We have shown that our approach is feasible, and works well in some cases. Our approach is most beneficial for models that involve a majority of non-synchronised statements and a few synchronised ones.

We plan to add further features to GRIP. While the tool currently does not support translation of steady-state properties, we have conducted an initial investigation using manual translation of this type of property (not included in this paper). Our investigation has shown us that adding this feature in the future would be feasible. Similarly, we aim to introduce support for specification and analysis of properties based on costs and rewards. This would involve translating both the reward structures present in a model and the properties themselves. Specifically, (1) the translated reward structures should not reference any individual module, and (2) GRIP and SPSL would need to be extended to support the translation of reward-based properties.

Acknowledgments.. Ivaylo Valkov was supported by the EPSRC Doctoral Training Partnership award EP/N007565/1 and by a grant from the UKRI Strategic Priorities Fund to the UKRI Research Node on Trustworthy Autonomous Systems Governance and Regulation [EP/V026607/1, 2020-2024]. Alastair Donaldson was supported by the EPSRC IRIS project (grant EP/R006865/1).

References

1. Prism - case studies. https://www.prismmodelchecker.org/casestudies/index.php. Accessed 15 Mar 2024
2. Bošnački, D., Dams, D., Holenderski, L.: Symmetric spin. In: Havelund, K., Penix, J., Visser, W. (eds.) SPIN 2000. LNCS, vol. 1885, pp. 1–19. Springer, Heidelberg (2000). https://doi.org/10.1007/10722468_1
3. Bryant, R.E.: Graph-based algorithms for boolean function manipulation. IEEE Trans. Comput. **35**(8), 677–691 (1986). https://doi.org/10.1109/TC.1986.1676819
4. Bryant, R.E.: Symbolic boolean manipulation with ordered binary-decision diagrams. ACM Comput. Surv. **24**(3), 293–318 (1992). https://doi.org/10.1145/136035.136043
5. Burch, J.R., Clarke, E.M., McMillan, K.L., Dill, D.L., Hwang, L.J.: Symbolic model checking: 10-20 states and beyond. Inf. Comput. **98**(2), 142–170 (1992). https://doi.org/10.1016/0890-5401(92)90017-A
6. Cachin, C., Kursawe, K., Shoup, V.: Random oracles in constantipole: practical asynchronous byzantine agreement using cryptography (extended abstract). In: Neiger, G. (ed.) Proceedings of the Nineteenth Annual ACM Symposium on Principles of Distributed Computing, 16–19 July 2000, Portland, Oregon, USA, pp. 123–132. ACM (2000). https://doi.org/10.1145/343477.343531
7. Cimatti, A., Clarke, E., Giunchiglia, F., Roveri, M.: NuSMV: a new symbolic model checker. STTT **2**, 410–425 (2000). https://doi.org/10.1007/s100090050046
8. Clarke, E.M., Emerson, E.A., Sistla, A.P.: Automatic verification of finite-state concurrent systems using temporal logic specifications. ACM Trans. Program. Lang. Syst. **8**(2), 244–263 (1986). https://doi.org/10.1145/5397.5399
9. Clarke, E.M., Grumberg, O., Kroening, D., Peled, D., Veith, H.: Model Checking, second edition. Cyber Physical Systems Series. MIT Press (2018). https://books.google.co.uk/books?id=qJl8DwAAQBAJ

10. Clarke, E.M., McMillan, K., Zhaor, X., Fujita, M., Yang, J.: Spectral transforms for large boolean functions with applications to technology mapping. In: Proceedings of the 30th ACM/IEEE Design Automation Conference, pp. 54–60. IEEE Computer Society Press (1993)

11. Clarke, E.M., Jha, S., Enders, R., Filkorn, T.: Exploiting symmetry in temporal logic model checking. Formal Methods Syst. Des. **9**, 77–104 (1996). https://api. semanticscholar.org/CorpusID:14472493

12. Donaldson, A.F., Miller, A.: Symmetry reduction for probabilistic model checking using generic representatives. In: Graf, S., Zhang, W. (eds.) ATVA 2006. LNCS, vol. 4218, pp. 9–23. Springer, Heidelberg (2006). https://doi.org/10.1007/11901914_4

13. Donaldson, A.F., Miller, A., Parker, D.: Language-level symmetry reduction for probabilistic model checking. In: QEST 2009, Sixth International Conference on the Quantitative Evaluation of Systems, pp. 289 – 298. IEEE Computer Society (2009). https://doi.org/10.1109/QEST.2009.21

14. Donaldson, A.F., Miller, A.: Exact and approximate strategies for symmetry reduction in model checking. In: Misra, J., Nipkow, T., Sekerinski, E. (eds.) FM 2006. LNCS, vol. 4085, pp. 541–556. Springer, Heidelberg (2006). https://doi.org/10. 1007/11813040_36

15. Donaldson, A.F., Miller, A., Parker, D.: GRIP: generic representatives in PRISM. In: Proceedings of the 4th International Conference on Quantitative Evaluation of Systems (QEST 2007), pp. 115–116. IEEE Computer Society (2007)

16. Emerson, E.A., Trefler, R.J.: From asymmetry to full symmetry: new techniques for symmetry reduction in model checking. In: Pierre, L., Kropf, T. (eds.) CHARME 1999. LNCS, vol. 1703, pp. 142–157. Springer, Heidelberg (1999). https://doi.org/ 10.1007/3-540-48153-2_12

17. Emerson, E.A., Wahl, T.: On combining symmetry reduction and symbolic representation for efficient model checking. In: Geist, D., Tronci, E. (eds.) CHARME 2003. LNCS, vol. 2860, pp. 216–230. Springer, Heidelberg (2003). https://doi.org/ 10.1007/978-3-540-39724-3_20

18. Hendriks, M., Behrmann, G., Larsen, K., Niebert, P., Vaandrager, F.: Adding symmetry reduction to UPPAAL. In: Larsen, K.G., Niebert, P. (eds.) FORMATS 2003. LNCS, vol. 2791, pp. 46–59. Springer, Heidelberg (2004). https://doi.org/10. 1007/978-3-540-40903-8_5

19. Hensel, C., Junges, S., Katoen, J., Quatmann, T., Volk, M.: The probabilistic model checker Storm. CoRR abs/2002.07080 (2020). https://arxiv.org/abs/2002. 07080

20. Holzmann, G.: The SPIN Model Checker: Primer and Reference Manual, 1st edn. Addison-Wesley Professional, Boston (2011)

21. Holzmann, G.J.: Explicit-state model checking. In: Clarke, E., Henzinger, T., Veith, H., Bloem, R. (eds.) Handbook of Model Checking, pp. 153–171. Springer, Cham (2018). https://doi.org/10.1007/978-3-319-10575-8_5

22. IEEE Computer Society: IEEE standard for information technology-telecommunications and information exchange between systems-local and metropolitan area networks-specific requirements part 3: Carrier sense multiple access with collision detection (CSMA/CD) access method and physical layer specifications. IEEE STD 802.3-2002 (Revision of IEEE STD 802.3, 2000 edn), pp. 1–1550 (2002). https://doi.org/10.1109/IEEESTD.2002.93570

23. Kwiatkowska, M., Norman, G., Parker, D.: Symmetry reduction for probabilistic model checking. In: Ball, T., Jones, R.B. (eds.) CAV 2006. LNCS, vol. 4144, pp. 234–248. Springer, Heidelberg (2006). https://doi.org/10.1007/11817963_23

24. Kwiatkowska, M., Norman, G., Parker, D.: PRISM 4.0: verification of probabilistic real-time systems. In: Gopalakrishnan, G., Qadeer, S. (eds.) CAV 2011. LNCS, vol. 6806, pp. 585–591. Springer, Heidelberg (2011). https://doi.org/10.1007/978-3-642-22110-1_47

25. Miller, A., Donaldson, A.F., Calder, M.: Symmetry in temporal logic model checking. ACM Comput. Surv. **38**(3) (2006). https://doi.org/10.1145/1132960.1132962. http://eprints.gla.ac.uk/3197/

26. Queille, J.P., Sifakis, J.: Specification and verification of concurrent systems in CESAR. In: Dezani-Ciancaglini, M., Montanari, U. (eds.) Programming 1982. LNCS, vol. 137, pp. 337–351. Springer, Heidelberg (1982). https://doi.org/10.1007/3-540-11494-7_22

27. Valkov, I.: Formal analysis of communication protocols for wireless sensor systems. Ph.D. thesis, University of Glasgow, Glasgow, UK (2024, to appear)

28. Wahl, T., Donaldson, A.F.: Replication and abstraction: symmetry in automated formal verification. Symmetry **2**(2), 799–847 (2010). https://doi.org/10.3390/SYM2020799

A Hypergraph-Based Formalization of Hierarchical Reactive Modules and a Compositional Verification Method

Daisuke Ishii$^{(\boxtimes)}$

Japan Advanced Institute of Science and Technology, Nomi, Ishikawa, Japan
dsksh@jaist.ac.jp

Abstract. The compositional approach is important for reasoning about large and complex systems. In this work, we address synchronous systems with hierarchical structures, which are often used to model cyber-physical systems. We revisit the theory of reactive modules and reformulate it based on hypergraphs to clarify the parallel composition and the hierarchical description of modules. Then, we propose an automatic verification method for hierarchical systems. Given a system description annotated with assume-guarantee contracts, the proposed method divides the system into modules and verifies them separately to show that the top-level system satisfies its contract. Our method allows an input to be a circular system in which submodules mutually depend on each other. Experimental result shows our method can be effectively implemented using an SMT-based model checker.

1 Introduction

Synchronous reactive systems are a basic model used in the design of *cyber-physical systems (CPSs)* [3], which are typically described as a feedback loop model with plant and digital controller modules. Although it is important for industrial CPS products to formally verify the safety of their models, the effort has not been sufficient. The scale and complexity of the models are hampered by the expertise and computational complexity required by formal method tools.

A divide-and-conquer approach could be the cure for scalability. The theory of reactive modules [3,6] provides a foundation of the compositional reasoning [16]. It formalizes a system as a set of *modules* (or agents or components) that behave synchronously on a sequence of rounds and enables to verify the *implementation* (or refinement) relation between composite modules. In the verification, the *assume-guarantee rule* [6] is crucial to reason about circular systems in which submodules depend on each other. Although it has been studied for decades, the theory is still underutilized in practice due to its discrepancy from the actual CPS descriptions and the lack of automated verification methods.

In this paper, we consider the verification of synchronous system models that are composed of reactive modules $M_1, \ldots, M_n$. Assuming that each module M_j is given a contract $(M_{a.j}, M_{g.j})$, consisting of assume (a) and guarantee (g)

T. Neele and A. Wijs (Eds.): SPIN 2024, LNCS 14624, pp. 67–84, 2025.
https://doi.org/10.1007/978-3-031-66149-5_4

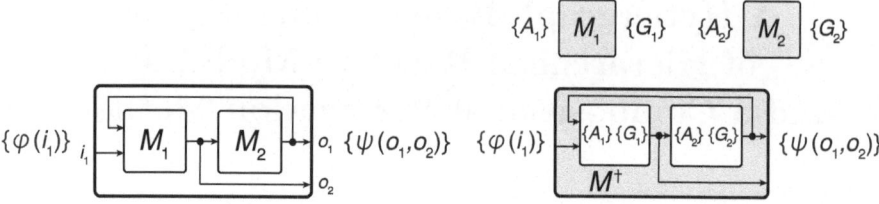

Fig. 1. Example system.   **Fig. 2.** Compositional verification.

properties, and satisfies it (we denote the fact by $M_j || M_{a.j} \preceq M_{g.j}$), we verify that the top-level system $M_1 || \cdots || M_n$ satisfies its contract (M_a, M_g). As is the case in [2,8,14], verification based on the assume-guarantee contracts requires an interactive proof process. An automatic method [11,12] has been proposed that abstracts submodules using their contracts and efficiently performs model checking. However, because of the abstraction, the method has the disadvantage of finding spurious counterexamples, especially when dealing with circular systems.

The objective of this research is to bridge the gap between the theory of reactive modules and the hierarchical design of practical systems, and to propose an automated compositional verification method based on the theory. Our contributions are summarized as follows:

1. We formalize the hierarchical structure commonly used in practical modeling languages, e.g. Lustre and Simulink, as a composition of reactive modules. Our formulation features the use of hypergraphs to describe hierarchical structures. We extend the definition of modules for hierarchization and show how it differs from parallel composition.
2. We propose a compositional verification method that transforms a hierarchical module $M[M_1, \ldots, M_n]$ into a form of a composition $M_1 || \cdots || M_n || M^\dagger$ and then checks that it satisfies a given contract (M_a, M_g). We show how to validate the implementation relation $M[M_1, \ldots, M_n] || M_a \preceq M_g$ automatically. The effectiveness of the method is confirmed by applying our implementation to several examples.

Example. Hierarchical modules can be illustrated as a flow diagram in Fig. 1 in which rectangles represent modules e.g. M_1; outer rectangle represents a hierarchical module $M[M_1, M_2]$. Each module is equipped with input, output, and hidden state variables, and is interpreted as a reaction relation between their values. We consider to verify that the module satisfies a contract $(\varphi(i_1), \psi(o_1, o_2))$, assuming the submodules are given sub-contracts (A_1, G_1) and (A_2, G_2). A compositional verification can be done in three ways:

- The method based on the reactive module theory regards $M[M_1, M_2]$ as a parallel composition $M_1 || M_2 || \cdots$ (if other modules are used, they should be composed together). Then, we deduce that the system satisfies the top-level contract by applying the inference rules to the assumptions, along with the composition structure. As the number of modules increases, the proof becomes more complex.

- The method based on abstraction [11] regards each submodule M_j as a reaction relation, e.g. $\mathsf{H}\,A_j \Rightarrow G_j$ described with a past fragment of LTL, and verifies $M[M_1, M_2]$ as a whole. Since there is a circular wiring in Fig. 1, this method may result in a spurious counterexample.
- The proposed method decomposes $M[M_1, M_2]$ as $M_1 \| M_2 \| M^\dagger$ (Fig. 2) where the module $M^\dagger$ represents the top-level description content equipped with interface variables with M_1 and M_2. We formalize $M^\dagger$ using hypergraphs and propose how to properly give sub-contracts to it. Then, we show that if the verification for the submodules M_1, M_2 and $M^\dagger$ succeeds, then the verification goal for $M[M_1, M_2]$ is also valid.

Paper Organization. Section 2 introduces the basics of the theory of reactive modules, which is reformulated using hypergraphs. Section 3 describes a formalization of hierarchical modules. Section 4 presents a proposed method that transforms hierarchical modules into decomposed forms. Section 5 describes a prototype implementation of the method and Sect. 6 reports an experimental result. Section 7 describes the related work.

Preliminaries. We assume a basic knowledge of *directed hypergraphs* (V, E) where V and E are sets of vertices and *(hyper)edges* (or hyperarcs), respectively. Each hyperedge consists of two lists of vertices called the *source* (or head) and the *target* (or tail), respectively. Given $e \in E$, we denote by $\mathrm{src}(e)$ and $\mathrm{tgt}(e)$ the sets of the vertices in the source and target. We call a vertex v *initial* if $\forall e \in E, v \notin \mathrm{tgt}(e)$ and *terminal* if $\forall e \in E, v \notin \mathrm{src}(e)$. For details of the hypergraph theory, see e.g. [9, 19].

2 Reactive Modules

This section is a run-through introduction to the basics of the reactive module theory [3, 6], which is modified for our purpose.

We consider *variables* typed as unit, bool, int, etc., referring to the *domains* $\mathcal{D}(\text{unit}) = \{()\}$, $\mathcal{D}(\text{bool}) = \{\top, \bot\}$, $\mathcal{D}(\text{int}) = \mathbb{Z}$, etc. Given a variable v of type t, we denote its *evaluation* by $[\![v]\!] \in \mathcal{D}(t)$. Given a set of variables $V = \{v_1, \ldots, v_n\}$ and a family of types $\{t_v\}_{v \in V}$, we denote by $\mathcal{D}(V)$ the *domain*,

$$\prod_{v \in V} \mathcal{D}(t_v) \quad \text{if } V \neq \emptyset, \qquad \{()\} \quad \text{if } V = \emptyset,$$

i.e. a Cartesian product of the family $\{\mathcal{D}(t_v)\}_{v \in V}$. For variable sets V ($\neq \emptyset$) and $W \subseteq V$, and a subdomain $D = \Pi_{v \in V} D_v \subseteq \mathcal{D}(V)$, $\pi_W(D)$ denotes the projection onto W i.e. $\Pi_{v \in W} D_v$.

2.1 Task Hypergraphs

As a unit to describe a composite system, we consider stateless tasks that are non-blocking and can be nondeterministic.

Definition 1. *Let R and W be finite and mutually disjoint sets of variables. A task e with a* read set R *and a* write set W *is represented by a total relation in $\mathcal{D}(R) \times \mathcal{D}(W)$ such that $\forall r \in \mathcal{D}(R), \exists w \in \mathcal{D}(W), (r, w) \in e$. We also denote a task by $e(R, W)$ to clarify its read and write sets.*

For example, a task $e(\emptyset, W)$ represents a task that outputs a constant or a nondeterministically chosen value in $\mathcal{D}(W)$. Given $e : \mathcal{D}(R) \to \mathcal{D}(W)$, it can represent a task that applies the function to the value of R and writes it to W.

In this paper, we propose to formalize a composite task description as a hypergraph representing a network of tasks connected via read and write variables. It aims to be an extension of task graphs in [3] to depict relations between both tasks and variables.

Definition 2. *A task (hyper)graph (TG) (V, E) is a directed hypergraph whose vertices represent variables and hyperedges represent tasks. We assume that (i) (V, E) is acyclic, (ii) no vertex is isolated, (iii) each vertex has at most one incoming edge, (iv) for a task $e \in E$ such that $e \in \mathcal{D}(R) \times \mathcal{D}(W)$ and a variable $v \in V$, $v \in \mathrm{src}(e)$ iff $v \in R$ and $v \in \mathrm{tgt}(e)$ iff $v \in W$.*

If it is clear from the context, we do not distinguish between vertices and variables, or hyperedges and tasks (relations in the variable domains), respectively. *Precedence relation* between tasks (denoted by $e \prec e'$ in [3]) and *await dependency* between variables ($v \succ v'$ in [3,6]) are represented by the existence of a path between the two in the graph. The condition (iii) prevents conflicts between writes to a variable by multiple tasks.

Example 3. Figure 4 illustrates an example TG. Each dot (with or without circle) represents a vertex in $\{i_1, i_2, o_1, o_2, s_1, s_1', l_1\}$ and each set of directed lines mediated by a numbered circle represents a hyperedge in $\{e_1, e_2, e_3\}$. For instance, $\mathrm{src}(e_2) = \{i_2, s_1\}$ and $\mathrm{tgt}(e_2) = \{l_1\}$. The hypergraph in Fig. 9b is not a TG since it contains a cycle.

TGs can be regarded as total relations i.e. tasks.

Definition 4. *Let (V, E) be a TG, R the set of initial vertices, and W the set of vertices such that $W \subseteq V \setminus R$. We consider a relation*

$$\exists v_1 \in \mathcal{D}(t_1), \cdots \exists v_m \in \mathcal{D}(t_m), \ e_1(R_1, W_1) \wedge \cdots \wedge e_n(R_n, W_n), \tag{1}$$

where $\{v_1, \ldots, v_m\} = V \setminus (R \cup W)$, $t_1, \ldots, t_m$ are their types, and $\{e_1, \ldots, e_n\} = E$. We denote the set of all such relations represented with a TG by $\mathcal{T}(R, W)$.

Note that variables in R may always be the initial in the TG, but those in W are not necessarily the terminal (they are shown as circled dots in the figures). Every variable in R_j or W_j ($j = 1, \ldots, n$) not included in $R \cup W$ are bound by a quantifier in Eq. (1).

Lemma 5. *Every relation in $\mathcal{T}(R, W)$ is total.*

Proof. The vertices in a TG can be partially ordered by the lengths of the longest paths from any initial vertex; we group the vertices according to the ordering. The initial vertices in R belong to the first group. The vertices written by tasks with the empty read set belong to the second group. Then, we check $\forall u_r \in \mathcal{D}(R), \exists u_w \in \mathcal{D}(W), \mathcal{R}(u_r, u_w)$ holds where $\mathcal{R}$ represents the relation (1). We check by induction that a value exists for vertices in every group to satisfy the relation. The first group is universally quantified, so any values can be assigned to the vertices. Assuming that the previous groups has been assigned values, the values for the next group are determined by the incoming tasks.   □

Although tasks in [3] are stateful, we formulate them as stateless. Modules defined later specify the state variables among the read/write set of tasks and properly manage the states. *Atoms* are used instead of tasks in [6] to represent the initialization of the state of a module when executed and the reactions in each round. We embed the initial conditions in modules and represent only the reactions with TGs.

2.2   Modules, Implementation Relation, and Parallel Composition

Reactive and synchronous systems are formalized as compositional modules (called *components* in [3]) executed in a series of rounds.

Definition 6 *([3,6]). A module is a tuple $(I, O, S, \mathcal{I}, \mathcal{R})$ where I, O and S are mutually disjoint sets of input, output and state variables. $\mathcal{I} \subseteq \mathcal{D}(S)$ is an initial condition, and $\mathcal{R}$ is a reaction relation that is a TG in $T(S \cup I, O \cup S')$, where S' represents a set of variables renamed from S. An execution of a module is a sequence of reactions*

$$s(-1) \xrightarrow{i(0)/o(0)} s(0) \xrightarrow{i(1)/o(1)} s(1) \cdots =$$
$$\{(s(j{-}1), i(j), o(j), s(j)) \in \mathcal{R} \mid j \in \{0\} \cup \mathbb{N}, s(-1) \in \mathcal{I}\}.$$

A trace of an execution is a sequence of values $(i(0), o(0))\,(i(1), o(1)) \cdots$, i.e. the projection onto $I \cup O$.

Note that $\mathcal{R}$ is interpreted as a set of quadruples of values, each of which is a family of values indexed by S, I, O or S'. We assume the state variables in S and S' always be the initial and terminal vertices of $\mathcal{R}$, respectively. Differently from the formalization in [3], which embeds state variables within tasks, modules must designate the vertices representing state variables from among the TG's initial and terminal vertices. Hereafter, for a module M_i with an identifier i, we denote its elements by e.g. I_i and $\mathcal{R}_i$.

If I, O or S of a module is empty, there may be an execution that involves the value (). For example, the module $M_\top$ defined by $(\emptyset, \emptyset, \emptyset, \{()\}, \{((), (), (), ())\})$ has an execution represented by a sequence of $((), (), (), ())$.

As a graphical language to describe modules (e.g. Simulink), we consider *signal flow diagrams* (SFDs). An SFD consists of rectangles and directed lines

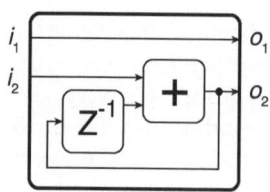

Fig. 3. A signal flow diagram describing the module $M_{\text{Ex.7}}$.

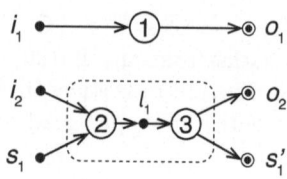

Fig. 4. The task hypergraph $\mathcal{R}_{\text{Ex.7}}$.

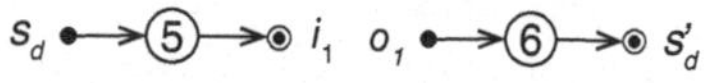

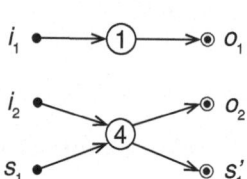

Fig. 5. Another TG for Example 7.

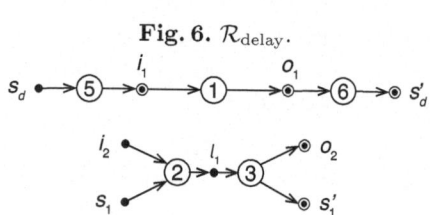

Fig. 6. $\mathcal{R}_{\text{delay}}$.

Fig. 7. The TG of $M_{\text{Ex.7}} \| M_{\text{delay}}$.

annotated with variable names and other labels. The rectangles represent modules that can be stateful and the lines represent synchronous communication between the processes. Input signals are connected to the left side and output signals are extracted from the right side of rectangles.

Example 7. The SFD of a counter constructed with an addition module and a delay module is shown in Fig. 3, which consists of $I_{\text{Ex.7}} = \{i_1, i_2\}$, $O_{\text{Ex.7}} = \{o_1, o_2\}$, $S_{\text{Ex.7}} = \{s_1\}$, $\mathcal{I}_{\text{Ex.7}}(s_1) \equiv (\llbracket s_1 \rrbracket = 0)$, and $\mathcal{R}_{\text{Ex.7}}$ represents a function $(\llbracket i_1 \rrbracket, \llbracket i_2 \rrbracket, \llbracket s_1 \rrbracket) \mapsto (\llbracket i_1 \rrbracket, \llbracket i_2 \rrbracket + \llbracket s_1 \rrbracket, \llbracket i_2 \rrbracket + \llbracket s_1 \rrbracket)$. Figure 4 shows $\mathcal{R}_{\text{Ex.7}}$ as a TG. The vertices in $O_{\text{Ex.7}}$ are shown as circled dots and the vertex l_1 represents a local variable that does not belong to $I_{\text{Ex.7}}$, $O_{\text{Ex.7}}$ or $S_{\text{Ex.7}}$. The hyperedges e_1, e_2 and e_3 represent the identity, addition and copy functions. An execution can be

$$0 \xrightarrow{(\top,1)/(\top,1)} 1 \xrightarrow{(\bot,0)/(\bot,1)} 1 \xrightarrow{(\top,2)/(\top,3)} 3 \cdots .$$

Example 8. There can be multiple TGs representing a reaction relation. Figure 5 shows another TG for $\mathcal{R}_{\text{Ex.7}}$ in which e_4 outputs two copies of the sum.

We also represent a safety property of a coexisting module as a module. A safety property $\varphi(v)$ involving a list of variables $v = (v_1, \ldots, v_n)$ is a module $M_{\varphi(v)}$ with $O_{\varphi(v)} = \{v_1, \ldots, v_n\}$, which nondeterministically outputs a signal satisfying $\varphi(v)$.

Example 9. We can consider an invariance property $\mathsf{G}\, o_2 \geq 0$ of $M_{\mathrm{Ex.7}}$ where G is the "always/globally" operator of LTL. Output signals represented by variable o_2 of the module $M_{\mathrm{Ex.7}}$ must be mimicked by $M_{\mathsf{G}\, o_2 \geq 0}$.

Next, we introduce the implementation (or refinement) relation and the composition mechanism for the modules.

Definition 10. *([6]) Let M_1 and M_2 be modules. We say M_1 implements M_2, denoted by $M_1 \preceq M_2$, if (i) $O_2 \subseteq O_1$, (ii) $I_2 \subseteq I_1 \cup O_1$, (iii) the await dependency of $y \in O_2$ on $x \in I_2 \cup O_2$ in $\mathcal{R}_2$ (i.e. $y \succ x$) is preserved in $\mathcal{R}_1$, and (iv) for every trace t of M_1, the projection of t onto $I_2 \cup O_2$ is a trace of M_2. We denote $M_1 \preceq M_2 \wedge M_2 \preceq M_1$ by $M_1 \cong M_2$.*

Lemma 11. *([6]) The implementation relation is a preorder.*

Definition 12. *([3,6]) We say modules M_1 and M_2 are compatible if (i) $O_1 \cap O_2 = S_1 \cap S_2 = \emptyset$ and (ii) $\mathcal{R}_1 \cup \mathcal{R}_2$ (the union as graphs) is acyclic. The parallel composition (PC) $M_1 \| M_2$ is a module $(I, O, S, \mathcal{I}, \mathcal{R})$, where $I = (I_1 \cup I_2) \setminus O$, $O = O_1 \cup O_2$, $S = S_1 \cup S_2$, $\mathcal{I} = \mathcal{I}_1 \times \mathcal{I}_2$ and $\mathcal{R} = \mathcal{R}_1 \cup \mathcal{R}_2$.*

Example 13. Consider composing the module in Example 7 with a module M_{delay} that represents a delay task whose TG is Fig. 6. The TG of PC $M_{\mathrm{Ex.7}} \| M_{\mathrm{delay}}$ is shown in Fig. 7.

Lemma 14 *([6]). The PC operation is associative, transitive and symmetric.*

To deal with extra output variables added by the PC operation, we introduce the hiding operator.

Definition 15. *([3,6]) Given a module M and a subset of output variables $O' \subseteq O$, hiding of O' in M, denoted $M \backslash O'$, is a module consisting of the same elements as M but excluding O' from O.*

2.3 Compositional Verification

We consider to verify that a module M fulfils an *assume-guarantee contract* (M_a, M_g), i.e. a pair of modules. For that purpose, we can show $M \| M_a \preceq M_g$ holds. In our experiment in Sect. 6, we consider contracts consisting of safety properties.

Example 16. $M_{\mathrm{Ex.7}} \| M_{\mathsf{G} i_2 \geq 0} \preceq M_{\mathsf{G} o_2 \geq 0}$ holds.

In this paper, we consider the verification of systems composed of n submodules. Here, we assume that each submodule satisfies a given contract, and aim at efficiently verifying the fact that the entire system fulfils the top-level contract by utilizing the assumptions.

Definition 17. *A compositional verification problem consists of n modules M_1, ..., M_n, a top-level contract (M_a, M_g) and n sub-contracts $(M_{a.j}, M_{g.j})$ where $j = 1, \ldots, n$; we assume $M_j \| M_{a.j} \preceq M_{g.j}$ for every j. The goal is a condition $M_1 \| \cdots \| M_n \| M_a \preceq M_g$.*

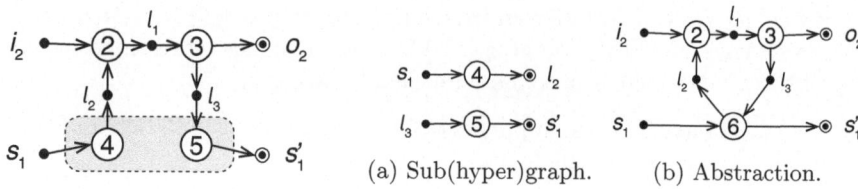

Fig. 8. A detailed TG for Example 7.   Fig. 9. Separation of a hypergraph.

The module M_T can be used to omit some elements of contracts. For arbitrary M, $M \preceq M_T$ and $M||M_T \cong M$ hold.

The following lemma provides two basic inference rules for the compositional reasoning on modules. The second rule allows a PC $M_1||M_2$ to be *circular* i.e. $I_1 \cap O_2$ and $I_2 \cap O_1$ are nonempty.

Lemma 18. *([6]) Let M_1, M_2, M_3 and M_4 be modules, where M_1 and M_2, and M_3 and M_4 are respectively compatible, and $I_3 \cup I_4 \subseteq I_1 \cup I_2 \cup O_1 \cup O_2$. (i) $M_1||M_2 \preceq M_1$. (ii) If $M_1||M_3 \preceq M_4$ and $M_2||M_4 \preceq M_3$, then $M_1||M_2 \preceq M_3||M_4$.*

Example 19. We consider a compositional verification problem of the system $M_{\text{Ex.7}}||M_{\text{delay}}$ in Example 13, which consists of

- submodules $M_{\text{Ex.7}}$ and M_{delay},
- top-level contract $(M_T, M_{\text{G}o_2 \geq 0})$, and
- sub-contracts $(M_{\text{G}i_2 \geq 0}, M_{\text{G}o_2 \geq 0})$ and $(M_{\text{G}o_2 \geq 0}, M_{\text{G}i_2 \geq 0})$.

The goal $M_{\text{Ex.7}}||M_{\text{delay}}||M_T \preceq M_{\text{G}o_2 \geq 0}$ is provable. First, we can deduce as follows.

$$\cfrac{\cfrac{M_{\text{Ex.7}}||M_{\text{G}i_2 \geq 0} \preceq M_{\text{G}o_2 \geq 0} \quad M_{\text{delay}}||M_{\text{G}o_2 \geq 0} \preceq M_{\text{G}i_2 \geq 0}}{M_{\text{Ex.7}}||M_{\text{delay}} \preceq M_{\text{G}i_2 \geq 0}||M_{\text{G}o_2 \geq 0}}\text{Lem.18(ii)} \quad \cfrac{M_{\text{G}i_2 \geq 0}||M_{\text{G}o_2 \geq 0} \preceq M_{\text{G}o_2 \geq 0}}{}\text{Lem.18(i)}}{M_{\text{Ex.7}}||M_{\text{delay}} \preceq M_{\text{G}o_2 \geq 0}}\text{Trans.}$$

Then, the goal follows from $M_{\text{Ex.7}}||M_{\text{delay}}||M_T \preceq M_{\text{Ex.7}}||M_{\text{delay}}$.

3   Hierarchical Reactive Modules

Practical languages for describing modules (e.g. Lustre and Simulink) tend to support hierarchical structures. We formalize such structures by allowing a TG to be hierarchical, separating a subhypergraph of the TG as a submodule.

Definition 20. *For a TG (V, E), its sub(hyper)graph is a TG such that $V' \subseteq V$ and $E' \subseteq E$. Assume a subgraph (V', E') of (V, E) is in $\mathcal{T}(R, W)$. Then, the abstraction of (V', E') in (V, E) is a hypergraph (V'', E'') where V'' is a set such that $V' \cup V'' = V$ and $E'' = (E \setminus E') \cup \{e\}$ where e is a fresh hyperedge such that $\mathrm{src}(e) = R$ and $\mathrm{tgt}(e) = W$.*

According to [9], subgraphs defined above are partial subhypergraphs without isolated vertex. Note that an abstraction of a TG may contain a cycle as shown in the following example.

Example 21. Consider a hypergraph in Fig. 8 that describes yet another TG (lower part) for Example 7. The graph in Fig. 9a is its subgraph that consists of the separated edges e_4 and e_5. The abstraction of the subgraph in Fig. 8 is shown in Fig. 9b, in which e_6 is associated with the subgraph.

Now, we consider the TGs of modules to be hierarchical. Intuitively, a hierarchical module is a module that separates subgraphs as submodules. To ensure the consistency among decomposed descriptions, and because of the proposed method, we assume several conditions.

Definition 22. *Let* $M_1, \ldots, M_n$ *be modules. A hierarchical module* $M[M_1, \ldots, M_n]$ *(also denoted by* M *or* $M[M_1..M_n]$*) is a module that satisfies the following conditions for each submodule* M_j*: (i)* $I \cap I_j = O \cap O_j = \emptyset$*, (ii)* $S_j \subseteq S$*, (iii)* $\pi_{S_j}(\mathcal{I}) \equiv \mathcal{I}_j$*, and (iv)* $\mathcal{R}_j$ *is a subgraph of* $\mathcal{R}$*.*

The condition (i) forces a submodule to handle its own input and output variables to facilitate the separation of the submodules in Sect. 4. The input and output variables must be copied to local variables before communicating with submodules. The conditions (ii) and (iii) make the state variables of a submodule shared with the parent and are maintained properly. A hierarchical module is interpreted as a *flattened* module whose TG embeds the submodules' TGs. In order for a hierarchical module $M[M_1..M_n]$ and its submodules $M_1, \ldots, M_n$ to be interpreted as modules properly, the TGs of the submodules must avoid a write conflict (Definition 2) and their states must be managed with separate variables, i.e., $\bigcap_{j=1,\ldots,n} O_j = \bigcap_{j=1,\ldots,n} S_j = \emptyset$ must hold.

Example 23. We consider a hierarchical module $M_{\text{Ex.23}}[M_1, M_2]$ that has two counter modules as submodules (Fig. 10). We assume they are the same as in Example 7, except for the variable names. Its TG is shown in Fig. 11 where the dashed frames enclose the hyperedges in the subgraphs. The state variables of the submodules are inherited to $M_{\text{Ex.23}}$ ($S_{\text{Ex.23}} = \{s_1, s_{1.1}, s_{2.1}\}$) and $\mathcal{I}_{\text{Ex.23}}$ is set as $\{\top\} \times \mathcal{I}_1 \times \mathcal{I}_2$. $M_{\text{Ex.23}} \| M_{\text{G}i_1 \geq 0} \preceq M_{\text{G}o_1 \geq 0}$ holds.

Abstraction of Hierarchical Modules. The hierarchization of modules can be viewed as an abstraction as shown in Example 21 (Fig. 9b). Naively, we can replace a fragment of the TG that belongs to a submodule with a hyperedge, which is then regarded as a complete graph between input and output vertices. Then, it may help to efficiently search for a counterexample that can be executed by the parent module. The Kind2 model checker [11] abstracts each submodule M_j with a hyperedge representing a property $\mathsf{H}\varphi(i_j) \Rightarrow \psi(o_j)$, where H is the pLTL "historically" modality, to exploit the contract given as a pair of safety properties $(\varphi(i_j), \psi(o_j))$ (where i_j/o_j is a variable list in I_j/O_j). Separately, the fact $M_j \| M_{\varphi(i_j)} \preceq M_{\psi(o_j)}$ has to be verified to check that the submodule M_j satisfies the contract.

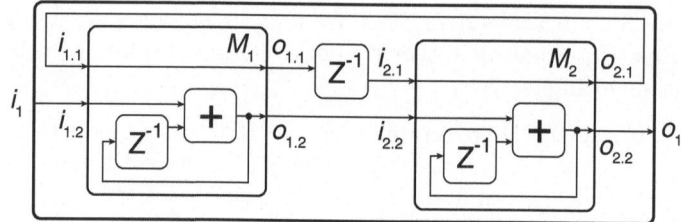

Fig. 10. A hierarchical module $M_{\text{Ex.23}}$.

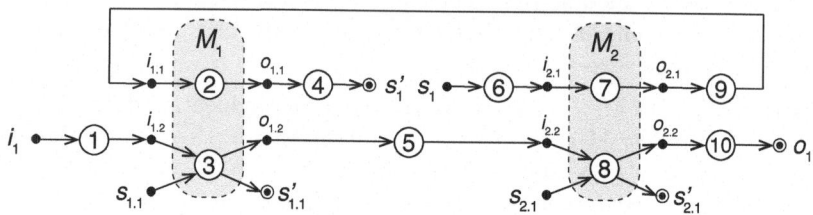

Fig. 11. The TG of $M_{\text{Ex.23}}$.

Example 24. Let $\varphi(x) \equiv \mathsf{G}\,x \geq 0$. For the submodules M_j of $M_{\text{Ex.23}}$ ($j = 1, 2$), $M_j \| M_{\varphi(i_{j.2})} \preceq M_{\varphi(o_{j.2})}$ holds. Then, the fact $M_{\text{Ex.23}} \| M_{\varphi(i_1)} \preceq M_{\varphi(o_1)}$ can be verified with an abstraction that replaces each submodule M_j with $M_{\mathsf{H}\,\varphi(i_{j.2}) \Rightarrow \varphi(o_{j.2})}$.

While the abstraction methods enable a verification process that leverages the sub-contracts, they cannot properly handle circular systems whose submodules assume the existence of other submodules. In such cases, the abstraction approach might not work; the process may detect spurious counterexamples.

Example 25. Consider the verification of $\varphi(x) \equiv \mathsf{G}\,x \geq 0$ on $M_{\text{Ex.23}}$ again. Let $\psi(x_1, x_2) \equiv \mathsf{G}(x_1 \wedge x_2 \geq 0)$. For the submodules M_j ($j = 1, 2$), $M_j \| M_{\psi(i_{j.1}, i_{j.2})} \preceq M_{\psi(o_{j.1}, o_{j.2})}$ holds. So, we can abstract the subgraph for M_j with $\mathsf{H}\,\psi(i_{j.1}, i_{j.2}) \Rightarrow \psi(o_{j.1}, o_{j.2})$ in the TG of $M_{\text{Ex.23}}$. Then, the verification of $M_{\text{Ex.23}} \| M_{\varphi(i_1)} \preceq M_{\varphi(o_1)}$ will result in a false counterexample such that $[\![i_{1.1}]\!] = [\![i_{2.1}]\!] = \bot$.

4   Compositional Verification of Hierarchical Modules

In this section, we propose a compositional verification method that can handle circular hierarchical modules. The method validates that a module $M[M_1..M_n]$ satisfies a contract (M_a, M_g). We describe how to validate the goal under the assumption that every submodule M_j ($j \in \{1, \ldots, n\}$) satisfies its contract $(M_{a.j}, M_{g.j})$. The key idea is to prepare a module $M^\dagger$ called *adapter* by extracting only the top-level part of the hierarchical TG. The subgraphs are separated from the top-level TG and the variables that correspond to the boundary vertices

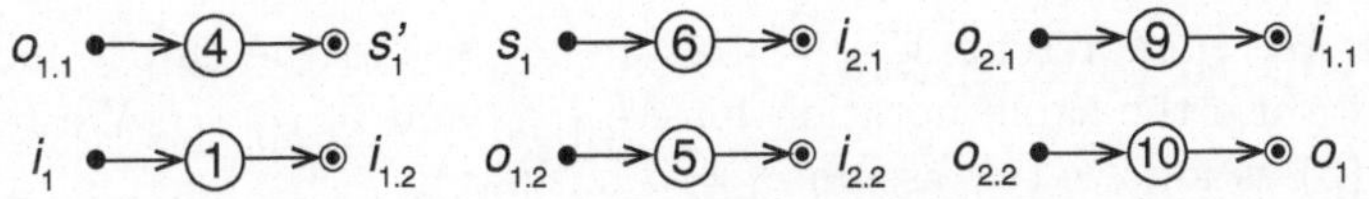

$$o_{1.1} \bullet\!\!\to\!\!\boxed{4}\!\!\to\!\!\circledcirc s_1' \quad s_1 \bullet\!\!\to\!\!\boxed{6}\!\!\to\!\!\circledcirc i_{2.1} \quad o_{2.1} \bullet\!\!\to\!\!\boxed{9}\!\!\to\!\!\circledcirc i_{1.1}$$

$$i_1 \bullet\!\!\to\!\!\boxed{1}\!\!\to\!\!\circledcirc i_{1.2} \quad o_{1.2} \bullet\!\!\to\!\!\boxed{5}\!\!\to\!\!\circledcirc i_{2.2} \quad o_{2.2} \bullet\!\!\to\!\!\boxed{10}\!\!\to\!\!\circledcirc o_1$$

Fig. 12. The TG of $M_{\mathrm{Ex.23}}^\dagger$.

between the top-level part and the subgraphs are set as input and output variables of the adapter module. The adapter $M^\dagger$ is prepared in a way $M[M_1..M_n]$ and $M_1||\cdots||M_n||M^\dagger$ be isomorphic, and thus yields a compositional verification problem (Definition 17).

Definition 26. *For sets* $V_1, \ldots, V_n$, *we denote their union by* $V_{\{1..n\}}$. *The adapter* $M^\dagger$ *of a hierarchical module* $M[M_1..M_n]$ *is a module with the components* $I^\dagger = I \cup O_{\{1..n\}}$, $O^\dagger = O \cup I_{\{1..n\}}$, $S^\dagger = S \setminus S_{\{1..n\}}$, $\mathcal{I}^\dagger = \pi_{S^\dagger}(\mathcal{I})$, *and* $\mathcal{R}^\dagger$ *obtained from* $\mathcal{R}$ *by removing the hyperedges* e_j *and the vertices corresponding with the variables in* S_j *for every* j.

Example 27. The TG of $M_{\mathrm{Ex.23}}^\dagger$ is with $R = \{i_1, o_{1.1}, o_{1.2}, o_{2.1}, o_{2.2}, s_1\}$ and $W = \{o_1, i_{1.1}, i_{1.2}, i_{2.1}, i_{2.2}, s_1'\}$. It is illustrated as Fig. 12.

A hierarchical module can now be regarded as a PC of decomposed modules (up to $\cong$ in Definition 10). However, since extra output variables of the submodules are added by the PC, they must be hidden (Definition 15).

Lemma 28. *Let* $M[M_1..M_n]$ *be a hierarchical module and* M' *be* $M_1||\cdots||M_n|| M^\dagger$. *Then,* $M[M_1..M_n] \cong M' \setminus (I_{\{1..n\}} \cup O_{\{1..n\}})$.

Proof. In Definition 26, the variable sets $I_{\{1..n\}}$ and $O_{\{1..n\}}$ are added to the adapter as initial and terminal vertices in $\mathcal{R}^\dagger$. Then, the PC of M_j and $M^\dagger$ merges $\mathcal{R}_j$ and $\mathcal{R}^\dagger$ by matching the vertices for the variables in $I_j \cup O_j$. Hence, the resulting TG is equivalent of replacing e_j with $\mathcal{R}_j$ in Definition 22 (ii). Each set of the variables is equal to that of $M[M_1..M_n]$ by the hiding operation. $\square$

Once a hierarchical module is decomposed into a PC of submodules and an adapter, it is possible to perform a compositional verification of facts about the module based on the theory of reactive modules. To do so, we can make a proof using the inference rules in Sect. 2, which may require a manual effort. However, the following theorem shows that such verification always succeeds if the conditions on the submodules and the adapter are valid.

Theorem 29. *Consider a goal* $M[M_1..M_n] || M_a \preceq M_g$. *If* (a) $M_j||M_{a.j} \preceq M_{g.j}$ *for every* j *and* (b) $M^\dagger||M_a||M_{g.1}||\cdots||M_{g.n} \preceq M_g||M_{a.1}||\cdots||M_{a.n}$, *then the goal holds.*

Proof. Let $M_{[j..k]}$ represent $M_j||\cdots||M_k$ if $j \leq k$ and the empty module $M_\top$ if $j > k$ (we use the same notation for $M_{a.[j..k]}$ and $M_{g.[j..k]}$). We rewrite the condition (b) as follows (we assume $j = n$ initially):

$$M_{[(j+1)..n]}||M^\dagger||M_a||M_{g.[1..j]} \preceq M_g||M_{a.[1..j]}. \tag{2}$$

From (a), $(M_j||M_{a.j})||(M_{a.[1..(j-1)]}||M_g) \preceq M_j||M_{a.j}$ (Lemma 18 (i)), and the transitivity of $\preceq$, we have

$$M_j||M_g||M_{a.[1..j]} \preceq M_{g.j}. \tag{3}$$

From (2), (3) and Lemma 18 (ii), we have

$$M_{[j..n]}||M^\dagger||M_a||M_{g.[1..(j-1)]} \preceq M_g||M_{a.[1..j]}||M_{g.j}.$$

The rhs can be simplified to $M_g||M_{a.[1..(j-1)]}$ (by Lemma 18 (i) and the transitivity). By repeating the above for $j = n-1, \ldots, 1$, we will obtain the fact, which is equivalent to the goal due to Lem. 28. □

The proposed method can be regarded as transferring the proof process along with top-level's compositional structure to the verification process of the condition (b) of Theorem 29. If the top-level can be viewed as a nested PC structure of submodules, then a proof of a hierarchical module may be obtained by applying the inference rules in Lemma 18 to the PCs. However, the proof process for a verification goal is non-trivial in general. Applying the inference rules requires a number of deductions, introduction of PCs of submodules and property modules, and adjusting the form of PC terms while checking module compatibility.

The proposed method provides an automated process for the compositional verification problem. As implemented in the next section, the separation of adapters can be automated, and the entire process can also be automated when combined with an automatic verifier for submodules.

When a system contains multiple hierarchies, we can simply repeat the process to check the two conditions in Theorem 29 in a bottom-up fashion to verify the whole system. Each process for a hierarchical module decomposes it after deriving an adapter, then either verifies the contract for the PC of the submodules or composes them with other submodules of the parent.

5   Implementation

We have implemented the proposed method by extending the Kind2 tool.

5.1   Lustre, CoCoSpec, and Kind2

Kind2 (version 1.6.0) [11] is an SMT-based model checking tool (we used Z3 4.12.4 as an SMT solver). Its input language is *Lustre* [10], a textual language for describing hierarchical synchronous modules, and the modules can be annotated contracts with the *CoCoSpec* language [12].

```
1   node Filter (in1 : bool; in2 : real)
2   returns (out1 : bool; out2 : real);
3   (*@contract
4     assume    in1;   assume    -1.0 <= in2 and in2 <= 1.0;
5     guarantee out1; guarantee -1.0 <= out2 and out2 <= 1.0;
6   *)
7   var sum, D1, D2: real;
8   let
9     out1 = in1;
10    sum = 0.0582*(if in1 then in2 else -in2) - (-1.49*D1) - 0.881*D2;
11    D1 = 0.0 -> pre sum;       D2 = 0.0 -> pre D1;
12    out2 = (sum - D2) / 1.25;
13  tel
14
15  node Toplevel (in : real) returns (out : real);
16  (*@contract
17    assume    -1.0 <= in and in <= 1.0;
18    guarantee -1.0 <= out and out <= 1.0;
19  *)
20  var b1, b2, pre_b2 : bool; s1 : real;
21  let
22    b1, s1 = Filter(b2, in);   pre_b1 = true -> pre b1;
23    b2, out = Filter(pre_b1, s1);
24    --%MAIN;
25  tel
```

Fig. 13. An example Lustre program annotated with CoCoSpec.

Example 30. Figure 13 shows an example described with Lustre and CoCoSpec. It is a module similar to Example 23 in which the counters are replaced with second-order digital filters. A Lustre node is defined by a section started with node, which is followed by

- the node name (e.g. Filter and Toplevel),
- the input variable list (in parentheses),
- the output variable list (with keyword returns),
- the contract annotation (described within a comment),
- the local variable list (with keyword var), and
- the body (enclosed in let and tel). The line "--%MAIN;" specifies that the node is a verification target.

We consider modules to be instances of Lustre nodes whose input and output variables are substituted by the arguments. The above program describes the verification conditions $M_{F0}||M_{F0.a} \preceq M_{F0.g}$ and $M_{T0}[M_{F1}, M_{F2}]||M_{a.T0} \preceq M_{g.T0}$ where M_{Ni} represents the ith instance of the node N and $M_{a.Ni}$ and $M_{g.Ni}$ represent the annotated properties (we abbreviate Filter and Toplevel as F and T).

When an input describes a goal (verification condition for the target module) $M[M_1..M_n]||M_a \preceq M_g$ and conditions for submodules $M_j||M_{a.j} \preceq M_{g.j}$ where $j = 1, \ldots, n$, Kind2 is able to verify its validity in three modes:

- *Monolithic mode* that interprets the target $M[M_1..M_n]$ as a module (cf. Definition 22) and verifies only the goal.
- *Modular mode* that verifies only the conditions for submodules $M_1, \ldots, M_n$.
- *Compositional mode* that verifies the goal with an abstraction as described in Sect. 3.

We used Kind2 with the default setting, which runs several model checking algorithms e.g. BMC, k-induction and PDR, in parallel.

5.2   Implementation of the Proposed Method

We have implemented the proposed method in OCaml by modifying Kind2.[1] The function we have added translates a hierarchical Lustre program to a program in which the hierarchical modules are replaced with adapter modules (here, we also refer to Lustre *nodes* as modules). The input is Lustre programs annotated with CoCoSpec contracts. It generates a list of reactive modules by applying the following processes:

1. *Instantiation of module definitions.* Because a Lustre node may be invoked several times by the parent module, we instantiate a node definition with the real argument of each call.
2. *Modification of the top-level into an adapter.* For each hierarchical module, we remove the submodule invocation statements and modify the variable list as described in Definition 26.
3. *Pretty printing.* The intermediate data will be printed as a decomposed and properly annotated Lustre program.

By feeding the output Lustre program to Kind2 with the modular mode, the satisfaction of the assume-guarantee contract by each module will be checked. The success of the process implies the validity of the annotated top-level module in the original Lustre program.

6   Experiment

To evaluate the effectiveness and the performance of the proposed method, we have conducted the verification of several examples using the implementation. The experiment was done with a MacBook Pro (10-core Apple M2 Pro chip and 32 GB RAM).

We prepared several circular hierarchical modules for the experiment.

---

[1] The artifact is available at https://doi.org/10.5281/zenodo.10559936 and the source code is available at https://github.com/dsksh/kind2.

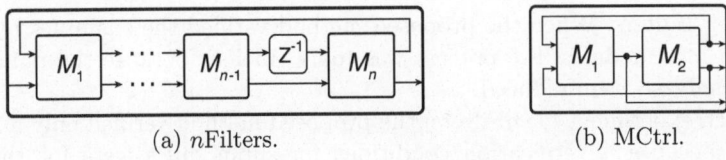

(a) nFilters.   (b) MCtrl.

Fig. 14. SFDs of the example modules.

Table 1. Execution result.

	Monolithic	Proposed method	
Example	Time	Time	#Guarantees
2Filters	204 s	14.9 s	7
3Filters	TO	14.9 s	9
36Filters	TO	15.4 s	75
MCtrl	TO	3.1 s	24

- *Feedback loop system containing n digital filters* (nFilters). This is an extended and parameterized version of Example 30. The Toplevel instance has n Filter instances, and they form a loop as illustrated in Fig. 14a. We gave the same assume-guarantee contracts to Filter and Toplevel as in Example 30 and verified that Toplevel satisfies the contract.
- *DC motor control (MCtrl).* This is a more practical and typical example in which a motor and a controller are described as submodules M_2 and M_1 and form a feedback loop as illustrated in Fig. 14b. We annotated the top-level module with 8 safety (guarantee) properties, and submodules with 3 to 5 assume/guarantee properties.

We first verified the target hierarchical modules with the monolithic mode of Kind2. Second, we decomposed the target modules into the PC forms using the proposed method, and then verified that the submodules and the adapters satisfied their contracts (i.e. the conditions in Theorem 18) using Kind2. Note that, if we verify any of the examples with the compositional mode of Kind2, it will result in a spurious counterexample as explained in Example 25.

Experimental Result. The result is shown in Table 1. Each column shows wall clock time for the monolithic process or the process with the proposed method, and the total number of guarantee properties verified with the proposed method ("#Guarantees"). "TO" means the process did not terminate within 600 s.

Discussions. Since the submodules i.e. the digital filters, motor and controller behave in a stateful manner, using Kind2 to check the safety properties requires analyzing the execution prefixes of certain lengths, which would be time-consuming. Therefore, the verification of the system at once resulted in timeouts except for 2Filters. When the proposed method verified the examples by dividing them into modules, the process was more efficient because the numbers of rounds analyzed were reduced.

In each experiment for nFilters, the proposed method verified only a `Filter` instance because n verification conditions for submodules were for the same Lustre node. Therefore, differences in the value of n should appear only in the verification of adapter module; the number of read/write variables of the module and the number of corresponding guarantee properties would increase. The execution time increased slightly; although this was due to the simplicity of the top-level content, we consider the overhead of our method was small since increasing the number of variables did not have much effect. The effectiveness of our method was also confirmed in MCtrl.

7   Related Work

The hierarchy with nesting parallel compositions has been considered in an implementation [2] and extensions e.g. [4] of the reactive module theory. The hierarchization we consider in this paper is slightly different from the nesting of composition operations; ours corresponds to the embedding operation in dataflow diagrams. Alur et al. [5] address two kinds of hierarchies by agents and modules, and propose to perform reasoning on them separately; they do not consider a transformation between the two unlike this work.

More recently, there has been work on compositional verification of hierarchical Simulink models [8,13,14,17]. As in this paper, these methods give contracts to subsystems and verify the models according to a hierarchical structure. Boström and Wiik [8] propose to convert both models and contracts to specific dataflow graphs, then to sequential programs, and perform program verification. Their method does not support models with algebraic loops, and it is not clear whether it can handle the circular systems we consider. Dragomir et al. [14,21] introduce QLTL properties and dedicated refinement calculus to perform verification on hierarchical models interactively on Isabelle. Notably, they handle liveness properties which we do not. Their verification method requires human support unlike ours. Since they do not seem to provide any inference rules that explicitly deal with circular cases, it is unclear whether our method is applicable to their framework.

The Kind2 tool [11] supports the modular and compositional model checking of hierarchical Lustre programs as described in Sect. 3 and Sect. 5. It is limited in handling circular programs. Murugesan et al. [17] also propose a compositional model checking method for Simulink based on a similar abstraction method.

Dragomir et al. [13] propose to analyze hierarchical models with composite predicate transformers (CPTs) that use several composition operators, e.g. serial and parallel compositions and feedbacks. In comparison, we use only one composition operator and formalize the hierarchical structure in a fixed way. Tripakis et al. [22] formalize hierarchical dataflow diagrams and propose a profiling method to assure the modularity; although the subject is similar, their purpose is different from ours. Bakirtzis et al. [7] propose a general framework for various compositional CPS models, which includes concepts equivalent to modules and hierarchies and a formalization of contracts; however, they do not discuss either a transformation between the two concepts or verification methods. Fong and Spivak [15] formalize various hierarchical models of reactive systems, but do not consider synchronous behavior or assume-guarantee verification.

Automation of contract generation has been studied, e.g. [1,18,20]. We assume that contracts are given; the combination of our method and contract generation is a future issue.

8   Conclusion

We have formalized hierarchical synchronous systems based on the theory of reactive modules. We have then proposed a verification method that decomposes a hierarchical module into non-hierarchical modules and checks each module separately to show that the whole system satisfies the contract. As the experimental results show, the proposed method can effectively verify the systems with circular structures, which are suitable for describing plant control CPSs.

In general, compositional reasoning requires proof of the consistency among the verification results of each module, but this is not necessary when a top-level module is decomposed with our method. The proof task to analyze the system structure is delegated to the implementation relation on the adapter, and then it is efficiently discharged using a tool like Kind2.

Future work includes integrating the proposed method with other compositional methods such as for triggered modules and different-rate modules. Also, cooperation with automatic contract generation methods remains an issue.

Acknowledgements. Tien Duc Ngo contributed to the preliminary experiment of this work. This work was supported by JST, CREST Grant Number JPMJCR23M1 and JSPS, KAKENHI Grant Number 23K11969.

References

1. Abd Elkader, K., Grumberg, O., Păsăreanu, C.S., Shoham, S.: Automated circular assume-guarantee reasoning. Formal Aspects Comput. **30**(5), 571–595 (2018). https://doi.org/10.1007/s00165-017-0436-0
2. Alur, R., Henzinger, T.A., Mang, F.Y.C., Qadeer, S., Rajamani, S.K., Tasiran, S.: MOCHA: modularity in model checking. In: Hu, A.J., Vardi, M.Y. (eds.) CAV 1998. LNCS, vol. 1427, pp. 521–525. Springer, Heidelberg (1998). https://doi.org/10.1007/BFb0028774
3. Alur, R.: Synchronous Model. In: Principles of Cyber-Physical Systems, pp. 13–64. MIT Press (2015)
4. Alur, R., Grosu, R.: Modular refinement of hierarchic reactive machines. In: POPL, pp. 390–402 (2000). https://doi.org/10.1145/325694.325746
5. Alur, R., Grosu, R., Lee, I., Sokolsky, O.: Compositional modeling and refinement for hierarchical hybrid systems. J. Logic Algebraic Program. **68**(1–2), 105–128 (2006). https://doi.org/10.1016/j.jlap.2005.10.004

6. Alur, R., Henzinger, T.A.: Reactive modules. Formal Methods Syst. Des. **15**(1), 7–48 (1999). https://doi.org/10.1023/A:1008739929481
7. Bakirtzis, G., Fleming, C.H., Vasilakopoulou, C.: Categorical semantics of cyber-physical systems theory. ACM Trans. Cyber-Phys. Syst. **5**(3), 1–32 (2021). https://doi.org/10.1145/3461669
8. Boström, P., Wiik, J.: Contract-based verification of discrete-time multi-rate Simulink models. Softw. Syst. Model. **15**(4), 1141–1161 (2016). https://doi.org/10.1007/s10270-015-0477-x
9. Bretto, A.: Hypergraph Theory: An Introduction. Mathematical Engineering, Springer, Cham (2013). https://doi.org/10.1007/978-3-319-00080-0
10. Caspi, P., Pilaud, D., Halbwachs, N., Plaice, J.A.: LUSTRE: a declarative language for programming synchronous systems. In: POPL, pp. 178–188 (1987)
11. Champion, A., Mebsout, A., Sticksel, C., Tinelli, C.: The KIND 2 model checker. In: Chaudhuri, S., Farzan, A. (eds.) CAV 2016. LNCS, vol. 9780, pp. 510–517. Springer, Cham (2016). https://doi.org/10.1007/978-3-319-41540-6_29
12. Champion, A., Gurfinkel, A., Kahsai, T., Tinelli, C.: CoCoSpec: a mode-aware contract language for reactive systems. In: De Nicola, R., Kühn, E. (eds.) SEFM 2016. LNCS, vol. 9763, pp. 347–366. Springer, Cham (2016). https://doi.org/10.1007/978-3-319-41591-8_24
13. Dragomir, I., Preoteasa, V., Tripakis, S.: Compositional semantics and analysis of hierarchical block diagrams. In: Bošnački, D., Wijs, A. (eds.) SPIN 2016. LNCS, vol. 9641, pp. 38–56. Springer, Cham (2016). https://doi.org/10.1007/978-3-319-32582-8_3
14. Dragomir, I., Preoteasa, V., Tripakis, S.: The refinement calculus of reactive systems toolset. Int. J. Softw. Tools Technol. Transfer **22**(6), 689–708 (2020). https://doi.org/10.1007/s10009-020-00561-4
15. Fong, B., Spivak, D.I.: An Invitation to Applied Category Theory: Seven Sketches in Compositionality. Cambridge University Press, Cambridge (2019)
16. Giannakopoulou, D., Namjoshi, K.S., Păsăreanu, C.S.: Compositional reasoning. In: Handbook of Model Checking, pp. 345–383. Springer, Cham (2018). https://doi.org/10.1007/978-3-319-10575-8_12
17. Murugesan, A., Whalen, M.W., Rayadurgam, S., Heimdahl, M.P.: Compositional verification of a medical device system. In: ACM SIGAda Annual Conference on High Integrity Language Technology, pp. 51–64 (2013). https://doi.org/10.1145/2527269.2527272
18. Neele, T., Sammartino, M.: Compositional automata learning of synchronous systems. In: Lambers, L., Uchitel, S. (eds.) FASE 2023. Lecture Notes in Computer Science, vol. 13991, pp. 47–66. Springer, Cham (2023). https://doi.org/10.1007/978-3-031-30826-0_3
19. Ouvrard, X.: Hypergraphs: an introduction and review. CoRR arXiv:2002.05014 (2020)
20. Păsăreanu, C.S., Giannakopoulou, D., Bobaru, M.G., Cobleigh, J.M., Barringer, H.: Learning to divide and conquer: applying the L* algorithm to automate assume-guarantee reasoning. Formal Meth. Syst. Des. **32**(3), 175–205 (2008). https://doi.org/10.1007/s10703-008-0049-6
21. Preoteasa, V., Dragomir, I., Tripakis, S.: The refinement calculus of reactive systems. Inf. Comput. **285**, 104819 (2022). https://doi.org/10.1016/j.ic.2021.104819
22. Tripakis, S., Lublinerman, R.: Modular code generation from synchronous block diagrams: interfaces, abstraction, compositionality. In: Lohstroh, M., Derler, P., Sirjani, M. (eds.) Principles of Modeling. LNCS, vol. 10760, pp. 449–477. Springer, Cham (2018). https://doi.org/10.1007/978-3-319-95246-8_26

Anniversary

Two Decades of Industrializing Formal Verification: The Reactis Story

Rance Cleaveland[1]([⊠])(iD), David Hansel[2], Steve Sims[2], and Scott A. Smolka[3](iD)

[1] Department of Computer Science, University Maryland, College Park, MD 20742, USA
[2] Reactive Systems Inc., 341 Kilmayne Dr. Suite 101, Cary, NC 27511, USA
{hansel,sims}@reactive-systems.com
[3] Department of Computer Science, Stony Brook University, Stony Brook, NY 11794, USA
sas@cs.stonybrook.edu

Abstract. Reactis® is a suite of tools produced by Reactive Systems, Inc. (RSI), for automated test generation from, and verification of, systems given in either the modeling languages MATLAB® / Simulink® / Stateflow® of The MathWorks, Inc., or ANSI C. RSI was founded by three of the authors of this paper in 1999, with the first release of Reactis coming in 2002; the tools are used in the testing and validation of embedded control systems in a variety of industries, including automotive and aerospace / defense. This paper traces the development of the Reactis tool suite from earlier research on model-checking tools undertaken by the authors and others, highlighting the importance of both the foundational basis of Reactis and the essential adaptations and extensions needed for a commercially successful product.

Keywords: Reactis · Model-based testing · MATLAB / Simulink / Stateflow · Guided simulation · Instrumentation-based specification and verification

This paper is dedicated to the loving memory of the first author, Rance Cleaveland, who passed away suddenly and unexpectedly on March 27, 2024. Rance was the best friend, collaborator, and business partner one could hope for. He was instrumental in launching Reactive Systems in 1999, and in the early 2000s, he took a sabbatical from his distinguished academic career to work full-time at the company. During that period, he spent much of his time visiting engineers at different companies and using his natural rapport with people and gift for explaining complicated topics clearly to convince engineering teams to try our tools. After stepping back from day-to-day management, Rance continued to chair the Board of Directors and help set the strategic direction of Reactive Systems. He is deeply missed by all who knew him.

Supported by Small Business Innovation Research (SBIR) funding from the US National Science Foundation and Office of Naval Research.
www.reactive-systems.com

T. Neele and A. Wijs (Eds.): SPIN 2024, LNCS 14624, pp. 87–105, 2025.
https://doi.org/10.1007/978-3-031-66149-5_5

1   Introduction

Reactis®[1] [43] is a suite of tools for the validation of embedded control systems. The tool suite provides two key functionalities:

1. automated test generation from The Mathworks Inc.'s MATLAB® / Simulink® / Stateflow® (hereafter referred to collectively as "Simulink") modeling environment[2] models and ANSI C code; and
2. verification of user-specified requirements against models / code.

Reactis is developed and sold by Reactive Systems., Inc., a company founded by three of the authors of this paper (Cleaveland, Smolka and Sims) in 1999; the fourth author (Hansel) was the first non-founder technical employee of the company. The tools' primary user base consists of engineers in the automotive and aerospace / defense industries, where the use of Simulink and C is ubiquitous in the design and development of embedded control systems. Customers include companies in over 20 countries world-wide.

While targeting testing-based-validation use cases, Reactis has its intellectual foundations squarely in the formal-verification research community, and more specifically, in model checking [2,11,27]. In particular, the techniques Reactis uses to generate thorough suites of test cases from models / code are heavily influenced by the state-space search methodologies used in *on-the-fly* model checking [7,12,26,48] (which is also sometimes referred to as local or tableau-based model checking). They also employ constraint-solving procedures from Satisfaction Modulo Theories (SMT) [3,10,36] tools. This foundational basis is fundamental to the quality of Reactis' performance, although other aspects of the tools' functionality have played essential roles in its success in the marketplace.

This paper traces the development of Reactis, from its formal-verification roots, to its initial market-driven design, to its subsequent and ongoing development. In particular, we highlight how the tools' grounding in formal methods has been a key factor in their eventual uptake, but this basis would not have been sufficient on its own; other concerns have also been critically important. The remainder of this paper is structured as follows. The next section describes the origins of the Reactis tool, and in particular the research efforts of the RSI founders before the company's launch that informed the design and development of Reactis. Section 3 reviews the launch of RSI, the initial vision of the founders for the company as a developer and purveyor of model-checking tools, and a crucial strategic pivot made early in the life of the company to focus on Simulink. The section following discusses the initial design decisions of Reactis and their relation to the founders' choice to focus on Simulink. Section 5 focuses on the commercial considerations inherent in obtaining customers for Reactis, and how these considerations influenced subsequent additions and modifications to the Reactis tool suite. Section 6 gives a snapshot of the current Reactis tools, while the final section offers our concluding remarks and perspectives on the future.

---

[1] Reactis® is a registered trademark of Reactive Systems, Inc.

[2] MATLAB ®, Simulink® and Stateflow®are registered trademarks of The Math-Works, Inc.

2   Origins of Reactis: Formal Methods / Model Checking

This section describes the intellectual foundations of Reactis, with a focus on research by the company founders (Cleaveland, Sims and Smolka) that laid the groundwork for the launch of Reactive Systems Inc. and the design and development of Reactive Systems' tools.

Cleaveland and Smolka have been active researchers in the formal methods community since the 1980s, both in collaboration with one another and with others. Smolka was an early pioneer in the development of verification algorithms based on bisimulation equivalence for finite-state systems [29,30]. This style of verification has its roots in the process-algebra [5] community; it features the use of a (higher-level) system models as specifications for (lower-level) system implementations, with semantic relations such as bisimulation equivalence used to check conformance between specification and implementation. Cleaveland and others showed how bisimulation-equivalence and simulation-preorder [25] algorithms could be adapted to compute a variety of other semantic-refinement relations in process algebra [15]. Both Cleaveland and Smolka have also worked on model-checking algorithms for finite-state systems and temporal logics, with special attention to the *modal mu-calculus* [31], an expressive logic based on fixpoints that can encode all known temporal logics. In model-checking-based verification, specifications are given as formulas in a temporal logic, such as the modal mu-calculus; a system is deemed to satisfy the formula according to the logic's semantics. In the case of finite-state systems and the modal mu-calculus (as well as other temporal logics), this check can be decided algorithmically. Cleaveland and Smolka have each contributed efficient algorithms for various fragments of the modal mu-calculus [21,33,47] and have also designed strategies for parallelizing model checking [49] and achieving improved practical performance via translations of model checking into logic programming [41].

Of special note for the purposes of this paper is Cleaveland's and Smolka's work on so-called *on-the-fly* [26] methods (these methods have also been called *local* [48] and *tableau-based* [12]) for model-checking the mu-calculus [8,12,22,32]. Traditional model-checkers compute the entire state space of a system, then check whether the states in the state space satisfy the temporal formula in question. In contrast, on-the-fly checkers generate states in a demand-driven fashion, based on the formula being checked, leading to more efficiency when not every state requires checking in order to determine the verification result. This approach also permits approximations to model checking to be performed by allowing some successor states to be omitted.

All four of the authors of this paper have developed research tools that implemented verification algorithms and other analyses for finite-state systems. Smolka and colleagues developed the Winston tool [34], which supported the graphical creation of hierarchical networks of processes and included capabilities for verification based on bisimulation and observational equivalence [35]. Cleaveland and colleagues created the Concurrency Workbench (CWB) [16], a tool for verifying systems described in Milner's Calculus of Communicating systems (CCS) [35]. A noteworthy feature of that tool was the compositional design of its analysis routines; a variety of semantic equivalences and preorders were implemented based on the combination of core equivalence routines combined with semantic transformations of the systems being analyzed. The tool also included a mu-calculus model checker. Cleaveland and Sims, whose PhD work was supervised by Cleaveland, redeveloped the Concurrency Workbench, yielding the Concurrency Workbench of North Carolina (CWB-NC) [18]. That tool refactored the design of the original CWB by decoupling the verification routines from the syntax and semantics of CCS. This enabled the development of a tool, the Process Algebra Compiler [13], that, given syntactic and semantic specifications of a modeling language, generates the necessary routines for building a CWB-NC that analyzes system descriptions given in that language. Sims and colleagues also worked on checking invariants of so-called Software Cost Reduction (SCR) tabular specifications of systems [24] (these specifications are sometimes called *Parnas* tables after their inventor [40]). The result of this work was the Salsa tool for checking invariants of SCR models [6]. Hansel's master's thesis was supervised by Cleaveland and Smolka, and focused on the generation of prototype implementations of distributed systems from validated formal specifications; this work is described in [23].

3 The Launch of Reactive Systems, and a Strategic Pivot

This section discusses the activities leading to the founding of Reactive Systems, and also describes a key turning-point in the company's strategy that led to the development of Reactis.

3.1 Lead-up to the Founding of Reactive Systems

As the previous section notes, the co-founders of Reactive Systems had extensive experience in the development of model-checking-based verifiers of finite-state systems, including both theory and tool development as well as distribution of these tools, primarily to other research groups. In the late 1990s, Cleaveland and Smolka began discussing the possibility of commercializing verification tools based on their research. The .com boom was in full swing, and both thought that, given the successful uptake of formal verification in the so-called Electronic Design Automation (EDA) marketplace for hardware design, and the dearth of verification tools targeting software design despite the evident need for improving software quality, there were market opportunities that could be pursued. With other colleagues they developed a hierarchical modeling notation and analysis tool, the Concurrency Factory [14], heavily influenced by the Winston tool [34], for the compositional modeling and verification of networks of finite-state machines. Their intention was to commercialize the Concurrency Factory; consequently, they applied for, and subsequently obtained, a patent for the concept [19]. They then began looking for investors for the following idea: a modeling and verification environment for concurrent / distributed software, with an initial focus on telecommunications companies. Also during this time, co-author Hansel arrived at Stony Brook University to work on his master's thesis under the supervision of Cleaveland and Smolka.

Meanwhile, after finishing his PhD studies at North Carolina State University Sims had taken a job as a research scientist at the US Naval Research Laboratory and was working on the Salsa tool [6] described above. Cleaveland and Smolka approached him to see if he might be interested in joining their commercialization effort; he agreed, and in 1999 the three incorporated Reactive Systems, Inc., still with the idea of developing and selling verification tools for concurrent and distributed systems based on the Concurrency Factory's bespoke modeling framework.

3.2 Early Reactive Systems: The Concurrency Factory

Once the company was launched, the founders needed capital to fund the development of the envisioned Concurrency Factory tool. In late 1999, they applied for a grant from the National Science Foundation (NSF) through the agency's Small-Business Innovation Research program, which provides research awards to small companies to help them commercialize basic research. They were successful in this endeavour; in June 2000, NSF awarded Reactis Systems enough funds beginning June 2000 for six months of exploratory research for the Concurrency Factory. The NSF grant also included a possibility for two years of additional follow-on funding if the results of the first six-month phase were promising.

An important meeting. The company also needed a business plan, which required market research, so the founders had begun attending industry-leaning conferences to gain insights into the prevailing commercial practices in software quality. At the 19th Digital Avionics Systems Conference (DASC) in October 2000, Sims met a key contact, an engineer from an automotive OEM[3]. The OEM engineer and Sims talked about the plans for Reactive Systems and the software-verification techniques of the Concurrency Factory, as well as about the software-validation techniques being used within the OEM. The engineer admitted that a tool like the Concurrency Factory would be a hard sell within his company, but that automated support for testing software against Simulink models would be of serious interest. The OEM at that time was using Simulink to design automotive subsystems, and there was an internal effort underway for using these models to guide the development of the embedded software to be run in these subsystems. Sims was intrigued, and the OEM engineer offered to send

---

[3] In the automotive, and other, industries, companies are referred to as *Original Equipment Manufacturers*, or OEMs, if they produce end-of-the-line products for customers outside the industry. Car companies such as BMW and Toyota are OEMs, for example. There is terminology for suppliers in these industries as well. Tier-1 suppliers sell their products to OEMs, while Tier-2 suppliers sell to Tier-1 suppliers, etc. As an example, Robert Bosch is a Tier-1 automotive supplier that sells components such as anti-lock braking subsystems to automotive OEMs. Companies selling e.g. electronic components to Bosch would then be considered Tier-2 suppliers.

Reactive Systems a couple of their Simulink models to see what our analyses might infer about them.

This interaction between Sims and the OEM engineer, and the subsequent collaborative pilot study that ensued, had a couple of profound long-term impacts on the technical and business strategy of Reactive Systems.

First, the collaboration gave us our first experience working with automotive engineers who were developing embedded software applications. In the early 2000s, the use of software in vehicles was just beginning its rapid growth towards its current state, in which each new vehicle model contain dozens of microprocessors and millions of lines of code. To this day, automotive customers make up the largest component by far of Reactive Systems' customer base.

Second, the work with the automotive pilot study introduced us to Simulink as a modeling notation for designing and implementing automotive applications. As a startup with limited resources, the decision of which notation(s) our analysis tools would support was one of the most important choices we had to make. We also were in the midst of our initial NSF funding, and we had to decide whether to devote these resources to continuing to develop an initial prototype of the Concurrency Factory, which was based on our home-grown modeling notation, or to work with our automotive OEM collaborator on the Simulink-model analysis study. In the end we gambled that developing a relationship with a potential reference customer was a better way forward than purely focusing on developing our own modeling technology. Consequently, even though our familiarity with Simulink was non-existent (we did know about MATLAB, however), we elected to devote the remaining time of our initial NSF funding to pursuing the Simulink pilot study with our OEM partner.

Simulink pilot study. After Sims' return from DASC he contacted the OEM engineer, and they defined the parameters of a pilot study involving one of the OEM's Simulink models (part of a powertrain[4] application). There was of course insufficient time to create any sort of analysis tool to run on the model. Instead, we agreed to apply the Salsa invariant checker to the Simulink model provided by the OEM [46], in order to see if there were any issues in the model the OEM should be aware of. The model contained a number of Simulink subsystems and Stateflow diagrams; the C code implementing the the model consisted of approximately a thousand lines. Our experience has been that models of interest to industry developers are rarely any smaller than this. In other words, any tool we wanted to develop would need to handle models with at least this much functionality.

To perform the analysis, we hand-translated the Simulink model into the SAL input notation supported by Salsa. Simulink is a large language with a complicated semantics, so this translation was the most time-consuming portion of the work. It also clarified that expecting a customer to perform such a hand-

---

[4] The *powertrain* of an automobile consists of all components responsible for delivering power to a vehicle's wheels. These include the engine, driveshaft, transmission, axles, and differentials.

translation, or to implement an automated translator, was clearly impractical: any tool would need to natively support a language, such as Simulink, directly used by embedded software developers.

After performing this hand translation, we then applied Salsa to the generated model and discovered several anomalies in the original Simulink model, including dead code (i.e. parts of the model that would never execute) and a number of violations of the OEM's modeling policies. These results caught the attention of the OEM team, since the Simulink model had previously undergone, and passed, a thorough manual review by OEM engineers. On the other hand, some limitations of our approach were also clear. First, as mentioned above, expecting engineers to translate Simulink models into a notation like SAL was a non-starter. Second, even if an automated translation from Simulink to SAL were implemented, we would still be constrained by the expressiveness of SAL, which supported only boolean and integer state variables and linear constraints. Based on the pilot study and subsequent discussions with the OEM team, we learned that most automotive applications, at least in the powertrain area, included both floating-point variables and non-linear constraints.

3.3   From the Concurrency Factory to Reactis

We received enthusiastic feedback from our OEM collaborators in response to the pilot study, and this became part of our application to NSF for a second, two-year round of funding. We planned to use this funding, which we were awarded in March 2001, to support the construction of the first commercial release of a tool that we could market to customers. The positive results from the pilot study led us to believe that the automotive industry would be a fertile one for our offerings, especially if our tools worked with Simulink. However, pursuing this strategy would be at odds with our initial vision of selling model-checking tools based on our own, Concurrency Factory, notation. What course should the company pursue?

Of course, from a current-day perspective, it is clear what decision we arrived at; we abandoned the Concurrency Factory and instead focused on developing testing and validation tools for Simulink, i.e. Reactis. While this strategic shift was initially driven in our minds by a market opening we perceived in a specific industry, namely, automotive, our decision also exposes important considerations regarding the commercialization of formal-verification tools. In academic formal-methods research, much work focuses on algorithms and analysis routines; less effort has traditionally been devoted to notational issues, such as modeling and specification languages. Reactive Systems' original focus on the Concurrency Factory was very much rooted in the founders' more foundational research; the expectation was that industrial users would see the benefits of formal verification and easily adapt their development processes to Reactive Systems' home-grown notations. Based on our pilot study, however, we realized that notational issues are crucially important in practice: companies want assurances that the notations they train their engineers in will be used in years to come, and they want an ecosystem of consultants and technical support to engage with on developing their design processes to accommodate new technologies. For all its intellectual rigor, the Concurrency Factory did not address these concerns. Simulink, on the other hand, had already seen significant uptake in the automotive industry, and there was substantial support for incorporating these notations into control-system design practice. Consequently, building validation and testing tools targeting Simulink removed a key barrier to developing a customer base that he Concurrency Factory would have almost certainly been unlikely to overcome.

4   Desiderata in the Design of Reactis

In order to achieve our goal of having a commercial version of Reactis ready before the second phase of our NSF funding ran out, we knew that additional technical talent would need to be added to the team. Consequently, in 2001, co-author David Hansel accepted our offer for him to join Reactive Systems.

With our technical team in place, we then needed to make a number of decisions about the design and development of Reactis. This section highlights some of the important choices that we took.

4.1   Supporting Simulink

Our experience during the pilot study with our OEM collaborators told us that supporting Simulink would be the promising route to obtaining customers in the automotive realm. The question was, how should we do this?

We decided to retain our original verification strategy, which was based on the state exploration / back-tracking approaches of on-the-fly model checking, but to re-purpose these ideas for validating Simulink diagrams. We explain the rationale for this choice more below; for now, however, we note that doing state-space exploration on Simulink diagrams of course requires the capability of generating states, computing transitions, and *backtracking*, i.e. restoring the system state to a previously constructed and stored state.

Our original hope was to use the Mathworks' simulation engine for these purposes; this would ensure that our state and transition computations would be fully consistent with those of Simulink, and also prevent us from having to implement these routines ourselves. However, the MathWorks APIs did not export the necessary functionality, and in our estimation it was quite unlikely that The MathWorks would modify these APIs to suit our purposes.

We opted instead to build our own Simulink interpreter. This approach gave us full access to all internals of states and transitions and in particular enabled us to retrieve model information necessary for state exploration with backtracking. There were significant on-going costs and challenges associated with this decision, however. To begin with, Simulink is a complex language, with a large number of operators (or "blocks") tuned for modeling control systems, and we would need to guarantee to our customers that our interpreter implemented the same semantics as The Mathworks, even as new versions of these notations were / are released. In addition, as a proprietary language, Simulink did not have a reference semantics. This lack of an independent definition of the behavior of Simulink has led to two different types of challenges over the years. The first is that corner cases in the meanings of blocks have to be uncovered via trial and error and extensive testing. The second is that customers sometimes misunderstand the intended semantics of the Simulink and will report bugs in Reactis that are, in fact, not bugs at all, but rather due to misunderstandings about the behavior of Simulink.

Our Simulink interpreter is built around a home-grown intermediate representation based on ideas from *synchronous languages* [4], of which Simulink is also an example, at least for the language's discrete (as opposed to continuous) semantics. Our simulator translates Simulink into this intermediate language; it then computes states and transitions for use in simulation on the intermediate representations. (We briefly considered using C as this intermediate notation, since there are tools for translating Simulink into C. However, these translators involve significant license fees, and for cost reasons we decided to avoid them.)

4.2   Supported Verification Approaches

Our original, Concurrency-Factory-inspired, vision featured the use of model checking to automatically determine if a given system model satisfied a formula in temporal logic. We initially considered adapting this paradigm to Simulink, but it did not fit well with our prospective customers design processes, which did not involve the formulation of such properties. However, our early discussions with industry professionals revealed interest in generic checks that should be passed by all models. For example, no model should have dead code or runtime errors such as divide-by-zero, out-of-bounds array indexing, or numerical overflows.

There was also another key role played by models in the development processes of the companies we spoke with: as specifications for the eventually deployed vehicular sub-systems (including software). For example, a model might be developed for a vehicle-transmission controller. The real-world controller hardware would contain a microprocessor on which software would be deployed to implement the behavior contained in the model. Confirming that the software source code, and the sub-system containing the deployed executable generated from the software, behaved consistently with the the model was of great interest because of the possibility of detecting and fixing errors before the sub-system was integrated into a larger vehicle design. Our industrial contacts, which had grown to included automotive OEMs and Tier-1 suppliers in the US, Europe and Japan, wondered if comprehensive suites of test cases could be automatically generated from models and used in downstream testing, with the model results being used as the oracle for judging the outcomes of software and system testing. This use of models in testing was, and is, routinely referred to as *Model-Based Testing* (MBT), and is part of a larger design paradigm called Model-Based Design (MBD) that The MathWorks markets to its customers.

We were aware of uses of model checkers for generating test cases [1, 42] in the research literature and decided to equip Reactis with support for MBT.

This Reactis feature allows users to decide on a level of "thoroughness" of the test suite by specifying structural coverage criteria, such as Condition coverage, Decision coverage, and Modified Condition / Decision Coverage (MC/DC) [9] of the model. Reactis then attempts to generate test suites from the model that achieve 100% coverage of the selected criteria. While these criteria have been criticized in the literature for being insufficiently rigorous, they have the benefit of being widely understood in industry and are included in various software safety standards for automotive [28] and aerospace [44,45].

The tool also needed support for user interaction to generate test cases, for a couple of reasons. First, users often had existing tests that they wanted incorporated into Reactis-generated test suites. Second, obtaining 100% coverage may be impossible to achieve in reasonable time, and users need to be able to manually add new tests to cover parts of the model that the Reactis test generator could not. Thus, we decided the tool needed an interactive test generator to help users with these tasks, in addition to a fully automatic test generator.

Some users also wanted to undertake model validation via testing. The general approach we decided to offer, which we call *instrumentation-based verification* [20], is based on ideas similar to the use of assertion-checking in software debugging. In traditional assertion-based debugging, users embed assertions, or boolean properties, at points in their code; at run time, these assertions are checked as the program executes, and violations are highlighted. In instrumentation-based verification, the idea is to allow users to define assertions in the form of *monitor models*, or small Simulink models with boolean outputs, that could be used as instrumentation for the main Simulink model. Generated test suites can then be run on the instrumented model, with any false outputs of monitor models flagged as errors.

Based on these considerations the earliest versions of Reactis, as well as current versions of software, included the following tools.

1. Reactis Tester generates a comprehensive yet compact test suite from a Simulink model.
2. Reactis Simulator allows users to interactively run and modify test suites on models, as well as debug models.
3. Reactis Validator checks if a Simulink model satisfies properties formulated as monitor models that have been instrumented into the model.

Reactis also uses static analysis to check for dead code, as well as a variety of run-time errors.

4.3   Verification Engine

With our desired verification functionality identified, we next needed to decide how best to implement it. In particular, the most important, and also most computationally intensive, operation was test-suite generation, and our experiences to that point led us to believe it would not be feasible to directly use existing model checkers or automatic theorem provers. The typical Simulink models that we had seen were too complex for state-of-the-art techniques then available; they often contain hundreds or even thousands of boolean, floating-point, and integer variables manipulated with the very rich (and continually growing) set of operators supported by Simulink.

Our solution to this problem was to develop (and patent) a technique, called *guided simulation* [17], for generating test cases that maximize coverage criteria. The rough idea is the following. A given coverage criterion for test suites defines a set of coverage targets that the suite must hit. For example, in Condition coverage, each atomic boolean expression in a model must be made both true and false by some test in the test suite in order for the test suite to fully cover the model. Each such expression therefore gives rise to two coverage targets: one corresponding to the expression being true, and one associated with it being false. In Reactis, a test case corresponds to a simulation run of the Simulink model being tested, with each step in the run specifying the values of the (unconnected) top-level inputs and also recording the (unconnected) top-level output values produced in response to these inputs. Thus, generating a test suite attaining 100% coverage of a given coverage criterion can be phrased as a search problem in the space of model simulation runs, with the objective of the search being to ensure that every coverage target is hit by some test case. This general search paradigm was inspired by our work with on-the-fly model-checking algorithms described in Section 2, which in effect search for proofs that a state in a model satisfies a given temporal property.

More specifically, our guided-simulation test-generation technique works by generating simulation steps, starting from an initial state of the model and tracking which coverage targets have been hit and which have not. From an intermediate state in a simulation run, Reactis selects the next inputs to provide in order to guide the next step in the simulation to the corresponding target next state, updating aggregate coverage-target data in the process. Input selection, which is a key mechanism (along with backtracking) for guiding the simulation being constructed, in turn involves two different strategies.

Constraint solving. A given coverage target can be viewed as a condition that must hold in the middle of the execution of a model. When the system is in a given state, we use techniques based on weakest preconditions to compute constraints on top level inputs whose solutions will guarantee coverage of the given target. We then attempt to solve these constraints using *Satisfaction Modulo Theories* (SMT) technology. The initial SMT solvers included in Reactis were home-grown and supported boolean constraints and linear constraints over reals (to approximate floating-point constraints). We later switched to use the Z3 solver developed by Microsoft [36].

Monte Carlo. Values for top-level inputs are generated randomly, the target state of the simulation step is computed, and coverage information is updated appropriately. These techniques have different strengths. When constraints can be formulated and solved, SMT-generated inputs are guaranteed to extend coverage. However, constraints may fall outside the theories supported by the constraint solver, or have solutions that are too computationally expensive to calculate. Indeed, for large models, even constructing the constraints can be infeasible. The Monte Carlo method on the other hand generates new inputs very quickly and works for any model supported by the interpreter, but at the cost of not guaranteeing growth in coverage.

Users can also place range and rate-of-change constraints on the top-level inputs of a model within Reactis. Reactis then ensures that these constraints are obeyed by any input values it generates, whether by constraint solving or randomly.

Our search strategy also uses a number of heuristics to improve test-suite coverage. These include, but are not limited to, ones for determining when a given state is a good candidate for being the source of potential future coverage and saving it; ones for when to abandon a given Monte Carlo run because coverage is not growing and backtracking to previous states that are promising; and ones for simplifying constraints to improve solver performance.

4.4   More on Coverage Metrics

The notion of coverage metrics / criteria for tests created from Simulink models derives from classical work done in the software engineering community for programming languages [37]. Some well-known metrics defined for (traditional imperative) programming languages include the following.

Statement coverage. Has every statement in a program been executed?

Decision coverage. Has every decision (i.e. maximal boolean-valued expression such as those appearing in loop guards and if-then-elese statements) in a program evaluated to both true and false?

Condition coverage. Has every atomic boolean-valued expression in a program evaluated to both true and false?

Reactis adapts these concepts to Simulink models. In addition, Reactis supports a number of metrics tailored to the Simulink notation. For example, the state-coverage criterion measures how many Stateflow states have been entered, while the transition-coverage measure tracks the number of Stateflow transitions that have fired.

The collection of metrics supported by Reactis has also expanded over time. One early evolution resulted from our work with aerospace customers. The DO-178B [44] (later updated to DO-178C [45]) safety standard for aviation software requires that the most safety-critical software components be tested to achieve 100% coverage for the modified condition/decision coverage (MC/DC) metric [9]. The ISO 26262 automotive safety standard [28] subsequently adopted

MC/DC as the level of testing required for safety critical automotive software; this created significant interest in this metric among our automotive customers, and we added support for it into Reactis.

Another driver for expanding the Reactis coverage metrics was due to a desire on the part of customers to have the coverage data for Reactis-generated tests from Simulink models match the coverage results obtained from running the test cases on (automatically generated) source code. For example, in 2009 we introduced so-called *multi-block MC/DC* coverage for Simulink into Reactis. This criterion allows for the aggregation of a collection of Simulink Logical Operator blocks into a single decision for the purpose of MC/DC tracking. This criterion closely mimics the decisions created in C code that is generated from such models, and which typically involves nested boolean expressions.

5 Selling Reactis

In June of 2002, we released the first commercial version of Reactis, and the business activities of marketing and selling the tool suite moved to the forefront of our concerns. There are many aspects of building a community of customers and users around a tool, and we will not describe all of these here. However, some features of building and maintaining a successful commercial tool have impacts on decisions regarding its technical development, and we describe several of these in this section.

Engineers in industry, especially those in a company's *business units* (as opposed to research and development), have numerous responsibilities related to a company's source of revenue — its products and services. For this reason, they typically have limited time to explore new tools and think about how they might evolve their development processes to accommodate them. In order to successfully sell Reactis, we needed to develop materials, including white papers, presentations and other marketing collateral, that explained not only the functionality of Reactis, but also how it could be deployed in different industrial scenarios. A substantial amount of our post-release resources were devoted to developing these materials, and to visiting prospective customers both to market our tool but also to learn how their internal model-based design processes were organized, so that we could offer tailored suggestions on how Reactis could be usefully deployed. We also offered training courses for customers at their sites.

In our experience, driving adoption of Reactis within companies has required extensive collaboration between Reactive Systems and the Reactis users within those companies. New users will need training, and users of all types will from time to time need advice on how to access and best use tool functionality. Users will also stumble across functionality they don't expect (either intended, or buggy), and want guidance from Reactive Systems on what to do. For its part, Reactive Systems gains direct insights into user experiences through its interactions with users, which give valuable feedback for refining Reactis to flatten the learning curve and best address specific needs in user-development processes. For this reason, the company has policies for responding quickly to all user requests.

These interactions consumes significant resources, but we view user support as a key component of our business strategy.

Tool users also often request new tool features. A case in point for Reactis was its treatment of C code embedded in Simulink diagrams. Simulink offers several ways to incorporate C code into a model, including S-Functions, C Caller blocks, and C code called from Stateflow. In each case, a portion of the model's functionality is determined by this C code. Early versions of Reactis treated C code in a model as a collection of black boxes. Reactis would compile the embedded C code and use it to compute the semantics of the given model component, thereby enabling simulation and test generation for these models. However, coverage was not tracked in the C code, and it was not possible to step into the C code for debugging (another use case of the Reactis Simulator). In response to user requests for the capability of tracking coverage information for this C code, in 2007 Reactive Systems launched the Reactis for C Plugin product, which enabled white-box testing for the C code parts of a model. With the C Plugin, coverage targets are identified within the C code, and Reactis Simulator is able to step into the C code during simulation. After positive feedback from customers on the C Plugin, in 2011 the company packaged this functionality for C into a separate tool suite, Reactis for C, which supports Reactis validation capabilities, including test generation, coverage tracking, and assertion verification, on ANSI C code. This tool reused significant parts of the original Reactis verification engine and drew upon experience we gained in developing customized versions of the Concurrency Workbench of North Carolina using the Process Algebra Compiler [18].

Of course, promising new functionality but not delivering it until much later can sour tool users on the tool in question. In the case of Reactis, we addressed this by scheduling new releases twice a year and also making beta versions of new functionality we were planning on releasing to users who had requested it.

The importance of engagement with, and responsiveness to, users cannot be overstated, and it highlights a crucial difference between traditional research tools and commercial ones: significant technical resources must be devoted to ensuring users can effectively and efficiently use tool functionality in order for the tools to be successful in the marketplace. In an academic setting, it can be difficult to justify these expenditures, since researchers and students naturally wish to focus on implementing new ideas of their own. In an industrial setting, on the other hand, without users there cannot be a tool!

6   Current State of Reactis

The most recent version, V2023.2, of Reactis was released in December of 2023. This release includes new versions of the following tools.

- Reactis Tester, Reactis Simulator and Reactis Validator, which target test-generation, model debugging and model validation for Simulink.

- Reactis for C Plugin, which is used in conjunction with the Reactis tools for Simulink to provide analytical capabilities C code that may be embedded in models.
- Reactis for C, which provides testing, simulation and testing capabilities for ANSI C code.

Each of the above categories of tools are licensed separately. That is, users would buy separate licenses to Reactis, which includes Tester, Simulator and Validator; to Reactis for C Plugin (which requires a license for Reactis); or to a license to Reactis for C (which is independent of the other tools).

In addition, the V2023.2 release includes licensing possibilities for two additional tools. The Reactis for EML Plugin, first released in 2015, supports white-box testing of the Embedded MATLAB (EML) portions of a model. EML is the subset of the MATLAB language that MathWorks supports for the generation of embedded C code. Simulink offers the capability to incorporate EML into a Simulink model, for example, in a MATLAB Function block or as a callable from Stateflow. With the EML Plugin, coverage targets are identified within the EML code, Reactis Tester attempts to exercise these targets, and Reactis Simulator steps into EML code for debugging and coverage visualization. To use the Reactis for EML Plugin, a user must also have a license for Reactis. Our Reactis Model Inspector tool enables users to navigate models. It does not include testing or validation functionality and is intended to support (read-only) model auditing and review.

7   Conclusions and Perspectives on the Future

In this paper we have given a brief history of the development of Reactis, a commercial tool suite we have developed over the past 20 years that uses ideas from formal methods to provide testing and validation support for Simulink and ANSI C. In the course of the paper, we have highlighted both the foundations of Reactis in model-checking research but also places where we needed to depart from the academic tool-development paradigm, and why. We are very strong believers in the value of formal methods. We also believe that adapting the techniques to industrial settings requires sustained effort and insight into company needs and concerns, so that engineers in these organizations can derive the benefits of this technology and verification-tool developers can maximize the impact of their offerings.

In terms of future plans for Reactis, at present, companies in over 20 different countries have acquired Reactis. Most users are in the automotive industry, although a significant number of users work in aerospace and defense companies. In general, different industries have different modeling cultures and use different notations as a result. While Simulink users are widespread, other modeling tools, such as LabVIEW [38], have strong user bases as well. Developing a "Reactis generator" for such modeling notations is something we see as a possibly fruitful direction for future development, as a way of expanding the Reactis market. Of course, such a development would need a business case, to include not only a significant body of users but also an industrial culture that is willing to pay for tools that support software design, development, and validation. We also are tracking developments in SMT tools with a view towards enhancing future versions of Reactis.

In terms of future prospects for industrial uptake of formal verification, we see reasons for optimism, provided the tools are appropriately targeted. For example, we do not see much of a market for "generic software-verification tools" that would be on par with, for example, compilers in terms of their widespread adoption. A primary reason for our pessimism on this front is two-fold. On the one hand, generic software-development tools are widely expected to be no-cost in the marketplace, which limits the financial return available to developers wishing to introduce formal-methods capabilities into them. On the other hand, much "generic software" has traditionally had relatively relaxed expectations in terms of functional correctness on the part of the software's users, meaning software companies have incentives to devote more development resources to new features and less to quality assurance. However, in many "non-software" industries whose products include significant software, expectations on the part of users, regulatory agencies, and insurance companies for correct functional behavior are much higher, and we see ongoing significant opportunities for formal methods in these sectors. The automotive and aerospace / defense sectors fall into this category; so do medical devices, household appliances, infrastructure (electricity, water, transportation, data communications, etc.), and banking, to name a few. Finally, consumers, companies, and governments are increasingly concerned about privacy and cybersecurity. While historically there may have been a certain tolerance for software functionality shortcomings, there is growing push-back against cybersecurity vulnerabilities. A 2024 report [39] by the US Office of the National Cybersecurity Director highlights the importance of formal methods as a tool for improving cybersecurity, and the cloud-services sector, for example, offers an interesting future opportunity for commercial applications of formal methods for this purpose, in our view.

References

1. Ammann, P., Black, P., Majurski, W.: Using model checking to generate tests from specifications. In: Second International Conference on Formal Engineering Methods, pp. 46–54 (1998). https://doi.org/10.1109/ICFEM.1998.730569
2. Baier, C., Katoen, J.P.: Principles of Model Checking. MIT Press (2008)
3. Barrett, C., et al.: CVC4. In: Gopalakrishnan, G., Qadeer, S. (eds.) CAV 2011. LNCS, vol. 6806, pp. 171–177. Springer, Heidelberg (2011). https://doi.org/10.1007/978-3-642-22110-1_14
4. Benveniste, A., Caspi, P., Edwards, S., Halbwachs, N., Le Guernic, P., De Simone, R.: The synchronous languages 12 years later. Proc. IEEE **91**(1), 64–83 (2003)
5. Bergstra, J.A., Ponse, A., Smolka, S.A.: Handbook of Process Algebra. Elsevier (2001)

6. Bharadwaj, R., Sims, S.: Salsa: combining constraint solvers with BDDS for automatic invariant checking. In: Graf, S., Schwartzbach, M. (eds.) Tools and Algorithms for the Construction and Analysis of Systems, pp. 378–395. Springer, Berlin, Heidelberg (2000). https://doi.org/10.1007/3-540-46419-0_26

7. Bhat, G., Cleaveland, R., Grumberg, O.: Efficient on-the-fly model checking for CTL. In: 10th Annual IEEE Symposium on Logic in Computer Science. pp. 388–397. IEEE (1995)

8. Bhat, G., Cleaveland, R.: Efficient local model-checking for fragments of the modal μ-calculus. In: Margaria, T., Steffen, B. (eds.) Tools and Algorithms for the Construction and Analysis of Systems, pp. 107–126. Springer, Berlin, Heidelberg (1996). https://doi.org/10.1007/3-540-61042-1_41

9. Chilenski, J., Miller, S.: Applicability of modified condition/decision coverage to software testing. Softw. Eng. J. 9(5), 193–200 (1994)

10. Cimatti, A., Griggio, A., Schaafsma, B.J., Sebastiani, R.: The MathSAT5 SMT solver. In: Piterman, N., Smolka, S.A. (eds.) Tools and Algorithms for the Construction and Analysis of Systems, pp. 93–107. Springer, Berlin, Heidelberg (2013). https://doi.org/10.1007/978-3-642-36742-7_7

11. Clarke, E., Emerson, E., Sistla, A.: Automatic verification of finite-state concurrent systems using temporal logic specifications. ACM Trans. Program. Lang. Syst. 8(2), 244–263 (1986)

12. Cleaveland, R.: Tableau-based model checking in the propositional μ-calculus. Acta Informatica 27, 725–747 (1990)

13. Cleaveland, R., Madelaine, E., Sims, S.: A front-end generator for verification tools. In: Brinksma, E., Cleaveland, W.R., Larsen, K.G., Margaria, T., Steffen, B. (eds.) TACAS 1995. LNCS, vol. 1019, pp. 153–173. Springer, Heidelberg (1995). https://doi.org/10.1007/3-540-60630-0_8

14. Cleaveland, R., Gada, J.N., Lewis, P.M., Smolka, S.A., Sokolsky, O., Zhang, S.: The concurrency factory - practical tools for specification, stimulation, verification, and implementation of concurrent systems. Specif. Parallel Algorithms 18, 75–90 (1994)

15. Cleaveland, R., Hennessy, M.: Testing equivalence as a bisimulation equivalence. Formal Aspects Comput. 5(1), 1–20 (1993). https://doi.org/10.1007/BF01211314

16. Cleaveland, R., Parrow, J., Steffen, B.: The concurrency workbench: a semantics-based tool for the verification of concurrent systems. ACM Trans. Program. Lang. Syst. 15(1), 36–72 (1993)

17. Cleaveland, R., Sims, S.T., Hansel, D.: System and method for automatic test-case generation for software, US Patent 7,644,398 (2010)

18. Cleaveland, R., Sims, S.T.: Generic tools for verifying concurrent systems, special issue on engineering automation for computer based systems. Sci. Comput. Program. 42(1), 39–47 (2002)

19. Cleaveland, R., Smolka, S.A., Lewis, P.M., Ramakrishna, Y.: Specification and verification for concurrent systems with graphical and textual editors, US Patent 6,385,765 (2002)

20. Cleaveland, R., Smolka, S.A., Sims, S.T.: An instrumentation-based approach to controller model validation. In: Broy, M., Krüger, I.H., Meisinger, M. (eds.) Model-Driven Development of Reliable Automotive Services, pp. 84–97. Springer, Berlin, Heidelberg (2008). https://doi.org/10.1007/978-3-540-70930-5_6

21. Cleaveland, R., Steffen, B.: A linear-time model-checking algorithm for the alternation-free modal μ-calculus. Formal Methods Syst. Des. 2, 121–147 (1993)

22. Du, X., Smolka, S.A., Cleaveland, R.: Local model checking and protocol analysis. Int. J. Softw. Tools Technol. Transf. 2, 219–241 (1999)

23. Hansel, D., Cleaveland, R., Smolka, S.A.: Distributed prototyping from validated specifications. J. Syst. Softw. **70**(3), 275–298 (2004)
24. Heitmeyer, C., Jeffords, R., Labaw, B.: Automated consistency checking of requirements specifications. ACM Trans. Softw. Eng. Methodol. **5**(3), 231–261 (1996)
25. Henzinger, M., Henzinger, T., Kopke, P.: Computing simulations on finite and infinite graphs. In: IEEE 36th Annual Foundations of Computer Science, pp. 453–462. IEEE (1995)
26. Holzmann, G.: On-the-fly model checking. ACM Comput. Surv. (CSUR) **28**(4es), 120–es (1996)
27. Holzmann, G.: The model checker SPIN. IEEE Trans. Softw. Eng. **23**(5), 279–295 (1997)
28. ISO 26262: Road vehicles - functional safety (2011)
29. Kanellakis, P., Smolka, S.: CCS expressions, finite state processes, and three problems of equivalence. In: Second Annual ACM Symposium on Principles of Distributed Computing, pp. 228–240 (1983)
30. Kanellakis, P., Smolka, S.: CCS expressions, finite state processes, and three problems of equivalence. Inf. Comput. **86**(1), 43–68 (1990)
31. Kozen, D.: Results on the propositional μ-calculus, special issue ninth international colloquium on automata, languages and programming (ICALP), Aarhus, Summer 1982. Theor. Comput. Sci. **27**(3), 333–354 (1983).https://doi.org/10.1016/0304-3975(82)90125-6
32. Liu, X., Ramakrishnan, C.R., Smolka, S.A.: Fully local and efficient evaluation of alternating fixed points. In: Steffen, B. (ed.) TACAS 1998. LNCS, vol. 1384, pp. 5–19. Springer, Heidelberg (1998). https://doi.org/10.1007/BFb0054161
33. Liu, X., Smolka, S.: Simple linear-time algorithms for minimal fixed points. In: Automata, Languages and Programming, pp. 53–66. Springer, Berlin, Heidelberg (1998). https://doi.org/10.1007/bfb0055040
34. Malhotra, J., Smolka, S., Giacalone, A., Shapiro, R.: Winston: a tool for hierarchical design and simulation of concurrent systems. In: Rattray, C. (ed.) Specification and Verification of Concurrent Systems, pp. 140–152. Springer, London (1990). https://doi.org/10.1007/978-1-4471-3534-0_7
35. Milner, R.: Communication and Concurrency. Prentice Hall, Englewood Cliffs, New Jersey (1989)
36. de Moura, L., Bjørner, N.: Z3: an efficient SMT solver. In: Ramakrishnan, C.R., Rehof, J. (eds.) Tools and Algorithms for the Construction and Analysis of Systems, pp. 337–340. Springer, Berlin, Heidelberg (2008). https://doi.org/10.1007/978-3-540-78800-3_24
37. Myers, G., Badgett, T., Thomas, T., Sandler, C.: The Art of Software Testing. Wiley, USA (2004)
38. National Instruments. https://www.ni.com/en/shop/labview.html
39. The While House, Office of the National Cybersecurity Director: Back to the Building Blocks: A Path Towards Secure and Measurable Software (2024)
40. Parnas, D.: Tabular representation of relations. Technical report 260, Communications Research Laboratory, McMaster University (1992)
41. Ramakrishna, Y.S., Ramakrishnan, C.R., Ramakrishnan, I.V., Smolka, S.A., Swift, T., Warren, D.S.: Efficient model checking using tabled resolution. In: Grumberg, O. (ed.) CAV 1997. LNCS, vol. 1254, pp. 143–154. Springer, Heidelberg (1997). https://doi.org/10.1007/3-540-63166-6_16

42. Rayadurgam, S., Heimdahl, M.: Coverage based test-case generation using model checkers. In: Eighth Annual IEEE International Conference and Workshop On the Engineering of Computer-Based Systems–ECBS 2001, pp. 83–91 (2001). https://doi.org/10.1109/ECBS.2001.922409
43. Reactive Systems Inc.: https://reactive-systems.com/
44. RTCA DO-178B (EUROCAE ED-12B), Software Considerations in Airborne Systems and Equipment Certification, 2nd Edition (1992)
45. RTCA DO-178C, Software Considerations in Airborne Systems and Equipment Certification (2011)
46. Sims, S., Cleaveland, R., Butts, K., Ranville, S.: Automated validation of software models. In: 16th Annual International Conference on Automated Software Engineering (ASE 2001), pp. 91–96. IEEE (2001)
47. Sokolsky, O.V., Smolka, S.A.: Incremental model checking in the modal μ-calculus. In: Dill, D.L. (ed.) CAV 1994. LNCS, vol. 818, pp. 351 363. Springer, Heidelberg (1994). https://doi.org/10.1007/3-540-58179-0_67
48. Stirling, C., Walker, D.: Local model checking in the modal μ-calculus. Theoret. Comput. Sci. **89**(1), 161–177 (1991)
49. Zhang, S., Sokolsky, O., Smolka, S.: On the parallel complexity of model checking in the modal μ-calculus. In: Ninth Annual IEEE Symposium on Logic in Computer Science, pp. 154–163. IEEE (1994)

Automated Reasoning in Quantum Circuit Compilation

Dimitrios Thanos[✉], Alejandro Villoria, Sebastiaan Brand, Arend-Jan Quist,
Jingyi Mei, Tim Coopmans, and Alfons Laarman

Leiden Institute of Advanced Computer Science (LIACS), Leiden University,
Leiden 2333, CA, The Netherlands
d.thanos@liacs.leidenuniv.nl

Abstract. Automated reasoning techniques have been proven of immense importance in classical applications like formal verification, circuit design and probabilistic inference. The domain of quantum computing poses new challenges of a different nature, such as the compilation of quantum circuits, which involves "quantum-hard" tasks such as the simulation, optimization, synthesis, and equivalence checking of quantum circuits. We ask the question of how effective the methods motivated by classical automated reasoning can be for quantum compilation. We assess their current applicability to this new domain by discussing the recent advances. In particular, we focus on three core automated reasoning approaches: decision diagrams, satisfiability and graphical calculus-based methods. In this survey, we explain in a manner accessible to those unfamiliar with quantum computing concepts how these prominent automated reasoning methods have found numerous applications in quantum circuit compilation. We find that surprisingly all considered reasoning methods, while originally developed for classical purposes, can excel at various compilation tasks for even universal quantum circuits.

1 Towards Quantum Supremacy

Quantum supremacy [137] is the question of obtaining the first example of a problem for which quantum computers provably surpass classical computers in theory and/or practice. There are various reasons to pursue this challenge [193].

First, technology advance has led to the miniaturization of classical computers, rendering them powerful and cost-effective. However, this trend has now extended into the micro-level where quantum phenomena come into play, causing insurmountable hurdles for further miniaturization. Alternatively, the embrace of quantum effects could lead to further miniaturization and innovation.

Second, quantum computing (QC) holds the promise of revolutionizing the field of computation by surpassing classical computers in terms of efficiency, particularly in tackling tasks that are deemed classically intractable [113,121]. Quantum algorithms can leverage quantum phenomena like entanglement and constructive interference to tackle problems beyond the reach of classical computers. For example, for period finding, Shor's algorithm features a performance exponentially faster than their best-known classical versions [121].

Third, theoretical computer science aims to understand the strengths and limitations of the most powerful computers that nature allows. So it makes sense to focus on studying the potential of quantum computers which are closer to the limit of our current understanding of nature. Physicists, chemists, and other scientists are constantly dealing with quantum-hard problems like Hamiltonian simulation, computing ground state energies, etc. This is where Feynman's conception of the quantum computer originated [104].

However, despite progress [6,89], quantum supremacy for *useful* problems remains elusive. A large-scale quantum computer with sufficient error-correction is yet to be built. On the theory side, one cannot rule out the possibility that the runtime of certain quantum algorithms gets matched by new classical algorithms. This has happened before. The quantum recommender algorithms were thought to be exponentially faster than their classical counterparts, however, Tang [163] showed (at the age of 18) that there exist classical algorithms with similar asymptotic performance.

Nonetheless, a good reason to remain optimistic about this kind of research is that challenges in quantum circuit compilation are of exactly the same nature as challenges with which physicists and quantum chemists struggle, as expressed in the third point above. (Not to mention that Tang's proof has led to a class of improved, quantum-inspired, classical algorithms.) Therefore, progress in QC is progress in fundamental research that has the potential to advance our understanding of the many quantum-hard problems that nature confronts us with.

Gate-based quantum computing, one of the most prevalent models of quantum computing, involves the utilization of a limited set of quantum gates, specifically reversible operators designed to manipulate qubits. These operators form quantum circuits, which are not necessarily unique, meaning that different circuits that implement the same computation can exist. In the current era of noisy intermediate-scale quantum computing (NISQ) [138], there are many challenges that we need to overcome when compiling quantum circuits into real-world devices. Such challenges are the high noise levels, the shallow depth of the circuits that can be practically implemented, and the various constraints (connectivity, topology, native gate sets, etc.) [46,65]. It is evident that circuit compilation problems form an important hurdle on the road to achieving quantum supremacy.

A promising range of techniques for addressing these questions exists within the field of automated reasoning. In the analysis of (classical) system behavior, computer scientists are often dealing with a combinatorial explosion. As a consequence, many powerful formalisms and approaches were developed to reason about such systems. For instance, decision diagrams [3,29], satisfiability [21] and theorem provers [19,24,99,115,130], offer rigorous techniques to verify the behavior of classical systems and ensure their accuracy and dependability.

The state of n quantum bits is generally represented as 2^n complex values [121]. Consequently, already the simulation of quantum circuits, a core task in circuit compilation, as we will see, must tackle a combinatorial explosion similar to that encountered when analyzing the behavior of classical systems. Hence, many of the techniques traditionally used for the analysis of classical systems in the field of automated reasoning and formal methods have been proven useful for quantum circuit compilation. For instance, decision diagrams have been established as efficient analysis and optimization tools for quantum circuits [187,199], model counting shows promise for simulation [107] and equivalence checking [108], satisfiability for analysis of non-universal quantum circuits [17,189,195], satisfiability modulo theories for circuit verification [12,39] and equivalence checking [4,5], and theorem-prover-style deductive approaches for simulation and equivalence checks [59,131,185]. In the context of quantum computing, the most prominent type of theorem provers is graphical calculi, so we will narrow the discussion about theorem provers to graphical calculi.

This survey aims to comprehensively document the progress of the transfer of 'classical' automated reasoning tools to the relatively new field of quantum computing. The following methodology was utilized to define its scope. We concentrate on the following three core automated reasoning approaches, focusing on some of the most prominent ones employed for quantum circuit compilation.

- Decision diagrams;
- Satisfiability (SAT / SMT / #SAT);
- Graphical calculus-based methods

For each, we used Google Scholar to identify the ten most relevant publications applying the particular method to a task in quantum compilation, as defined in Sect. 3 to include the simulation, synthesis, optimization, and equivalence checking of quantum circuits. We use the search terms "decision diagrams" / "satisfiability / "SAT" / "SMT" / "Graphical calculi" and "quantum circuit". The publications outside the scope of our paper are discarded. Starting from these works, we did a literature review following the trace of citations. We explicitly exclude formal verification, such as model checking and Hoare logic-based deduction, from our queries, as we are mainly interested in how automated reasoning methods transfer onto progressing the quantum supremacy challenge and not how formal verification can be lifted to the quantum domain. In this, our survey differs substantially from earlier surveys [38,100], which focus on correctness verification of quantum circuits and algorithms.

The text is structured as follows. We first introduce quantum computing in more detail in Sect. 2. We then define what we mean by quantum circuit compilation by discussing the challenging tasks it encompasses in Sect. 3. In Sect. 4-6, we discuss in detail how the three automated reasoning approaches found applications in quantum circuit compilation. We also report on automated reasoning that superseded the particular version tailored to quantum computing, finding for instance an interesting parallel in the early invention of decision diagrams representing pseudo-Boolean functions (i.e., quantum states) and their later development in quantum computing.

In Sect. 7, we also discuss other methods that we could identify as used for quantum circuit compilation (some drawn from physics, such as tensor networks or path integral-based methods). We conclude in Sect. 8 that automated reasoning methods originally developed for classical problems also excel in various compilation tasks and will likely find more applications in quantum computing.

2 Fundamental Concepts of Quantum Computing

In this section we briefly explain the core concepts of quantum computing. The goal is not to give a complete overview of all of the underlying mathematics, but rather to familiarize the reader with some of the concepts and terms used. For a comprehensive overview of quantum computing, we refer the reader to [121].

Quantum States. Where classical states are described by bits, quantum states are described by quantum bits (qubits). A qubit, just like a classical bit, can be in a state 0 or 1, described by vectors $|0\rangle = \begin{bmatrix} 1 & 0 \end{bmatrix}^{\mathsf{T}}$ and $|1\rangle = \begin{bmatrix} 0 & 1 \end{bmatrix}^{\mathsf{T}}$. Unlike classical bits, qubits can be in *superposition*, described by a linear combination of the $|0\rangle$ and $|1\rangle$ states. Concretely, the state of a qubit is given by $|\psi\rangle = \alpha_0|0\rangle + \alpha_1|1\rangle = \begin{bmatrix} \alpha_0 & \alpha_1 \end{bmatrix}^{\mathsf{T}}$, with $\alpha_0, \alpha_1 \in \mathbb{C}$ and $|\psi\rangle$ a unit vector (i.e. $|\alpha_0|^2 + |\alpha_1|^2 = 1$).

Another key difference between classical bits and qubits is how multi-(qu)bit systems are composed. Whereas as n-bit state can simply be described by a bit string $b \in \{0,1\}^n$, multi-qubit states are composed from single-qubit states by means of the tensor product. Intuitively, this tensor product is similar to the Cartesian product of sets. Formally, the tensor product of an $r_a \times c_a$ matrix A and an $r_b \times c_b$ matrix B yields a matrix $A \otimes B$ of dimension $r_a r_b \times c_a c_b$ equal to

$$A \otimes B = \begin{bmatrix} A_{11}B & A_{12}B & \cdots & A_{1c_a}B \\ \vdots & \vdots & \ddots & \vdots \\ A_{r_a 1}B & A_{r_a 2}B & \cdots & A_{r_a c_a}B \end{bmatrix}.$$

For example, for two single-qubit states $|\psi_A\rangle = \begin{bmatrix} \alpha_0 & \alpha_1 \end{bmatrix}^{\mathsf{T}}$ and $|\psi_B\rangle = \begin{bmatrix} \beta_0 & \beta_1 \end{bmatrix}^{\mathsf{T}}$, we get that $|\psi_A\rangle \otimes |\psi_B\rangle = \begin{bmatrix} \alpha_0\beta_0 & \alpha_0\beta_1 & \alpha_1\beta_0 & \alpha_1\beta_1 \end{bmatrix}^{\mathsf{T}}$. In general, the state of n qubits is given by a 2^n-dimensional vector. For convenience, we define $|b\rangle = |b_1\rangle \otimes |b_2\rangle \otimes \cdots \otimes |b_n\rangle$ for $|b\rangle \in \{0,1\}^n$. Observe that $|b\rangle = e_b$, i.e., a 2^n-length vector with only index $b \in \{0,1\}^n$ set to 1 and the other entries to 0.

Similar to the conjugate of a complex number, a quantum state $|\varphi\rangle = \begin{bmatrix} \alpha_0 & \alpha_1 & \ldots & \alpha_{2^n} \end{bmatrix}^{\mathsf{T}}$ has an 'adjoint' $\langle\varphi| = (|\varphi\rangle^*)^{\mathsf{T}} = \begin{bmatrix} \alpha_0^* & \alpha_1^* & \ldots & \alpha_{2^n}^* \end{bmatrix}$, i.e., its transpose, a row vector, with all complex entries conjugated. Observe that $\langle b|\varphi\rangle = \alpha_b$. All states must have unit length, which we can now formulate as $\langle\varphi|\varphi\rangle = 1$.

It is important to note that some joint states cannot be decomposed as a tensor product of smaller states. Such states are *entangled*. For example, states $|\psi_3\rangle = 1/\sqrt{2}(|00\rangle + |11\rangle) = \begin{bmatrix} 1/\sqrt{2} & 0 & 0 & 1/\sqrt{2} \end{bmatrix}^{\mathsf{T}}$ and $|\psi_4\rangle$ from Fig. 1 are entangled.

Quantum Operations. There are two types of operations on quantum states: gates, and measurements. Quantum gates are linear maps that are information-preserving (i.e., reversible) and norm-preserving. This makes it so that an n-qubit quantum gate U is given by an $2^n \times 2^n$ unitary matrix, where unitarity is defined as $UU^\dagger = U^\dagger U = I$ with $U^\dagger = (U^*)^\mathsf{T}$ denoting the conjugate transpose of U.

The effect of a gate (matrix) on a qubit state (vector) is computed through matrix-vector multiplication. A sequence of quantum gates acting on a state is typically visualized in a quantum circuit (Fig. 1). Some examples of quantum gates are the Pauli gates (X, Y, Z) and Clifford gates (H, S, CNOT).

$$X = \begin{bmatrix} 0 & 1 \\ 1 & 0 \end{bmatrix} \qquad Z = \begin{bmatrix} 1 & 0 \\ 0 & -1 \end{bmatrix} \qquad Y = \begin{bmatrix} 0 & -i \\ i & 0 \end{bmatrix}$$

$$H = \frac{1}{\sqrt{2}} \begin{bmatrix} 1 & 1 \\ 1 & -1 \end{bmatrix} \qquad S = \begin{bmatrix} 1 & 0 \\ 0 & -i \end{bmatrix} \qquad \mathrm{CNOT} = \begin{bmatrix} 1 & 0 & 0 & 0 \\ 0 & 1 & 0 & 0 \\ 0 & 0 & 0 & 1 \\ 0 & 0 & 1 & 0 \end{bmatrix}$$

To give an intuition, the X gate acts as a classical bit-flip and the Z and S gates as a 'phase-flip' on $|1\rangle$, yielding $-|1\rangle$ and $i|1\rangle$ respectively, the Hadamard (H) gate brings a qubit from the $|0\rangle$ state into a *uniform superposition* of $|0\rangle$ and $|1\rangle$, and the two-qubit controlled-not gate (CNOT) acts as a bit flip on a 'target' only when the control qubit is one. Combined with the Hadamard, the CNOT gate can produce entangled states (as in Fig. 1).

We say that two unitaries U and V are equivalent up to 'global phase' if $U = cV$, where $c \in \mathbb{C}$. The term 'global phase' refers to the complex factor c, which does not affect any observable properties of unitaries [121].

A (standard) measurement of a qubit in the state $|\psi\rangle = \alpha_0|0\rangle + \alpha_1|1\rangle$ collapses the qubit onto the $|0\rangle$ ($|1\rangle$) state, with probability equal to $|\alpha_0|^2$ ($|\alpha_1|^2$). For measuring a single qubit within a multi-qubit state $|\varphi\rangle$, we note that it can always be written as $|\varphi\rangle = \alpha_0|0\rangle \otimes |\varphi_0\rangle + \alpha_1|1\rangle \otimes |\varphi_1\rangle$, where $|\alpha_0|^2$ and $|\alpha_1|^2$ again correspond to the probabilities of collapsing to the $|0\rangle$ and $|1\rangle$ states.

The Stabilizer Formalism. The final concept we introduce is the stabilizer formalism [69], which describes a class of quantum circuits and corresponding quantum states that are known to be classically simulatable in polynomial time.

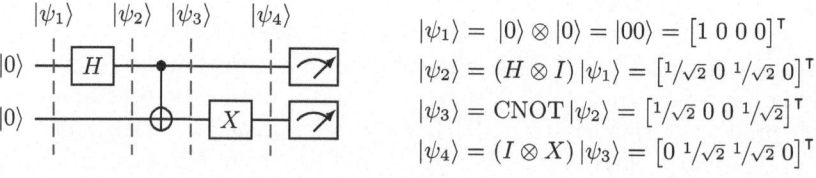

$$|\psi_1\rangle = |0\rangle \otimes |0\rangle = |00\rangle = [1\ 0\ 0\ 0]^\mathsf{T}$$
$$|\psi_2\rangle = (H \otimes I)\,|\psi_1\rangle = [1/\sqrt{2}\ 0\ 1/\sqrt{2}\ 0]^\mathsf{T}$$
$$|\psi_3\rangle = \mathrm{CNOT}\,|\psi_2\rangle = [1/\sqrt{2}\ 0\ 0\ 1/\sqrt{2}]^\mathsf{T}$$
$$|\psi_4\rangle = (I \otimes X)\,|\psi_3\rangle = [0\ 1/\sqrt{2}\ 1/\sqrt{2}\ 0]^\mathsf{T}$$

Fig. 1. An example 2-qubit quantum circuit. Each qubit is represented by a horizontal wire, and operations are applied from left to right. As is common, we write $|xy\rangle$ as shorthand for $|x\rangle \otimes |y\rangle$. Measuring both qubits after obtaining $|\psi_4\rangle = 1/\sqrt{2}(|01\rangle + |10\rangle)$ gives $|01\rangle$ or $|10\rangle$ each with probability $|1/\sqrt{2}|^2 = 1/2$.

The circuits in question are those composed of Clifford gates, which are all quantum gates that can be generated (under multiplication and the tensor product) from H, S, and CNOT. The stabilizer states are all states that can be produced by a Clifford circuit initialized to the all-zero state $|0\rangle^{\otimes n}$. Any stabilizer state can be represented by a set of n "stabilizers". In this context, a matrix $P \in \{\pm P_1 \otimes \cdots \otimes P_n \mid P_i \in \{X, Y, Z, I\}\}$ is a stabilizer of an n-qubit state $|\sigma\rangle$ if $|\sigma\rangle$ is an eigenvector of P, i.e. $P|\sigma\rangle = \pm|\sigma\rangle$. Tracking these n stabilizers (including the $\pm$ sign) can be done with $2n^2 + n$ Boolean values, which are typically encoded in a so-called stabilizer tableau.

Moreover, this tableau can be efficiently updated to implement the semantics of any Clifford gate. The fact that stabilizer states are representable by a succinct and easy-to-update representation makes Clifford circuits classically simulatable.

It is important to note that while stabilizer states play an important role in quantum computing, such as in quantum error correction [16,78,98,164] and measurement-based quantum computing [141], they are by themselves not sufficient for universal quantum computation. Generalizations of the stabilizer formalism further allow a complete discretization of the (universal) quantum state space as we shall see in subsequent sections.

By adding a single non-Clifford gate, like a T gate or any arbitrary rotation gate R_X, R_Y, R_Z, to the Clifford gate set, we obtain universal quantum computing [120]. Stabilizer-rank-based methods [28] allow (classical) fixed-parameter tractable simulation in the number of T gates in the circuit.

3   Challenges of Quantum Circuit Compilation

Quantum circuit compilation is the process of mapping an instance of a quantum algorithm to a specific quantum hardware, i.e., a specific quantum processing unit (QPU). A more efficient mapping can bring us closer to quantum supremacy, especially when we aim to identify *useful* problems at which quantum computers excel, since they are likely to require structured circuits instead of randomized ones [6]. As instances, we will consider circuits, and not programming languages (which can be translated to circuits), since we are interested in suing automated reasoning for hard problems. There are multiple problems crucial to quantum circuit compilation. For this text, we focus on the following tasks.

1. **Circuit Simulation**: Classical simulation of quantum circuits enables basic analyses, but, as we shall see, is also often a sub-task in other compilation tasks. Better simulation methods arguably translate into improvement in other compilation tasks. We can distinguish two types of simulation.
   (a) **Strong Circuit Simulation**: Given a quantum circuit, and a computational basis state, compute the probability of measuring that state.
   (b) **Weak Circuit Simulation**: Given a quantum circuit, sample from the probability distribution of its measurement outcomes.[1]

---

[1] Measurements can always be deferred until the end of the circuit [121, Sect. 4.4].

2. **Circuit Optimization**: Given a quantum circuit and a set of hardware constraints, produce another circuit that represents an equivalent computation satisfying the constraints. The new circuit should match the hard constraints posed by the QPU, such as topology (i.e., connectivity between qubits), as well as soft constraints that allow the reduction of execution time and noise exposure introduced by imperfections in the QPU. For example, certain gates might be more expensive in terms of noise or entangling certain pairs of qubits could be more error-prone [10]. Tasks such as explicitly reducing the number of gates in a circuit fall in this category [158], as well as other tasks such as layout synthesis and qubit mapping and routing [112], which consist of mapping a logical circuit into a hardware-aware one that satisfies the physical qubit-connectivity constraints.

3. **Circuit Synthesis**: Given a specification, produce a quantum circuit that adheres to it [58]. This specification can be relational, e.g., another circuit or a Hoare logic expression [196], or simply an input and output state, in which case, we speak more specifically of **quantum state preparation**. The specification can also be a unitary, in which case we speak of 'decomposition'. Synthesis is often combined with optimization. In that case, the specification is extended with the soft and hard constraints discussed above.

4. **Circuit Equivalence Checking**: Given two circuits, decide whether they represent equivalent unitaries. Since circuit compilation tasks modify circuits, having a method for checking the correctness of the result is essential.

Although we have categorized the tasks into four main categories, it is important to note that they come in a whole spectrum. For example, there are many hardware constraints under which one can optimize and trade-offs come into play. Moreover, a heuristic solution is sometimes good enough for soft constraints. While we are mainly interested in sound and complete methods, since they match the capabilities of the considered automated reasoning techniques, we also include heuristic solutions built on automated reasoning.

Each compilation task can be defined as an exact or approximate problem (i.e., with bounded error like in the classical class BPP). For instance, exact equivalence checking requires that the unitaries are equal up to global phase (see Sect. 2), while approximate checking merely requires that the two unitaries always produce the same states up to a certain *fidelity* [82,103] (a measure for 'closeness' of two quantum states [121]). The complexity classes for exact problems depend strongly on the gate set, because the gate set determines the reachable states [67]. On the other hand, the bounded-error complexity classes, where BQPis the quantum analog of P and QMA the quantum analog of NP [94], are invariant under the gate set (a motivation behind their definition). The exact version is harder, e.g., exact strong simulation is already #P-complete [88,120]. Nonetheless, exact reasoning can be appealing because it allows discretization.[2] Surprisingly, exact reasoning methods are even used to compute the approximate versions of these compilation tasks. For example, [82,184] even solve the "quantum NP"-hard circuit equivalence, which is QMA-hard [84] to approximate and

---

[2] With a ring discretizing all complex numbers calculable in a given gate set [67,92].

NQP-hard [162] to compute exactly. Exact reasoning also allows linear #SAT encodings of simulation and equivalence checking [107,108] (see Sect. 5) and is used extensively in ZX-calculus (see Sect. 6).

In Sect. 4–6, we detail the applications of three main automated reasoning approaches in quantum circuit compilation. We focus on the historical context and the technical aspects that have influenced the adaptation and evolution of these automated reasoning methods in this new application area. This illustrates how quantum circuit compilation benefited from automated reasoning techniques originally developed for reasoning about classical systems.

4 Decision Diagrams

To point out parallels in the early development of decision diagrams and their later use in quantum computing, we order this section mostly chronologically.

Historical Background. Akers [3] defined the binary decision diagram (BDD) for compactly representing a Boolean function $f : \{0,1\}^n \to \{0,1\}$. Bryant [29] established the importance of BDDs by giving efficient manipulation algorithms

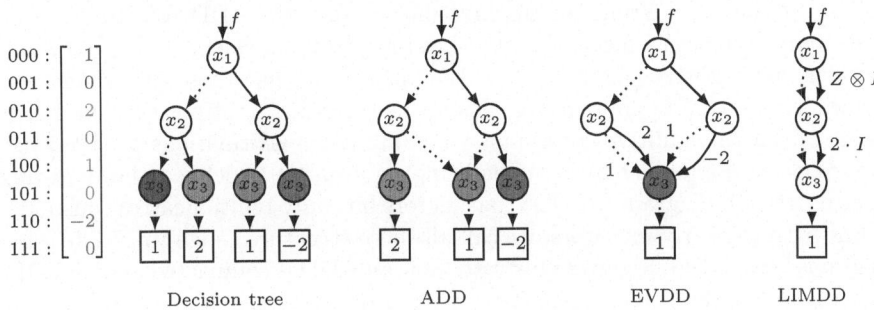

000 :	1
001 :	0
010 :	2
011 :	0
100 :	1
101 :	0
110 :	-2
111 :	0

Decision tree   ADD   EVDD   LIMDD

Fig. 2. A pseudo-Boolean function $f(x_1, x_2, x_3) = \overline{x_3} \cdot (2 \cdot x_2 \cdot (\overline{x_1} - x_1) + \overline{x_2})$ represented as (non-normalized quantum state) vector and as an (exponentially-sized) decision tree. A node labeled with variable x represents a Shannon decomposition on x. An outgoing dashed edge represents the sub functions where the variable x is restricted to false (0) and a solid edge where the variable is restricted to true (1). The boxes represent leaves. We omit zero leaves. From a decision tree to algebraic decision diagram (ADD): The red node is merged into the green as both represent the same sub-function $f_{00}(x_3) = f_{10}(x_3) = 1 - x_3$. From algebraic decision diagram to edge-valued decision diagram (EVDD): The purple, green, and blue nodes represent sub-functions that are equivalent up to a constant factor of 2, 1, -2 respectively. By putting the factors on the edges, these equivalent nodes can be merged. Unlabeled edges have implicit factor 1. From edge-valued decision diagram (EVDD) to LIMDD: The Pauli LIM $Z \otimes I$ maps f_0 to f_1 (as vectors, i.e. $(Z \otimes I) \cdot [\, f_{100}\ f_{101}\ f_{110}\ f_{111} \,]^T = [\, f_{100}\ f_{101}\ -f_{110}\ -f_{111} \,]^T = f_0$). Both x_2 nodes can thus be merged by putting this map on the (high) edge. Unlabeled edges to nodes x_i have implicit map $1 \cdot I^{\otimes i}$.

that can compute $F \circ G$ where F, G are any functions represented as BDDs and $\circ$ is any Boolean operator. Using Bryant's work as a basis, operations have been invented for quantification [106], matrix multiplication [56,66], and even reachability [25]. With the multi-terminal [40] or algebraic decision diagram (ADD) [9], BDDs were generalized to support the representation of pseudo-Boolean function $f \colon \{0,1\}^n \to D$ for any (ring-structured) domain $D \in \{\mathbb{R}, \mathbb{C}, \dots\}$.

The main idea that led to the construction of decision diagrams can be traced back to Boole and his identity formula $f = (x \wedge f_x) \vee (\overline{x} \wedge f_{\overline{x}})$ [22], where $f_x, f_{\overline{x}}$ are the sub-functions with variable x restricted to $1, 0$ respectively. This is the expansion identity (also known as decomposition identity or Shannon decomposition) and is often attributed to Shannon since its appearance in his paper on classical circuits [152]. This decomposition can be viewed as "peeling off" a single variable x. By iterating this process on sub-functions, we can build the typical exponential binary tree ending with $0, 1$ leaves as illustrated in Fig. 2 for a pseudo-Boolean function. Note that this results in a total variable order along a path in the decision diagram: All decision diagrams discussed here fix a single variable order along all paths as this crucially enables the efficient manipulation operations given by Bryant [29]. When decision diagram nodes in this binary tree represent the same sub-function, e.g., two leaf nodes with same value, they are merged. For many practical functions, such as those arising in verification problems, this node merging can result in significant size reductions.

Lai, Pedram and Vrudhula [97] first showed that the ADD structure can be made more succinct by merging nodes that represent equivalent sub-functions up to a constant additive factor, which is then placed on the edges. Tatershofer and Pedram [159,160,179] improved on this by also allowing multiplicative constants p apart from the additive constants a, i.e., affine transformations $p \cdot f + a$. By requiring the domain D to be a semi-ring, Wilson [191] defined the semi-ring labeled decision diagram (SLDD) that factors out only multiplicative constants. In line with the early inventors, we call all these structure "edge-valued decision diagrams" (EVDDs). Figure 2 illustrates how an ADD is compacted as an EVDD, with the factored out multiplicative constants on the edges. Finally, Sanner and McAllester [147] developed affine ADDs (AADDs), and Fargier, Marquis and Schmidt [62] generalized and related various of the above decisions diagrams.

As in all data structure design, there is a trade-off between succinctness and efficiency of operations. Darwiche [52] first mapped these trade-offs for different data structures representing Boolean functions. His 'knowledge compilation map' shows, inter alia, that, for BDD, basic Boolean operations are efficient, but more complicated operations like unbounded quantification are not, which was already known from the fact that poly-time matrix multiplication with BDDs would imply $\mathsf{P} = \mathsf{NP}$ [106] and that reachability using BDDs is PSPACE-hard [63].

Later, Fargier et al. [62] showed that more or less similar results hold for manipulation operations on ADDs, replacing Boolean operations for point-wise addition, multiplication, and min/max computations in the pseudo-Boolean domain. However, they also show that basic operations, such as point-wise addition, become intractable for EVDD and AADD. This affects the implementation of the Hadamard gate, as we discuss in the next section. On the other

Table 1. Various decision diagrams (DDs) used in the literature (extended from [177]). The column "node merge" lists the conditions under which two decision diagram nodes, representing functions f and g, are merged. Here, $p, a \in \mathbb{C}$ are complex constants, $P = P_1 \otimes \cdots \otimes P_n$ a sequence of single-qubit Pauli gates P_i, and $f + a$ means the function $f(\vec{x}) + a$ for all $\vec{x}$. All DDs, except QDD, have complex scalars as terminals and their internal nodes v, w thus represent functions $f, g \colon \{0,1\}^n \to \mathbb{C}$. In QDD, the terminal node represents the $|0\rangle$ vector, and thus QDD can be seen as functions $f, g \colon \{0,1\}^{n-1} \to \mathbb{C}^2$. Here R_X is a single-qubit rotation operator. All these decision diagrams use a fixed total variable order. CFLOBDD is not in the table as it processes variables via recursive bisection.

(Quantum) decision diagrams (and variants)	Node merge
Decision Tree	(no merging)
MTBDD (1993) [40], ADD (1997) [9], QuiDD (2003) [169]	$f = g$
SLDD$_\times$ (2004) [191], QMDD (2006) [111], XQDD (2008) [183], TDD (2021) [83]	$f = p \cdot g$
EVBDD (1994) [97], SLDD$_+$ (2005) [62]	$f = g + a$
FEVBDD (1994) [159,160,179], SLDD$_{+,\times}$ (2005) [62], AADD (2005) [147]	$f = p \cdot g + a$
QDD (2006) [2]	$f = R_X \cdot g$
LIMDD (2023) [176–178]	$f = p \cdot P \cdot g$

hand, this worst-case analysis obfuscates the fact that there are no functions for which AADD or EVDD is slower on point-wise addition than ADD [147]. This is because the intractability for AADD/EVDD addition only occurs when the corresponding ADD would already be large for representing the input functions.

The structural differences between these different decision diagrams are summarized in Table 1, which also summarizes the chronology, includes references and the quantum-inspired decision-diagram versions discussed next.

Decision Diagrams in Quantum Circuit Compilation. Decision diagrams have been pioneered for the efficient representation and manipulation of quantum states by Viamontes in the form of QuiDDs [169], which are essentially ADDs with a complex domain $D = \mathbb{C}$. The basic insight is that a quantum state $|\varphi\rangle$ can be viewed as a pseudo-Boolean function $f_\varphi(\vec{x}) = (\vec{x})\,\varphi$, from computational basis states $\vec{x}$ to complex amplitudes $(\vec{x})\,\varphi$. QuiDDs have been used for simulation of quantum computing [169–171].

QuiDDs were succeeded by QMDDs [111], which can be viewed as edge-valued decision diagrams on the domain of complex numbers, transferring the compactness of EVDD to the application of quantum circuit compilation. Here it should be noted that the (pointwise) addition operation is required for the implementation of the Hadamard gate, which can result in a possible blowup of the diagram, which nonetheless does not exceed the size of the corresponding QuiDDs, as discussed above for ADD and EVDD. The tractability of gate implementations for QMDD is presented in [176,177], showing that all other primitive gate operations are tractable for QMDD, except for the swap gate.

QMDD has been applied to many quantum circuit compilation tasks: quantum circuit simulation [199], equivalence checks [31–33, 122], including approximate equivalence checks [184], and synthesis [123, 198]. QMDDs have also been used to simulate Hamiltonians [80, 144] and circuits with noise [72–74].

QDD [2] offers an interesting variation on QMDD, as Table 1 illustrates. Further, while QMDD represents unitaries with quaternary DD nodes, TDD [83] clones each qubit variable as was traditionally done [66, 106, 159].

Interestingly, certain families of stabilizer states yield an exponentially-sized representation in QMDD as shown in [176]. This is surprising given the importance of stabilizer states and that simulation (and representation) of stabilizer states is tractable as discussed in Sect. 2. LIMDD [176, 178] generalizes QMDD to solve this, reducing the required space in worst case from $\Omega(2^{\sqrt{n}})$ to $\mathcal{O}(n^2)$ for all stabilizer states. LIMDD achieves this feat by merging nodes that are equivalent up to tensor products of single-qubit Pauli operators (as well as constant complex factors, such as in QMDDs). Consequently, where a QMDD node has decomposition $|0\rangle \otimes \alpha_0|\varphi_0\rangle + |1\rangle \otimes \alpha_1|\varphi_1\rangle$ pointing to a node φ_0 (φ_1) with the low (high) edge which is labelled with complex number α_0 (α_1), a LIMDD replaces the complex numbers α_0, α_1 by tensor products of Pauli matrices $P = P_1 \otimes \cdots \otimes P_n$ times a complex number. An example is given in Fig. 2, showing that some nodes which could not be merged in QMDD are equivalent in LIMDD.

The price paid for its succinctness is however that LIMDD must compute the stabilizer group for each node to achieve canonicity, causing cubic factor overhead in practice [176]. Asymptotic analysis [177] also shows that computing fidelity is also intractable for LIMDD (under common-place complexity-theoretic assump-

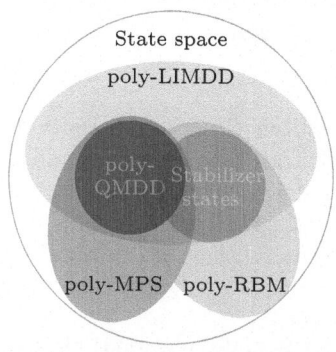

Fig. 3. The comparative succinctness of LIMDD, QMDD, MPS, and RBM (see Sect. 7) following from the asymptotic analysis in [177]. All these data structures, except the stabilizer formalism, can represent any quantum state (they are universal). The Venn diagram however shows families of states that can be represented in polynomial size with the different data structures. Consequently, MPS, RBM, and LIMDD are incomparable in terms of representation power: they each have their strengths and weaknesses. LIMDD is the only structure that can represent all stabilizer states and all (families of) states supported by polynomial QMDD. (Poly-)ADD and QuiDD are not shown but are strictly contained in (Poly-)QMDD.

tions), while tractable in QMDD. However, Vinkhuijzen et al. [177] produced a "knowledge compilation map" for quantum information (see Fig. 3) showing that only LIMDD can be more succinct than MPS and RBM (see Sect. 7).

Another type of decision diagram that is used for simulation is CFLOBDD. This decision diagram fundamentally differs from the decision diagrams described above, as it generalizes the linear variable order into a recursive bisection of the variables (which can be viewed as a Shannon decomposition over multiple variables at once): Every level recursively 'peels off' half of the variables instead of one variable. This is why CFLOBDD does not fit in Table 1. Moreover, a CFLOBDD has variable sharing, such that a node representing a function $f(x_1, \ldots, x_k)$ can also be used for different variables like e.g. $f(x_{k+1}, \ldots, x_{2k})$. This recycling, combined with handling half of the variables in every level, results in the best case in an exponential compression with respect to BDD. For instance, the multi-qubit Hadamard matrix $H^{\otimes m}$ which can be represented by CFLOBDD in only logarithmic space. With these reductions, highly-structured quantum circuits with very many qubits can be simulated efficiently [156,157]. The treatment of multiple variables at a time aligns with SDD [51] and 'hierarchical set decision diagram' [166], while the idea of variable shifting also was used for SDD [118]. To our knowledge, there does not yet exist a "knowledge compilation map" comparing CFLOBDD to other similar data structures.

In terms of performance, QMDD is usually among the fastest method for compilation tasks, outperforming QuiDD [199], array-based methods [82,199], tensor networks [83,157] (see Sect. 7) and ZX calculus (see Sect. 6) in some cases [132]. Variants like QDD and TDD have related performance [2,83], while CFLOBDD was shown to be competitive to QMDD, SDD and tensor network [155,157]. LIMDD, finally, proved somewhat impractical in empirical evaluation [178]. In practice, we therefore see QMDD used more, e.g. [34,187].

There are multiple tools for the different types of decision diagrams for quantum, e.g. [30,82,155,178,186–188,197]. It should however be pointed out that all of these use floating points by default to represent real weights on the edges. In practice, this can cause rounding errors to rapidly propagate in the discrete data structure, resulting in numerical instability [124,197].

5   SAT/SMT-Based Methods

While decision diagrams have been an indispensable data structure in many automated reasoning applications, both within as well as outside of quantum computing, they require (at worst) an exponential amount of space because they represent all satisfying assignments. While many relevant problems are known to be computationally hard in terms of time complexity, almost all of them can be solved in polynomial space. Boolean satisfiability (SAT) is the problem of deciding whether there exists a satisfying assignment to a given Boolean formula, i.e., an assignment for which the formula evaluates to true, i.e., is "satisfiable". The SAT problem is the first problem that was shown to be NP-complete [45]. Satisfiability modulo theories (SMT) generalizes SAT to other domains than Boolean, such as expressions containing bit vectors, integers or real numbers. SAT and SMT solvers are tools that tackle computationally hard problems with polynomial-space algorithms [21]. It has become common practice to solve other NP-complete problems by reducing them to SAT and then using a highly optimized solver to solve the SAT instance, the solution of which can then be translated back to the domain of the original problem. SAT-solving algorithms such as DPLL [53] and CDCL [154] traverse the exponentially large search space step-wise (thus taking only polynomial space omitting caches e.g. learned clauses), and use clever pruning heuristics to obtain good performance on many practical instances. Finally, model counting is the problem of counting the number of satisfying assignments of a Boolean formula and model counters are tools for solving it [37,128,145].

SAT-based approaches have been used to tackle different types of quantum circuit synthesis problems. While SMT allows for the encoding of problems with a continuous search space [12,39], in practice there tends to be a bias towards discretizing the problem as explained in Sect. 3. In quantum circuit synthesis and circuit optimization, we can identify two types of discrete problems to which SAT-based methods have been applied: layout synthesis (i.e. taking an existing circuit and remapping the qubits according to constraints) and the synthesis of discrete circuits (such as Clifford circuits and transformations of so-called "graph states"). We start with the use of SAT for simulation and equivalence checking.

Initial attempts have been made to harness the strengths of satisfiability solvers for the simulation of quantum circuits. For instance, [17] implements a simulator for Clifford circuits based on a SAT encoding. The authors also discuss a SAT encoding for simulating universal circuits [17,189] of exponential size, making it impractical. Subsequently, [107] achieved the strong simulation of universal quantum circuits by a linear encoding as a (weighted) model counting problem. The crux of the method is a generalization of the stabilizer formalism to universal quantum computing which effectively discretizes the state space. Empirical evaluation shows that the method is competitive to state-of-the-art methods based on DD and ZX calculus (see Sect. 6).

SAT-based solvers have also been used for the reversible simulation of irreversible (i.e. classical) circuits [110,190], which is relevant for the construction of space-efficient quantum oracles, e.g. for Grover's algorithm [71] and quantum backtracking algorithms [114,142]. This was later improved by the spooky pebble game [96], which is a model to trade off classical and quantum space. Using SAT solvers, the spooky pebble game has been used to study trade-offs between classical space, quantum space, and circuit depth of a computation, optimizing quantum circuits within hardware constraints [139,140]. SAT has also been used for equivalence checking of Clifford circuits, using the stabilizer formalism [17], though the problem is in P [165]. For universal quantum circuits, equivalence checking can be achieved via a Turing reduction to model counting, as proved by [108] based on the simulation through model counting approach mentioned above [107] and an equivalence checking approach [165].

Clifford circuit synthesis deals with the problem of finding a Clifford circuit, which, from the $|0 \dots 0\rangle$ state, generates a particular target stabilizer state (a form of state preparation). SAT-based methods have been applied to this problem, both with additional optimality constraints [55,57,102,116,125,148,153] and without [17]. The work [148] uses both a MaxSAT solver [21] and a regular SAT solver in combination with binary search. Additional constraints (such as searching for the shortest circuit) are important because without these, the problem is known to be in P [1]. And while SAT solvers are good at hard problems, they are impractical for tractable problems, as demonstrated in [165].

A different application of SAT solvers has been in optimal layout synthesis, where swap operations are inserted into the circuit to ensure that multi-qubit gates are only executed on connected physical qubits. Minimizing the number of swaps is NP-hard [105] and thus the problem was tackled with SMT [27,75,161] and MaxSAT-based methods [112].

Related is the synthesis of circuits that transform one "graph state" into another using only single-qubit operations. Graph states are a special type of stabilizer states that play a crucial role in quantum networking as well as measurement-based quantum computing [78,141]. An n-qubit graph state is a quantum state that can be described by an undirected graph $G = (V, E)$ with n vertices V and edges $E \subseteq V \times V$. Intuitively, the edges of the graph capture information about the way the qubits (corresponding to vertices in the graph) are entangled. Formally, a graph state $|G\rangle$ is the state $(\prod_{(v,w) \in E} CZ_{v,w}) H^{\otimes n} |0 \dots 0\rangle$, where $CZ_{v,w}$ is a two qubit quantum gate equal to $H_w \text{CNOT}_{v,w} H_w$.

An important property of graph states is that several important operations, i.e. single- and two-qubit Clifford gates and single-qubit standard measurements, can be expressed in terms of graph transformations. This allows for reasoning about transformations of graph states entirely in terms of transformations of graphs. A recurring problem [50,77,87] is finding transformations between graph states using only local (i.e. single-qubit) operations, a problem which is NP-complete in general [49]. By reducing the problem to SAT, this problem has been successfully solved up to 17 qubits [26].

6 Graphical Calculus-Based Methods

Another notable technique for quantum circuit compilation is the use of graphical calculi. These techniques consist of using a set of composable graphical generators that interact and can be rewritten via a set of equations. In recent years there have been many proposals for representing quantum systems as graphical languages. For example, there exist graphical languages with a focus on linear optical circuits [42,64,79], fermionic quantum systems [54], open quantum systems [194], Gaussian pure states [109], qudits [101], and more general systems [18,44]. Even more so, there has recently been a diagrammatic axiomatization of quantum circuits [41]. Still, the most prominent graphical language for quantum circuit compilation is the *ZX-calculus* [43], (the focus of this section) and its multiple variations and extensions [7,76,136,173,181].

Initially developed by Coecke and Duncan [43], the ZX-calculus is a graphical language used for reasoning about quantum systems and quantum computations. Inspired by Penrose's tensor network notation [134] and graphical languages stemming from category theory [150], ZX-diagrams consist of a set of graphical generators that semantically represent linear maps $f \colon \mathbb{C}^{2^n} \to \mathbb{C}^{2^m}$ between Hilbert spaces. We can compose these generators both sequentially and parallelly (corresponding to multiplication and Kronecker product of the linear maps they represent, respectively) to create more complex quantum systems. In fact, ZX-diagrams are *universal* in the sense that they can represent any linear map between qubit Hilbert spaces, not necessarily limited to quantum operations (and interestingly, it was shown in [175] how to do a conversion between QMDDs and ZH-diagrams, which in turn can be converted into ZX-diagrams). This makes ZX-diagrams more flexible than standard circuit notation. ZX-diagrams are also equipped with a set of rewrite rules, forming the ZX-calculus. These rewrite rules identify semantic identities between diagrams, and they are the basis for ZX-calculus-based algorithms, for which we are going to give an overview now. Figure 4 shows an example correspondence between a quantum circuit and equivalent ZX-diagram (up to scalar factor).

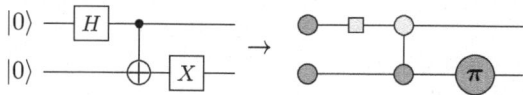

Fig. 4. A quantum circuit and corresponding ZX-diagram.

When it comes to simulation of quantum circuits with ZX-calculus, most approaches [35,92,93,95] opt for calculating individual amplitudes of the state vector using variations of the following set of steps. We start with a quantum circuit consisting of an input state, a sequence of gates, and an effect $\langle x |$ that we turn into a ZX-diagram. By applying some specific rewrite rules, we turn our diagram into a *graph-like* ZX-diagram [59]. We then contract the resulting diagram via a specific set of rewrite rules that strictly decrease the number of nodes (to ensure termination), and then split the reduced diagram into smaller ones by identifying sub-diagrams where we can apply a stabilizer decomposition. This results in a linear combination of ZX-diagrams that can be recursively simplified by again contracting and decomposing each diagram, resulting in a linear combination of smaller ZX-diagrams that we then semantically translate to calculate the resulting amplitude. Adjacent to the simulation of quantum circuits, recent techniques allow for differentiation and integration of ZX-diagrams with applications in quantum machine learning [85,167,182].

When the task is optimizing a circuit with ZX-calculus [13,81,90,91,158], we also proceed by translating the input circuit into a ZX-diagram, then to a graph-like diagram, followed by similar simplification rules until termination, and then extracting a circuit from the resulting simplified diagram. Since

ZX-diagrams are more general than quantum circuits, the process of extracting a circuit from a unitary ZX-diagram is #P-hard [14]. Fortunately, there are sufficient conditions that have been identified for circuit extraction to be done efficiently [8,59,60,185]. These are graph-theoretical *flow* conditions on the diagrams ensuring that a circuit can be extracted step-wise from them. The aforementioned circuit optimization technique can be augmented for quantum-classical circuits (i.e. quantum circuits that can interact with classical circuits) [23] for an extension of the ZX-calculus that allows the discarding of quantum systems [36].

When it comes to circuit synthesis (and also very much related to circuit optimization), most works revolve around *Pauli exponentials* [47,185]. Exponentiated Paulis are operations of the form $e^{-i\alpha P_1 \otimes P_2 \otimes \cdots \otimes P_n}$ and arise naturally when performing quantum chemistry simulations [47]. These operations have a compact representation as ZX-diagrams (referred to as *Pauli gadgets*) and their interactions (e.g. commutation rules between themselves and other quantum gates) can be intuitively reasoned about as ZX-diagram rewrite rules. Given a computation in the form of a composition of Pauli exponentials, the task of ZX-calculus based synthesis aims for a succinct representation of these operations in the form of a sequence of gates (that is, a circuit-like ZX-diagram), oftentimes taking into account device constraints such as qubit connectivity when choosing the layout of multi-qubit gates [48,68,70,192].

Lastly, we want to mention the task of circuit equivalence checking using ZX-calculus. As a language, the ZX-calculus is *complete* in the sense that there are sets of rewrite rules [174] that can provably transform any two semantically equivalent ZX-diagrams into one another. This does not mean that the procedure to do so (e.g., via normal forms) is optimal, so for equivalence a different approach [90,91,131,133] exists: Given two quantum circuits U, U' (in the form of ZX-diagrams), composing one with the dagger of the other in order to verify $U^\dagger U' = I$ (up to global phase). This would mean that after reducing the composed circuits we should arrive to the identity diagram, which is a collection of bare wires. The simplification algorithm [91] also boils down to turning a diagram to graph-like form and applying strictly reducing rewrites until no more nodes can be simplified. It was shown in [131] that this technique does not arrive to the identity diagram on certain (desirable) cases, such as when the two circuits differ by small errors or when one of them uses an ancillary qubit.

7 Other Quantum Circuit Compilation Methods

Without the intention to be complete, we present other important automated tools used in quantum circuit compilation in this section.

Bisimulation has been used in computer science for inter alia process analysis [146] but can also be used in quantum simulation. This line of research was introduced by Jimenez et al. [86], showing that this technique complements quantum simulation methods with decision diagrams.

Shaik and van der Pol [151] tackled circuit topology optimization with modern planning tools. Interestingly, the resulting optimizer is shown to outperform methods based on satisfiability. Venturelli et al. [168] also use planning and constraint programming to realize a quantum circuit optimization tool.

Quantum circuit optimization problems have also been solved with (mixed integer) linear programming. The tackled problems include finding optimal quantum circuits under restrictions on the topology (qubit connectivity) [119, 180] or under restrictions on the gate set [11, 117].

Finally, tensor networks are used in physics for simulation of physical quantum systems [127, 194] and have found various applications in quantum computing [129, 149, 172]. Tensor networks represent quantum information and operations in similar way as graph calculi (see Sect. 6). In fact, ZX-diagrams are tensor networks [185], but with an additional set of rewrite rules to find simplification strategies or to carry out proofs in purely diagrammatic form. A matrix product states (MPS) [135] is a linear tensor network with a variable order much like in decision diagrams (see Sect. 4). In fact, Vinkhuijzen et al. [177] found that MPS can polynomially simulate EVDD (QMDD), but not vice versa. Like quantum circuits [15, 61], tensor networks have also been used in machine learning [143]. A historical overview can be found in [126].

Finally, we point out that tensor networks are used in similar ways as decision diagrams and satisfiability. In formal verification, for instance, sets of states and transitions are represented in DD, before fixpoints are computed using matrix-vector operations [25, 56, 66, 106]. Tensor networks also treat single quantum states and operations that way (see MPS and MPO [127]). Similarly, the SAT-based approaches identified in Sect. 5 essentially use a bounded model checking [20] encoding of circuits. Moreover, in knowledge compilation in AI, information is first compiled in a succinct representation before analyzing it using queries, which is how tensor networks are also used in physics [126].

8   Conclusion

This survey found various types of decision diagram-, SAT- and graph-calculus-based approaches are used extensively in quantum circuit compilation. The result illustrates how quantum circuit compilation benefited from automated reasoning techniques originally developed for reasoning about classical systems. Perhaps surprising, is that not only universal quantum simulation has been efficiently handled with decision diagrams, model counting and ZX-calculus-based approaches, but also harder problems like (universal) equivalence checking. In fact, together these methods make up the state-of-the-art in quantum circuit simulation, often with different performance characteristics (e.g., model counting can be slower than decision diagrams on structured circuits but better on more unstructured circuits, while ZX-calculus can sometimes outperform both methods). Similarly, optimization and synthesis tasks have been shown to be solvable with DD, SAT/SMT and ZX-calculus for non-universal cases (e.g., for Clifford circuits) and for universal circuits, which are however mostly handled with heuristic approaches. In the latter case, ZX-calculus seems to excel.

From these successes, and recent progress in this direction [80, 144], we conclude that 'classical' automated reasoning tools will likely come to play a larger role in other quantum computing and physics applications, such as finding ground states, phase transitions and quantum error correction, where currently methods like tensor networks are often applied.

Acknowledgments. This work was supported by the Dutch National Growth Fund, as part of the Quantum Delta NL program. This work was funded by the European Union under the NEASQC project (Grant Agreement No. 951821) and the EQUALITY project (Grant Agreement No. 101080142). This publication is part of the project Divide & Quantum (with project number 1389.20.241) of the research program NWA-ORC which is (partly) financed by the Dutch Research Council (NWO).

References

1. Aaronson, S., Gottesman, D.: Improved simulation of stabilizer circuits. Phys. Rev. A **70**(5) (2004). https://doi.org/10.1103/PhysRevA.70.052328
2. Abdollahi, A., Pedram, M.: Analysis and synthesis of quantum circuits by using quantum decision diagrams. In: Proceedings of the Design Automation and Test in Europe Conference, vol. 1, pp. 1–6. IEEE (2006)
3. Akers. Binary decision diagrams. IEEE Trans. Comput. C **27**(6), 509–516 (1978)
4. Amy, M.: Towards large-scale functional verification of universal quantum circuits. arXiv preprint arXiv:1805.06908 (2018)
5. Amy, M.: Formal methods in quantum circuit design (PhD thesis). Ph.D. thesis, University of Waterloo (2019)
6. Arute, F., et al.: Quantum supremacy using a programmable superconducting processor. Nature **574**(7779), 505–510 (2019)
7. Backens, M., Kissinger, A.: ZH: A complete graphical calculus for quantum computations involving classical non-linearity. Electron. Proc. Theor. Comput. Sci. **287**, 23–42 (2019). https://doi.org/10.4204/EPTCS.287.2
8. Backens, M., Miller-Bakewell, H., de Felice, G., Lobski, L., van de Wetering, J.: There and back again: a circuit extraction tale. Quantum **5**, 421 (2021). https://doi.org/10.22331/q-2021-03-25-421
9. Bahar, R.I., et al.: Algebraic decision diagrams and their applications. Formal Methods Syst. Des. **10**(2–3), 171–206 (1997)
10. Barends, R., et al.: Superconducting quantum circuits at the surface code threshold for fault tolerance. Nature **508**(7497), 500–503 (2014). https://doi.org/10.1038/nature13171
11. Baßler, P., et al.: Synthesis of and compilation with time-optimal multi-qubit gates. Quantum **7**, 984 (2023). https://doi.org/10.22331/q-2023-04-20-984
12. Bauer-Marquart, F., Leue, S., Schilling, C.: symQV: automated symbolic verification of quantum programs. In: Formal Methods: 25th International Symposium, FM 2023, Lübeck, 6–10 March 2023, Proceedings, pp. 181–198. Springer, Heidelberg (2023)
13. de Beaudrap, N., Bian, X., Wang, Q.: Fast and effective techniques for T-count reduction via spider nest identities. In: Flammia, S.T. (ed.) 15th Conference on the Theory of Quantum Computation, Communication and Cryptography (TQC 2020). Leibniz International Proceedings in Informatics (LIPIcs), vol. 158, pp. 11:1–11:23. Schloss Dagstuhl – Leibniz-Zentrum für Informatik, Dagstuhl, Germany (2020). https://drops-dev.dagstuhl.de/entities/document/10.4230/LIPIcs.TQC.2020.11

14. de Beaudrap, N., Kissinger, A., van de Wetering, J.: Circuit extraction for ZX-diagrams can be #P-hard. In: Bojańczyk, M., Merelli, E., Woodruff, D.P. (eds.) 49th International Colloquium on Automata, Languages, and Programming (ICALP 2022). Schloss Dagstuhl - Leibniz-Zentrum für Informatik (2022). https://drops.dagstuhl.de/entities/document/10.4230/LIPIcs.ICALP.2022.119

15. Benedetti, M., Lloyd, E., Sack, S., Fiorentini, M.: Parameterized quantum circuits as machine learning models. Quant. Sci. Technol. **4**(4), 043001 (2019). https://doi.org/10.1088/2058-9565/ab4eb5

16. Berent, L., Burgholzer, L., Derks, P.J.H., Eisert, J., Wille, R.: Decoding quantum color codes with MaxSAT. arXiv preprint arXiv:2303.14237 (2023). https://doi.org/10.48550/arXiv.2303.14237

17. Berent, L., Burgholzer, L., Wille, R.: Towards a SAT encoding for quantum circuits: a journey from classical circuits to Clifford circuits and beyond. In: Meel, K.S., Strichman, O. (eds.) 25th International Conference on Theory and Applications of Satisfiability Testing (SAT 2022). Leibniz International Proceedings in Informatics (LIPIcs), vol. 236, pp. 18:1–18:17. Schloss Dagstuhl – Leibniz-Zentrum für Informatik, Dagstuhl, Germany (2022). https://drops-dev.dagstuhl.de/entities/document/10.4230/LIPIcs.SAT.2022.18

18. Bergholm, V., Biamonte, J.D.: Categorical quantum circuits. J. Phys. A: Math. Theor. **44**(24), 245304 (2011)

19. Bertot, Y., Castéran, P.: Interactive Theorem Proving and Program Development: Coq'Art: The Calculus of Inductive Constructions. Springer, Heidelberg (2013)

20. Biere, A., Cimatti, A., Clarke, E.M., Strichman, O., Zhu, Y.: Bounded model checking. Handb. Satisfiabil. **185**(99), 457–481 (2009)

21. Biere, A., Heule, M., van Maaren, H.: Handbook of Satisfiability, vol. 185. IOS Press (2009)

22. Boole, G.: An investigation of the laws of thought, on which are founded the mathematical theories of logic and probabilities. Walton and Maberly (1854)

23. Borgna, A., Perdrix, S., Valiron, B.: Hybrid quantum-classical circuit simplification with the zx-calculus. In: Oh, H. (ed.) Programming Languages and Systems, pp. 121–139. Springer, Cham (2021)

24. Bove, A., Dybjer, P., Norell, U.: A brief overview of Agda – a functional language with dependent types. In: Theorem Proving in Higher Order Logics: 22nd International Conference, TPHOLs 2009, Munich, 17–20 August 2009. Proceedings 22, pp. 73–78. Springer, Cham (2009)

25. Brand, S., Bäck, T., Laarman, A.: A decision diagram operation for reachability. In: International Symposium on Formal Methods, pp. 514–532. Springer, Cham (2023)

26. Brand, S., Coopmans, T., Laarman, A.: Quantum graph-state synthesis with SAT. Proceedings of the 14th International Workshop on Pragmatics of SAT (2023)

27. Brandhofer, S., Kim, J., Niu, S., Bronn, N.T.: SAT-based quantum circuit adaptation. In: 2023 Design, Automation and Test in Europe Conference and Exhibition (DATE), pp. 1–6. IEEE (2023)

28. Bravyi, S., Browne, D., Calpin, P., Campbell, E., Gosset, D., Howard, M.: Simulation of quantum circuits by low-rank stabilizer decompositions. Quantum **3**, 181 (2019). https://doi.org/10.22331/q-2019-09-02-181

29. Bryant, R.E.: Symbolic Boolean manipulation with ordered binary-decision diagrams. ACM Comput. Surv. **24**(3), 293–318 (1992)

30. Burgholzer, L., Bauer, H., Wille, R.: Hybrid Schrödinger-Feynman simulation of quantum circuits with decision diagrams. In: 2021 IEEE International Conference on Quantum Computing and Engineering (QCE), pp. 199–206 (2021)

31. Burgholzer, L., Kueng, R., Wille, R.: Random stimuli generation for the verification of quantum circuits. In: Proceedings of the 26th Asia and South Pacific Design Automation Conference, pp. 767–772 (2021)
32. Burgholzer, L., Wille, R.: Advanced equivalence checking for quantum circuits. IEEE Trans. Comput. Aided Des. Integr. Circuits Syst. **40**(9), 1810–1824 (2020)
33. Burgholzer, L., Wille, R.: Improved DD-based equivalence checking of quantum circuits. In: 25th Asia and South Pacific Design Automation Conference (ASP-DAC), pp. 127–132 (2020)
34. Burgholzer, L., Wille, R.: QCEC: A JKQ tool for quantum circuit equivalence checking. Software Impacts **7**, 100051 (2021)
35. Cam, T., Martiel, S.: Speeding up quantum circuits simulation using ZX-calculus. arXiv preprint arXiv:2305.02669 (2023)
36. Carette, T., Jeandel, E., Perdrix, S., Vilmart, R.: Completeness of graphical languages for mixed state quantum mechanics. ACM Trans. Quant. Comput. **2**(4), 1–28 (2021). https://doi.org/10.1145/3464693
37. Chakraborty, S., Fremont, D., Meel, K., Seshia, S., Vardi, M.: Distribution-aware sampling and weighted model counting for sat. In: Proceedings of the AAAI Conference on Artificial Intelligence, vol. 28 (2014)
38. Chareton, C., Bardin, S., Lee, D., Valiron, B., Vilmart, R., Xu, Z.: Formal methods for quantum programs: a survey. arXiv preprint arXiv:2109.06493 (2021)
39. Chen, Y.-F., Rümmer, P., Tsai, W.-L.: A theory of Cartesian arrays (with applications in quantum circuit verification). In: Pientka, B., Tinelli, C. (eds.) CADE 29, pp. 170–189. Springer, Cham (2023). https://doi.org/10.1007/978-3-031-38499-8_10
40. Clarke, E.M., McMillan, K.L., Zhao, X., Fujita, M., Yang, J.: Spectral transforms for large Boolean functions with applications to technology mapping. In: Proceedings of the 30th international Design Automation Conference, pp. 54–60 (1993)
41. Clément, A., Delorme, N., Perdrix, S., Vilmart, R.: Quantum circuit completeness: extensions and simplifications. In: Murano, A., Silva, A. (eds.) 32nd EACSL Annual Conference on Computer Science Logic (CSL 2024). Leibniz International Proceedings in Informatics (LIPIcs), vol. 288, pp. 20:1–20:23. Schloss Dagstuhl – Leibniz-Zentrum für Informatik, Dagstuhl (2024). https://drops-dev.dagstuhl.de/entities/document/10.4230/LIPIcs.CSL.2024.20
42. Clément, A., Heurtel, N., Mansfield, S., Perdrix, S., Valiron, B.: LOv-calculus: a graphical language for linear optical quantum circuits. In: Szeider, S., Ganian, R., Silva, A. (eds.) 47th International Symposium on Mathematical Foundations of Computer Science (MFCS 2022). Leibniz International Proceedings in Informatics (LIPIcs), vol. 241, pp. 35:1–35:16. Schloss Dagstuhl – Leibniz-Zentrum für Informatik, Dagstuhl (2022). https://drops-dev.dagstuhl.de/entities/document/10.4230/LIPIcs.MFCS.2022.35
43. Coecke, B., Duncan, R.: Interacting quantum observables: categorical algebra and diagrammatics. New J. Phys. **13**(4), 043016 (2011). arXiv:0906.4725 [quant-ph]
44. Coecke, B., Kissinger, A.: Picturing Quantum Processes: A First Course in Quantum Theory and Diagrammatic Reasoning. Cambridge University Press (2017)
45. Cook, S.A.: The complexity of theorem-proving procedures. In: Proceedings of the Third Annual ACM Symposium on Theory of Computing (STOC 1971), pp. 151–158. Association for Computing Machinery, New York (1971). https://doi.org/10.1145/800157.805047
46. Córcoles, A.D., et al.: Challenges and opportunities of near-term quantum computing systems. arXiv preprint arXiv:1910.02894 (2019)

47. Cowtan, A., Dilkes, S., Duncan, R., Simmons, W., Sivarajah, S.: Phase gadget synthesis for shallow circuits. Electron. Proc. Theor. Comput. Sci. **318**, 213–228 (2020). https://doi.org/10.4204/EPTCS.318.13

48. Cowtan, A., Simmons, W., Duncan, R.: A generic compilation strategy for the unitary coupled cluster ansatz. arXiv preprint arXiv:2007.10515 (2020)

49. Dahlberg, A., Helsen, J., Wehner, S.: How to transform graph states using single-qubit operations: computational complexity and algorithms. Quant. Sci. Technol. **5**(4), 045016 (2020)

50. Dahlberg, A., Wehner, S.: Transforming graph states using single-qubit operations. Philos. Trans. Roy. Soc. A: Math. Phys. Eng. Sci. **376**(2123), 20170325 (2018)

51. Darwiche, A.: SDD: a new canonical representation of propositional knowledge bases. In: Twenty-Second International Joint Conference on Artificial Intelligence (2011)

52. Darwiche, A., Marquis, P.: A knowledge compilation map. J. Artif. Intell. Res. **17**, 229–264 (2002)

53. Davis, M., Logemann, G., Loveland, D.: A machine program for theorem-proving. Commun. ACM **5**(7), 394–397 (1962)

54. De Felice, G., Hadzihasanovic, A., Ng, K.F.: A diagrammatic calculus of fermionic quantum circuits. Logic. Methods Comput. Sci. **15** (2019)

55. Deng, H., Tao, R., Peng, Y., Wu, X.: A case for synthesis of recursive quantum unitary programs. Proc. ACM Program. Lang. **8**(POPL), 1759–1788 (2024). https://doi.org/10.1145/3632901

56. van Dijk, T., Laarman, A., van de Pol, J.: Multi-core BDD operations for symbolic reachability. ENTCS **296**, 127–143 (2013)

57. Ding, J., Yamashita, S.: Exact synthesis of nearest neighbor compliant quantum circuits in 2-D architecture and its application to large-scale circuits. IEEE Trans. Comput. Aided Des. Integr. Circuits Syst. **39**(5), 1045–1058 (2019)

58. Ding, Y., Chong, F.T.: Circuit synthesis and compilation. In: Ding, Y., Chong, F.T. (eds.) Quantum Computer Systems: Research for Noisy Intermediate-Scale Quantum Computers, pp. 91–125. Springer, Cham (2020). https://doi.org/10.1007/978-3-031-01765-0_6

59. Duncan, R., Kissinger, A., Perdrix, S., van de Wetering, J.: Graph-theoretic simplification of quantum circuits with the ZX-calculus. Quantum **4**, 279 (2020). https://doi.org/10.22331/q-2020-06-04-279

60. Duncan, R., Perdrix, S.: Rewriting measurement-based quantum computations with generalised flow. In: Abramsky, S., Gavoille, C., Kirchner, C., Meyer auf der Heide, F., Spirakis, P.G. (eds.) Automata, Languages and Programming, pp. 285–296. Springer, Heidelberg (2010). https://doi.org/10.1007/978-3-642-14162-1_24

61. Dunjko, V., Briegel, H.J.: Machine learning and artificial intelligence in the quantum domain (2017)

62. Fargier, H., Marquis, P., Schmidt, N.: Semiring labelled decision diagrams, revisited: canonicity and spatial efficiency issues. In: IJCAI, pp. 884–890 (2013)

63. Feigenbaum, J., Kannan, S., Vardi, M.Y., Viswanathan, M.: Complexity of problems on graphs represented as OBDDs. In: STACS, pp. 216–226. Springer, Heidelberg (1998)

64. de Felice, G., Coecke, B.: Quantum linear optics via string diagrams. Electron. Proc. Theor. Comput. Sci. **394**, 83–100 (2023). https://doi.org/10.4204/EPTCS.394.6

65. Finigan, W., Cubeddu, M., Lively, T., Flick, J., Narang, P.: Qubit allocation for noisy intermediate-scale quantum computers. arXiv prepirnt arXiv:1810.08291 (2018)
66. Fujita, M., McGeer, P.C., Yang, J.Y.: Multi-terminal binary decision diagrams: an efficient data structure for matrix representation. FMSD **10**(2–3), 149–169 (1997)
67. Giles, B., Selinger, P.: Exact synthesis of multiqubit Clifford+T circuits. Phys. Rev. A **87**(3), 032332 (2013)
68. Gogioso, S., Yeung, R.: Annealing optimisation of mixed ZX phase circuits. Electron. Proc. Theor. Comput. Sci. **394**, 415–431 (2023). https://doi.org/10.4204/EPTCS.394.20
69. Gottesman, D.: Stabilizer codes and quantum error correction. arXiv preprint arXiv:quant-ph/9705052 (1997)
70. Meijer-van de Griend, A., Duncan, R.: Architecture-aware synthesis of phase polynomials for NISQ devices. Electron. Proce. Theor. Comput. Sci. **394**, 116–140 (2023). https://doi.org/10.4204/EPTCS.394.8
71. Grover, L.K.: A fast quantum mechanical algorithm for database search. In: Proceedings of the Twenty-Eighth Annual ACM Symposium on Theory of Computing, pp. 212–219 (1996)
72. Grurl, T., Fuß, J., Wille, R.: Considering decoherence errors in the simulation of quantum circuits using decision diagrams. In: Proceedings of the 39th International Conference on Computer-Aided Design, pp. 1–7 (2020)
73. Grurl, T., Fuß, J., Wille, R.: Noise-aware quantum circuit simulation with decision diagrams. IEEE Trans. Comput. Aided Des. Integr. Circuits Syst. **42**(3), 860–873 (2022)
74. Grurl, T., Kueng, R., Fuß, J., Wille, R.: Stochastic quantum circuit simulation using decision diagrams. In: 2021 Design, Automation and Test in Europe Conference and Exhibition (DATE), pp. 194–199. IEEE (2021)
75. Guo, Z.H., Wang, T.C.: SMT-based layout synthesis approaches for quantum circuits. In: Proceedings of the 2024 International Symposium on Physical Design (ISPD 2024), pp. 235–243. Association for Computing Machinery, New York (2024). https://doi.org/10.1145/3626184.3633316
76. Hadzihasanovic, A.: A diagrammatic axiomatisation for qubit entanglement (2015)
77. Hahn, F., Pappa, A., Eisert, J.: Quantum network routing and local complementation. NPJ Quant. Inf. **5**(1), 76 (2019)
78. Hein, M., Dür, W., Eisert, J., Raussendorf, R., Nest, M., Briegel, H.J.: Entanglement in graph states and its applications. arXiv preprint arXiv:quant-ph/0602096 (2006)
79. Heurtel, N.: A complete graphical language for linear optical circuits with finite-photon-number sources and detectors (2024)
80. Hillmich, S., Hadfield, C., Raymond, R., Mezzacapo, A., Wille, R.: Decision diagrams for quantum measurements with shallow circuits. In: 2021 IEEE International Conference on Quantum Computing and Engineering (QCE), pp. 24–34. IEEE (2021)
81. Holker, C.: Causal flow preserving optimisation of quantum circuits in the ZX-calculus. arXiv preprint arXiv:2312.02793 (2023)
82. Hong, X., Ying, M., Feng, Y., Zhou, X., Li, S.: Approximate equivalence checking of noisy quantum circuits. In: 2021 58th ACM/IEEE Design Automation Conference (DAC), pp. 637–642 (2021)

83. Hong, X., Zhou, X., Li, S., Feng, Y., Ying, M.: A tensor network based decision diagram for representation of quantum circuits. ACM Trans. Design Automat. Electron. Syst. **27**(6), 1–30 (2022)

84. Janzing, D., Wocjan, P., Beth, T.: "Non-identity-check" is QMA-complete. Int. J. Quant. Inf. **3**(03), 463–473 (2005)

85. Jeandel, E., Perdrix, S., Veshchezerova, M.: Addition and differentiation of ZX-diagrams. In: Felty, A.P. (ed.) 7th International Conference on Formal Structures for Computation and Deduction (FSCD 2022). Leibniz International Proceedings in Informatics (LIPIcs), vol. 228, pp. 13:1–13:19. Schloss Dagstuhl – Leibniz-Zentrum für Informatik, Dagstuhl (2022). https://drops-dev.dagstuhl.de/entities/document/10.4230/LIPIcs.FSCD.2022.13

86. Jiménez-Pastor, A., Larsen, K.G., Tribastone, M., Tschaikowski, M.: Efficient simulation of quantum circuits by model order reduction. arXiv preprint arXiv:2308.09510 (2023)

87. de Jong, J., Hahn, F., Tcholtchev, N., Hauswirth, M., Pappa, A.: Extracting maximal entanglement from linear cluster states. arXiv preprint arXiv:2211.16758 (2022)

88. Jozsa, R., van den Nest, M.: Classical simulation complexity of extended clifford circuits. Quant. Inf. Comput. **14**(7–8), 633–648 (2014). https://doi.org/10.26421/QIC14.7-8-7

89. Kim, Y., et al.: Evidence for the utility of quantum computing before fault tolerance. Nature **618**(7965), 500–505 (2023)

90. Kissinger, A., van de Wetering, J.: PyZX: large scale automated diagrammatic reasoning. In: QPL (2019). https://api.semanticscholar.org/CorpusID:104292461

91. Kissinger, A., van de Wetering, J.: Reducing the number of non-Clifford gates in quantum circuits. Phys. Rev. A **102**(2) (2020). https://doi.org/10.1103/PhysRevA.102.022406

92. Kissinger, A., van de Wetering, J.: Simulating quantum circuits with ZX-calculus reduced stabiliser decompositions. Quant. Sci. Technol. **7**(4), 044001 (2022). https://doi.org/10.1088/2058-9565/ac5d20

93. Kissinger, A., van de Wetering, J., Vilmart, R.: Classical simulation of quantum circuits with partial and graphical stabiliser decompositions. In: Le Gall, F., Morimae, T. (eds.) 17th Conference on the Theory of Quantum Computation, Communication and Cryptography (TQC 2022). Leibniz International Proceedings in Informatics (LIPIcs), vol. 232, pp. 5:1–5:13. Schloss Dagstuhl – Leibniz-Zentrum für Informatik, Dagstuhl (2022). https://drops-dev.dagstuhl.de/entities/document/10.4230/LIPIcs.TQC.2022.5

94. Kitaev, A.Y., Shen, A., Vyalyi, M.N.: Classical and Quantum Computation. American Mathematical Soc. (2002)

95. Koch, M., Yeung, R., Wang, Q.: Speedy contraction of ZX diagrams with triangles via stabiliser decompositions. arXiv preprint arXiv:2307.01803 (2023)

96. Kornerup, N., Sadun, J., Soloveichik, D.: The spooky pebble game. arXiv preprint arXiv:2110.08973 (2021)

97. Lai, Y.T., Pedram, M., Vrudhula, S.B.: EVBDD-based algorithms for integer linear programming, spectral transformation, and function decomposition. IEEE Trans. Comput. Aided Des. Integr. Circuits Syst. **13**(8), 959–975 (1994)

98. Landahl, A.J., Anderson, J.T., Rice, P.R.: Fault-tolerant quantum computing with color codes. arXiv preprint arXiv:1108.5738 (2011)

99. Leino, K.R.M.: Dafny: an automatic program verifier for functional correctness. In: International Conference on Logic for Programming Artificial Intelligence and Reasoning, pp. 348–370. Springer, Heidelberg (2010)

100. Lewis, M., Soudjani, S., Zuliani, P.: Formal verification of quantum programs: theory, tools, and challenges. ACM Trans. Quant. Comput. **5**(1), 1–35 (2023)
101. Lin, R.: A graphical calculus for quantum computing with multiple qudits using generalized Clifford algebras (2023)
102. Lin, S.W., Chen, S.H., Wang, T.F., Chen, Y.R.: A quantum SMT solver for bit-vector theory. arXiv preprint arXiv:2303.09353 (2023)
103. Linden, N., de Wolf, R.: Lightweight detection of a small number of large errors in a quantum circuit. Quantum **5**, 436 (2021)
104. Lloyd, S.: Universal quantum simulators. Science 273(5278), 1073–1078 (1996). https://doi.org/10.1126/science.273.5278.1073
105. Maslov, D., Falconer, S.M., Mosca, M.: Quantum circuit placement. IEEE Trans. Comput. Aided Des. Integr. Circuits Syst. **27**(4), 752–763 (2008)
106. McMillan, K.L.: Symbolic model checking: an approach to the state explosion problem. Ph.D. thesis, Carnegie Mellon University (1992)
107. Mei, J., Bonsangue, M., Laarman, A.: Simulating quantum circuits by model counting. arXiv preprint arXiv:2403.07197 (2024)
108. Mei, J., Coopmans, T., Bonsangue, M., Laarman, A.: Equivalence checking of quantum circuits by model counting. arXiv preprint (to appear) (2024)
109. Menicucci, N.C., Flammia, S.T., van Loock, P.: Graphical calculus for Gaussian pure states. Phys. Rev. A **83**(4), 042335 (2011)
110. Meuli, G., Soeken, M., De Micheli, G.: SAT-based CNOT, T quantum circuit synthesis. In: Kari, J., Ulidowski, I. (eds.) Reversible Computation, pp. 175–188. Springer, Cham (2018). https://doi.org/10.1007/978-3-319-99498-7_12
111. Miller, D.M., Thornton, M.A.: QMDD: A decision diagram structure for reversible and quantum circuits. In: 36th International Symposium on Multiple-Valued Logic (ISMVL 2006), p. 30 (2006)
112. Molavi, A., Xu, A., Diges, M., Pick, L., Tannu, S., Albarghouthi, A.: Qubit mapping and routing via MaxSAT. In: 2022 55th IEEE/ACM International Symposium on Microarchitecture (MICRO), pp. 1078–1091. IEEE (2022)
113. Montanaro, A.: Quantum algorithms: an overview. NPJ Quant. Inf. **2**(1), 15023 (2016). https://doi.org/10.1038/npjqi.2015.23
114. Montanaro, A.: Quantum-walk speedup of backtracking algorithms. Theory Comput. **14**(1), 1–24 (2018)
115. de Moura, L., Kong, S., Avigad, J., Van Doorn, F., von Raumer, J.: The Lean theorem prover (system description). In: Automated Deduction-CADE-25: 25th International Conference on Automated Deduction, Berlin, 1–7 August 2015, Proceedings 25, pp. 378–388. Springer, Heidelberg (2015)
116. Murali, P., Javadi-Abhari, A., Chong, F.T., Martonosi, M.: Formal constraint-based compilation for noisy intermediate-scale quantum systems. Microprocess. Microsyst. **66**, 102–112 (2019)
117. Nagarajan, H., Lockwood, O., Coffrin, C.: QuantumCircuitOpt: an open-source framework for provably optimal quantum circuit design. In: 2021 IEEE/ACM Second International Workshop on Quantum Computing Software (QCS), pp. 55–63. IEEE (2021)
118. Nakamura, K., Denzumi, S., Nishino, M.: Variable shift SDD: a more succinct sentential decision diagram. In: Faro, S., Cantone, D. (eds.) 18th International Symposium on Experimental Algorithms (SEA 2020). Leibniz International Proceedings in Informatics (LIPIcs), vol. 160, pp. 22:1–22:13. Schloss Dagstuhl – Leibniz-Zentrum für Informatik, Dagstuhl (2020). https://drops-dev.dagstuhl.de/entities/document/10.4230/LIPIcs.SEA.2020.22

119. Nannicini, G., Bishop, L.S., Günlük, O., Jurcevic, P.: Optimal qubit assignment and routing via integer programming. ACM Trans. Quant. Comput. **4**(1), 1–31 (2022)

120. van den Nest, M.: Classical simulation of quantum computation, the Gottesman-Knill theorem, and slightly beyond. Quant. Inf. Comput. **10**(3), 258–271 (2010)

121. Nielsen, M.A., Chuang, I.L.: Quantum Information and Quantum Computation, vol. 2, no. 8, p. 23. Cambridge University Press, Cambridge (2000)

122. Niemann, P., Wille, R., Drechsler, R.: Equivalence checking in multi-level quantum systems. In: Reversible Computation: 6th International Conference, RC 2014, Kyoto, 10–11 July 2014. Proceedings 6, pp. 201–215. Springer, Heidelberg (2014)

123. Niemann, P., Wille, R., Drechsler, R.: Advanced exact synthesis of Clifford+T circuits. Quant. Inf. Process. **19**, 1–23 (2020)

124. Niemann, P., Zulehner, A., Drechsler, R., Wille, R.: Overcoming the tradeoff between accuracy and compactness in decision diagrams for quantum computation. IEEE Trans. Comput. Aided Des. Integr. Circuits Syst. **39**(12), 4657–4668 (2020)

125. Oliveira Oliveira, M.D.: On the satisfiability of quantum circuits of small treewidth. Theory Comput. Syst. **61**, 656–688 (2017)

126. Orús, R.: Tensor networks for complex quantum systems. Nat. Rev. Phys. **1**(9), 538–550 (2019)

127. Orús, R.: A practical introduction to tensor networks: matrix product states and projected entangled pair states. Annals Phys. **349**, 117–158 (2014). https://www.sciencedirect.com/science/article/pii/S0003491614001596

128. Oztok, U., Darwiche, A.: A top-down compiler for sentential decision diagrams. In: IJCAI (IJCAI 2015), pp. 3141–3148. AAAI Press (2015)

129. Pan, F., Zhang, P.: Simulation of quantum circuits using the big-batch tensor network method. Phys. Rev. Lett. **128**(3), 030501 (2022)

130. Paulson, L.C. (ed.): Isabelle. LNCS, vol. 828. Springer, Heidelberg (1994). https://doi.org/10.1007/BFb0030541

131. Peham, T., Burgholzer, L., Wille, R.: Equivalence checking of quantum circuits with the ZX-calculus. IEEE J. Emerg. Select. Topics Circuits Syst. **12**(3), 662–675 (2022)

132. Peham, T., Burgholzer, L., Wille, R.: Equivalence checking of quantum circuits with the zx-calculus. IEEE J. Emerg. Select. Topics Circuits Syst. **12**(3), 662–675 (2022)

133. Peham, T., Burgholzer, L., Wille, R.: Equivalence checking of parameterized quantum circuits: Verifying the compilation of variational quantum algorithms. In: 2023 28th Asia and South Pacific Design Automation Conference (ASP-DAC), pp. 702–708 (2023)

134. Penrose, R.: Applications of negative dimensional tensors. In: Combinatorial Mathematics and its Applications. Academic Press (1971)

135. Perez-Garcia, D., Verstraete, F., Wolf, M., Cirac, J.: Matrix product state representations. Quant. Inf. Comput. **7**(5), 401–430 (2007). https://doi.org/10.5555/2011832.2011833

136. Poór, B., Wang, Q., Shaikh, R.A., Yeh, L., Yeung, R., Coecke, B.: Completeness for arbitrary finite dimensions of zxw-calculus, a unifying calculus. In: 2023 38th Annual ACM/IEEE Symposium on Logic in Computer Science (LICS). IEEE (2023). https://doi.org/10.1109/LICS56636.2023.10175672

137. Preskill, J.: Quantum computing and the entanglement frontier. Bull. Am. Phys. Soc. **58** (2013)

138. Preskill, J.: Quantum computing in the NISQ era and beyond. Quantum **2**, 79 (2018). https://doi.org/10.22331/q-2018-08-06-79
139. Quist, A.J., Laarman, A.: Optimizing quantum space using spooky pebble games. In: International Conference on Reversible Computation, pp. 134–149. Springer, Heidelberg (2023)
140. Quist, A.J., Laarman, A.: Trade-offs between classical and quantum space using spooky pebbling. arXiv preprint arXiv:2401.10579 (2024)
141. Raussendorf, R., Briegel, H.J.: A one-way quantum computer. Phys. Rev. Lett. **86**, 5188–5191 (2001). https://doi.org/10.1103/PhysRevLett.86.5188
142. Rennela, M., Brand, S., Laarman, A., Dunjko, V.: Hybrid divide-and-conquer approach for tree search algorithms. Quantum **7**, 959 (2023)
143. Rieser, H.M., Köster, F., Raulf, A.P.: Tensor networks for quantum machine learning. Proc. Roy. Soc. A: Math. Phys. Eng. Sci. **479**(2275), 20230218 (2023)
144. Sander, A., Burgholzer, L., Wille, R.: Towards Hamiltonian simulation with decision diagrams. In: 2023 IEEE International Conference on Quantum Computing and Engineering (QCE), vol. 1, pp. 283–294. IEEE (2023)
145. Sang, T., Bacchus, F., Beame, P., Kautz, H.A., Pitassi, T.: Combining component caching and clause learning for effective model counting. In: International Conference on Theory and Applications of Satisfiability Testing (2004). https://api.semanticscholar.org/CorpusID:52027
146. Sangiorgi, D.: Introduction to Bisimulation and Coinduction. Cambridge University Press (2011)
147. Sanner, S., McAllester, D.: Affine algebraic decision diagrams (AADDs) and their application to structured probabilistic inference. In: IJCAI, vol. 2005, pp. 1384–1390 (2005). https://doi.org/10.5555/1642293.1642513
148. Schneider, S., Burgholzer, L., Wille, R.: A SAT encoding for optimal Clifford circuit synthesis. In: Proceedings of the 28th Asia and South Pacific Design Automation Conference, pp. 190–195 (2023)
149. Seitz, P., Medina, I., Cruz, E., Huang, Q., Mendl, C.B.: Simulating quantum circuits using tree tensor networks. Quantum **7**, 964 (2023)
150. Selinger, P.: A survey of graphical languages for monoidal categories. In: Coecke, B. (ed.) New Structures for Physics, pp. 289–355. Springer, Heidelberg (2011). https://doi.org/10.1007/978-3-642-12821-9_4
151. Shaik, I., van de Pol, J.: Optimal layout synthesis for quantum circuits as classical planning. arXiv preprint arXiv:2304.12014 (2023)
152. Shannon, C.E.: The synthesis of two-terminal switching circuits. Bell Syst. Tech. J. **28**(1), 59–98 (1949)
153. Shutty, N., Chamberland, C.: Decoding merged color-surface codes and finding fault-tolerant Clifford circuits using solvers for satisfiability modulo theories. Phys. Rev. Appl. **18**(1), 014072 (2022)
154. Silva, J.M., Sakallah, K.A.: GRASP-a new search algorithm for satisfiability. In: Proceedings of International Conference on Computer Aided Design, pp. 220–227. IEEE (1996)
155. Sistla, M., Chaudhuri, S., Reps, T.: Symbolic quantum simulation with Quasimodo. In: International Conference on Computer Aided Verification, pp. 213–225. Springer, Heidelberg (2023)
156. Sistla, M., Chaudhuri, S., Reps, T.: Weighted context-free-language ordered binary decision diagrams. arXiv preprint arXiv:2305.13610 (2023)
157. Sistla, M.A., Chaudhuri, S., Reps, T.: CFLOBDDs: context-free-language ordered binary decision diagrams. ACM Trans. Program. Lang. Syst. (2023)

158. Staudacher, K., Guggemos, T., Grundner-Culemann, S., Gehrke, W.: Reducing 2-qubit gate count for ZX-calculus based quantum circuit optimization. Electron. Proc. Theor. Comput. Sci. **394**, 29–45 (2023). https://doi.org/10.4204/EPTCS. 394.3

159. Tafertshofer, P., Pedram, M.: Factored EVBDDs and their application to matrix representation and manipulation. Tech. rep., CENG Technical Report 94-27, Department of EE-Systems, University of Southern California (1994)

160. Tafertshofer, P., Pedram, M.: Factored edge-valued binary decision diagrams. Formal Methods Syst. Des. **10**(2), 243–270 (1997)

161. Tan, B., Cong, J.: Optimal layout synthesis for quantum computing. In: Proceedings of the 39th International Conference on Computer-Aided Design, pp. 1–9 (2020)

162. Tanaka, Y.: Exact non-identity check is NQP-complete. Int. J. Quant. Inf. **8**(05), 807–819 (2010)

163. Tang, E.: A quantum-inspired classical algorithm for recommendation systems. In: Proceedings of the 51st Annual ACM SIGACT Symposium on Theory of Computing, pp. 217–228 (2019)

164. Terhal, B.M.: Quantum error correction for quantum memories. Rev. Mod. Phys. **87**, 307–346 (2015). https://doi.org/10.1103/RevModPhys.87.307

165. Thanos, D., Coopmans, T., Laarman, A.: Fast equivalence checking of quantum circuits of Clifford gates. In: André, É., Sun, J. (eds.) Automated Technology for Verification and Analysis, pp. 199–216. Springer, Cham (2023)

166. Thierry-Mieg, Y., Poitrenaud, D., Hamez, A., Kordon, F.: Hierarchical set decision diagrams and regular models. In: TACAS 2009, ETAPS 2009, pp. 1–15. Springer, Heidelberg (2009)

167. Toumi, A., Yeung, R., Felice, G.: Diagrammatic differentiation for quantum machine learning. Electron. Proc. Theor. Comput. Sci. **343**, 132–144 (2021)

168. Venturelli, D., et al.: Quantum circuit compilation: an emerging application for automated reasoning. In: Scheduling and Planning Applications Workshop (2019). https://openreview.net/forum?id=S1eEBO3nFE

169. Viamontes, G.F., Markov, I.L., Hayes, J.P.: Improving gate-level simulation of quantum circuits. Quantum Inf. Process. **2**(5), 347–380 (2003)

170. Viamontes, G.F., Markov, I.L., Hayes, J.P.: Quantum Circuit Simulation. Springer, Cham (2009)

171. Viamontes, G., Markov, I., Hayes, J.: High-performance QuIDD-based simulation of quantum circuits. In: Proceedings Design, Automation and Test in Europe Conference and Exhibition, vol. 2, pp. 1354–1355 (2004)

172. Villalonga, B., et al.: A flexible high-performance simulator for verifying and benchmarking quantum circuits implemented on real hardware. NPJ Quant. Inf. **5**(1), 86 (2019)

173. Villoria, A., Basold, H., Laarman, A.: Enriching diagrams with algebraic operations. arXiv preprint arXiv:2310.11288 (2023)

174. Vilmart, R.: A near-optimal axiomatisation of ZX-calculus for pure qubit quantum mechanics. arXiv preprint arXiv:1812.09114 (2018)

175. Vilmart, R.: Quantum multiple-valued decision diagrams in graphical calculi. In: Bonchi, F., Puglisi, S.J. (eds.) 46th International Symposium on Mathematical Foundations of Computer Science (MFCS 2021). Leibniz International Proceedings in Informatics (LIPIcs), vol. 202, pp. 89:1–89:15. Schloss Dagstuhl – Leibniz-Zentrum für Informatik, Dagstuhl (2021). https://drops-dev.dagstuhl. de/entities/document/10.4230/LIPIcs.MFCS.2021.89

176. Vinkhuijzen, L., Coopmans, T., Elkouss, D., Dunjko, V., Laarman, A.: LIMDD: a decision diagram for simulation of quantum computing including stabilizer states. Quantum **7**, 1108 (2023). https://doi.org/10.22331/q-2023-09-11-1108

177. Vinkhuijzen, L., Coopmans, T., Laarman, A.: A knowledge compilation map for quantum information. arXiv preprint arXiv:2401.01322 (2024)

178. Vinkhuijzen, L., Grurl, T., Hillmich, S., Brand, S., Wille, R., Laarman, A.: Efficient implementation of LIMDDs for quantum circuit simulation. In: International Symposium on Model Checking of Software (SPIN) (2023)

179. Vrudhula, S.B.K., Pedram, M., Lai, Y.T.: Edge Valued Binary Decision Diagrams, pp. 109–132. Springer, New York (1996)

180. Wagner, F., Bärmann, A., Liers, F., Weissenbäck, M.: Improving quantum computation by optimized qubit routing. J. Optim. Theory Appl. **197**(3), 1161–1194 (2023)

181. Wang, Q.: An algebraic axiomatisation of ZX-calculus. Electron. Proc. Theor. Comput. Sci. **340**, 303–332 (2021). https://doi.org/10.4204/EPTCS.340.16

182. Wang, Q., Yeung, R., Koch, M.: Differentiating and integrating ZX diagrams with applications to quantum machine learning (2022)

183. Wang, S.A., Lu, C.Y., Tsai, I.M., Kuo, S.Y.: An XQDD-based verification method for quantum circuits. IEICE Trans. Fundam. Electron. Commun. Comput. Sci. **91**(2), 584–594 (2008)

184. Wei, C.Y., Tsai, Y.H., Jhang, C.S., Jiang, J.H.R.: Accurate BDD-based unitary operator manipulation for scalable and robust quantum circuit verification. In: Proceedings of the 59th ACM/IEEE Design Automation Conference, pp. 523–528 (2022)

185. van de Wetering, J.: ZX-calculus for the working quantum computer scientist. arXiv preprint arXiv:2012.13966 (2020)

186. Wille, R., Burgholzer, L., Artner, M.: Visualizing decision diagrams for quantum computing (special session summary). In: 2021 Design, Automation and Test in Europe Conference and Exhibition, pp. 768–773. IEEE (2021)

187. Wille, R., Burgholzer, L., Hillmich, S., Grurl, T., Ploier, A., Peham, T.: The basis of design tools for quantum computing: arrays, decision diagrams, tensor networks, and ZX-calculus. In: Proceedings of the 59th ACM/IEEE Design Automation Conference (DAC 2022), pp. 1367–1370. Association for Computing Machinery, New York (2022). https://doi.org/10.1145/3489517.3530627

188. Wille, R., Hillmich, S., Burgholzer, L.: Tools for quantum computing based on decision diagrams. ACM Trans. Quant. Comput. **3**(3), 1–17 (2022)

189. Wille, R., Przigoda, N., Drechsler, R.: A compact and efficient SAT encoding for quantum circuits. In: 2013 Africon, pp. 1–6. IEEE (2013)

190. Wille, R., Zhang, H., Drechsler, R.: ATPG for reversible circuits using simulation, Boolean satisfiability, and pseudo Boolean optimization. In: 2011 IEEE Computer Society Annual Symposium on VLSI, pp. 120–125 (2011)

191. Wilson, N.: Decision diagrams for the computation of semiring valuations. In: Proceedings of the 19th International Joint Conference on Artificial Intelligence, pp. 331–336 (2005)

192. Winderl, D., Huang, Q., Mendl, C.B.: A recursively partitioned approach to architecture-aware ZX polynomial synthesis and optimization. In: 2023 IEEE International Conference on Quantum Computing and Engineering (QCE). IEEE (2023). https://doi.org/10.1109/QCE57702.2023.00098

193. de Wolf, R.: Quantum computing: lecture notes. arXiv preprint arXiv:1907.09415 (2019)

194. Wood, C.J., Biamonte, J.D., Cory, D.G.: Tensor networks and graphical calculus for open quantum systems. Quant. Info. Comput. **15**(9–10), 759–811 (2015)
195. Yamashita, S., Markov, I.L.: Fast equivalence-checking for quantum circuits. In: 2010 IEEE/ACM International Symposium on Nanoscale Architectures, pp. 23–28. IEEE (2010)
196. Ying, M.: Floyd-Hoare logic for quantum programs. ACM Trans. Program. Lang. Syst. (TOPLAS) **33**(6), 1–49 (2012)
197. Zulehner, A., Hillmich, S., Wille, R.: How to efficiently handle complex values? implementing decision diagrams for quantum computing. In: 2019 IEEE/ACM International Conference on Computer-Aided Design (ICCAD), pp. 1–7. IEEE (2019)
198. Zulehner, A., Wille, R.: Improving synthesis of reversible circuits: exploiting redundancies in paths and nodes of QMDDs. In: Reversible Computation: 9th International Conference, RC 2017, Kolkata, 6–7 July 2017, Proceedings 9, pp. 232–247. Springer, Heidelberg (2017)
199. Zulehner, A., Wille, R.: Advanced simulation of quantum computations. IEEE Trans. Comput. Aided Des. Integr. Circuits Syst. **38**(5), 848–859 (2019)

Automated Reasoning

Random Access on Narrow Decision Diagrams in External Memory

Steffan Christ Sølvsten[(✉)][iD], Casper Moldrup Rysgaard[iD],
and Jaco Van de Pol[iD]

Aarhus University, Aarhus, Denmark
{soelvsten,rysgaard,jaco}@cs.au.dk

Abstract. The external memory BDD package Adiar can manipulate
Binary Decision Diagrams (BDDs) larger than the RAM of the machine.
To do so, it uses one or more priority queues to defer processing each
recursion until the relevant nodes are encountered in a sequential scan.
We outline how to improve the performance of Adiar's algorithms if the
BDD width of one of its inputs is small enough to fit into main memory. In
this case, one of the algorithms' priority queues can entirely be replaced
with (levelised) random access to the nodes of the narrow BDD. This
preserves the I/O efficiency of the original algorithm, is applicable to
other types of decision diagrams, and significantly improves performance
for many larger BDD computations.

Keywords: Binary Decision Diagrams · External Memory Algorithms

1 Introduction

Based on the work of Lars Arge [4,5], Adiar[1] [24] is an implementation of
Binary Decision Diagrams (BDD) [7] capable of handling BDDs larger than the
machine's random access memory (RAM). To achieve this, it uses time-forward
processing [3,8,15] to replace the conventional depth-first recursion stack with
one (or more) priority queue(s) that are synchronised with a sequential iteration
through the input BDD(s).

The high performance of conventional BDD implementations is the result
of several decades of research. Especially the unique node table and its layout
has been vital [12,14,16,19]. Yet, these and other ideas are not applicable to
time-forward processing. Hence, new ideas are needed to make Adiar achieve a
satisfactory performance. This has motivated the introduction of its levelised priority queue [23], its equality checking algorithm [24], and the concept of levelised
cuts [21]. Common to all these optimisations is the use of some meta information
about the BDD graph to substantially improve performance.

Adiar's performance was evaluated [22,24] on various combinatorial benchmarks. Each of these benchmarks accumulates a set of constraints, each of which

---

[1] github.com/ssoelvsten/adiar.

T. Neele and A. Wijs (Eds.): SPIN 2024, LNCS 14624, pp. 137–145, 2025.
https://doi.org/10.1007/978-3-031-66149-5_7

is a very narrow BDD, into one BDD whose size quickly grows large. Certain instances of symbolic model checking or symbolic SCC computation are quite similar. Here, the BDDs that represent transition relations in deterministic finite automata [9,11] or in asynchronous models of concurrency [10], e.g. Petri Nets [13,18], are narrow while the one for the accumulated state space is large.

1.1 Contributions

In the same vein as the prior optimisations in [21,24], we show in Sect. 3 how Adiar can exploit the width of the input BDDs (defined in Sect. 2). In this case, the product construction algorithm in [24] can omit the use of one of its priority queues in favour of per-level using random access directly on the narrow BDD. Our experiments in Sect. 4 show that this considerably improves performance for the larger instances of both the motivating use case, i.e. when computing on at least one narrow BDD, and also average use cases.

1.2 Related Work

Prior to this work, levelised cuts [21] improves Adiar's performance by soundly upper bounding the size of its priority queues. If it is smaller than main memory, then the external memory priority queue is replaced by a simpler and faster priority queue that only works in internal memory. This has been vital for Adiar's performance on BDDs that do fit into the RAM. In this work, we instead improve Adiar's performance by changing the algorithms' logic. Hence, in constrast to the prior work, this optimisation targets the entire spectrum of BDDs. It especially is of benefit to some larger instances.

CAL [20] (based on [6,17]) is to the best of our knowledge the only other BDD package to compute on BDDs that exceed main memory. To do so, it stores all BDD nodes in a unique node table and uses queues to execute its algorithms in the breadth-first manner. These node tables and queues can be offloaded to the disk via the operating system's swap memory. Yet, this is only efficient, if every level fits into memory. In general, CAL is not I/O-efficient [5], whereas Adiar is [24].

2 Preliminaries

2.1 I/O Model

Aggarwal and Vitter [1] designed the I/O-model to analyse the data transfers between two levels of a memory hierarchy. Here, the internal memory, e.g. the RAM, has a finite size of M and data exceeding its capacity needs to be transferred in blocks of size B to/from the external memory, e.g. the Disk.

The number of block data transfers (I/Os) needed to sequentially read and write N amounts of data is $\mathrm{scan}(N) \triangleq N/B$. To sort N amounts of data one

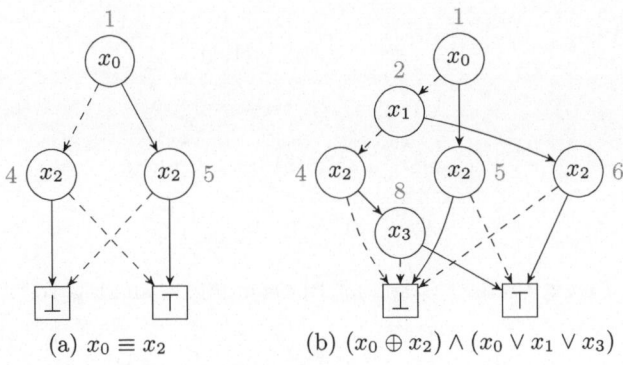

(a) $x_0 \equiv x_2$ (b) $(x_0 \oplus x_2) \wedge (x_0 \vee x_1 \vee x_3)$

Fig. 1. Examples of Reduced and Ordered Binary Decision Diagrams. Terminals are drawn as boxes while internal nodes are drawn as circles with its decision variable. The *then* and *else* edges are respectively drawn solid and dashed.

needs to use $\text{sort}(N) \triangleq N/B \cdot \log_{M/B}(N/B)$ I/Os. Furthermore, one can also design a priority queue capable of inserting and extracting N elements in the optimal $\Theta(\text{sort}(N))$ number of I/Os [3]. For all realistic values of N, M, and B, both $\text{scan}(N)$ and $\text{sort}(N)$ are several magnitudes smaller than N itself.

2.2 Binary Decision Diagrams

Binary Decision Diagrams (BDD) [7] provide a concise representation of Boolean functions $\mathbb{B}^n \to \mathbb{B}$ as a singly-rooted directed acyclic graph (DAG). As shown in Fig. 1, a BDD has two terminals with the Boolean values $\mathbb{B} = \{\bot, \top\}$ as the function's output whereas each internal BDD node provides an if-then-else decision on one of the n input variables, x_i.

What are colloquially referred to as BDDs are in fact *Reduced* and *Ordered* BDDs (ROBDDs). A BDD is ordered if the decision variables only occur once on each path from the root to a terminal and always following the same order. This induces a *levelisation* with level x_i only containing nodes with the said variable. The *width* of a BDD is the size of its largest level. A BDD is reduced, if there are (1) no *duplicate* nodes and (2) no *redundant* nodes. A node is a duplicate if it represents the same if-then-else. In conventional BDDs, a node is redundant if it has two identical children.

Fundamental to BDDs is the *Apply* operation, which, given BDDs f and g and a binary operator $\odot$, constructs the BDD for $f \odot g$. This is done via a product construction of both input BDDs and applying $\odot$ when arriving at a pair of terminals. As an example, Fig. 2 shows the product of Fig. 1a and 1b.

Here, we only provide a high-level description of Adiar's Apply algorithm that includes the details needed for Sect. 3; we refer to [24] for a detailed explanation. To make it I/O-efficient, Adiar imposes a total order on its BDD nodes such that the BDD is sorted level by level. Specifically, each node is associated with a numeric *time point* that they are encountered in the input (grey indices in

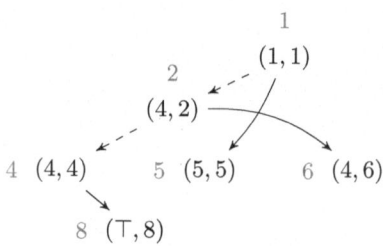

Fig. 2. Product Graph for BDDs in Fig. 1a and Fig. 1b.

Fig. 1). This ensures that each BDD node will always come after its parent during a sequential scan. To identify the pair of children of $(n_f, n_g) \in f \times g$ in a product construction, the ordering implies one needs the first to-be seen node, $\min(n_f, n_g)$, and possibly also the second one, $\max(n_f, n_g)$. Hence, a priority queue can make the recursion to target (n_f, n_g) match the sequential scan of the inputs by sorting it on the time point $\min(n_f, n_g)$. If $\max(n_f, n_g)$ is also needed then the BDD node $\min(n_f, n_g)$ is further forwarded to $\max(n_f, n_g)$ with a second priority queue. To do so, the second priority queue sorts its elements on the time point $\max(n_f, n_g)$. To guarantee a polynomial running time, recursions to the same target are grouped together. This is done by resolving ties in both priority queues' ordering via a lexicographical ordering of the recursion targets. For example, the product graph of the BDDs in Fig. 1a and Fig. 1b is resolved in [24] in the order depicted in Fig. 2.

This Apply algorithm only uses $\mathcal{O}(\text{scan}(N_f) + \text{scan}(N_g) + \text{sort}(T))$ I/Os unlike the $\mathcal{O}(T)$ I/Os used by conventional recursive implementations [5, 12], where N_f, N_g are the number of internal BDD nodes in f and g, respectively, and T is the number of BDD nodes in the output.

3   Using Random Access for Narrow Decision Diagrams

Without loss of generality, assume that the second input, g, to the Apply operation is *narrow*, i.e. the width of g is smaller than some threshold $\theta < M/2$. In this case, we can completely omit the second priority queue.

To do so, when processing level x_i, we load all BDD nodes of g at level x_i from external memory. This provides immediate random access to the entire level x_i of g. Hence, the second priority queue can be omitted if the ordering of the first priority queue is changed accordingly. Specifically, the first priority queue now solely has to respect the levels of the recursive calls and synchronise them with the sequential scan through f. Hence, the recursion target (n_f, n_g) should first be sorted on its level to not miss any requests where the level of n_f is below the one of n_g. Secondly, it is sorted lexicographically to respect the sequential scan through f. Futhermore, lexicographical sorting also groups recursive calls for the same target together, which preserves the polynomial running time and I/Os.

Doing so only affects the order in which all recursions are resolved; the output is still isomorphic to what is produced by the algorithm in [24]. For example in Fig. 2, if random access is used on the BDD from Fig. 1b then $(4, 6)$ is resolved prior to $(5, 5)$. In the previous algorithm [24], both the node at time point 4 in Fig. 1a and the one at 6 in Fig. 1b had to be visited in-order to resolve the product $(4, 6)$; hence, this product was resolved after $(5, 5)$. Instead, with random access the node at time point 6 in Fig. 1b is immediately available when reading the one at 4 in Fig. 1a; hence, $(4, 6)$ is resolved before $(5, 5)$.

Proposition 1. *The Apply algorithm with random access on a narrow BDD saves $\mathcal{O}(sort(T))$ I/Os in comparison to the prior algorithm from [24].*

Proof. Since the input BDDs are already sorted based on their level, loading the levels of g top-down is possible in a single sequential scan. Yet, the algorithm in [24] also needs to scan through g. Hence, loading the nodes of g for random access does not cost any additional I/Os. Yet, it completely removes the $\mathcal{O}(sort(T))$ I/Os incured by the second priority queue.

Note that this is only a constant improvement over the algorithm in [24]. Furthermore, it is only an $\mathcal{O}(sort(T))$ rather than a $\Theta(sort(T))$ improvement, since recursion requests do not necessarily need to be moved into the second priority queue.

4   Experimental Evaluation

We have implemented the modified algorithm of Sect. 3 and run the benchmarks from [22,24] with threshold $\theta = 0$, B (2 MiB), and ∞. Using $\theta = 0$ essentially turns the random access optimisation off and provides a baseline. On the other hand, using $\theta = \infty$ entirely replaces the previous Apply algorithm. Finally, $\theta = B$ provides a small value which covers the motivating use cases while also leaving more of the internal memory to the remaining priority queue.

The benchmarks of [22,24] consist of two categories. First, the *Combinatorial Counting* problems, e.g. the Queens problem, primarily involve the accumulation of lots of narrow decision diagrams. On the other hand, the *EPFL* [2] *Circuit Verification* provides a typical use case for decision diagrams.

As in [21,22,24], we have run all experiments on the CSCAA *Grendel* cluster where Adiar is initialised with $M = 300$ GiB. For each value of θ and each benchmark instance, the running time has been measured between 3 and 15 times (11.2 times on average) depending on its expected running time. We consider a measurement to be significant if the difference between the mean running time of $\theta = 0$ and $\theta = B, \infty$ is larger than twice their largest standard deviation. Figure 3 shows the speed-up in the mean running time for all 113 benchmark instances. Table 1 provides a summary of all significant instances.

Both $\theta = B$ and $\theta = \infty$ provide a significant performance increase in performance for the larger combinatorial benchmarks compared to the $\theta = 0$ baseline. Of the 64 combinatorial instances, 14 (21.9% of all of these instances) had

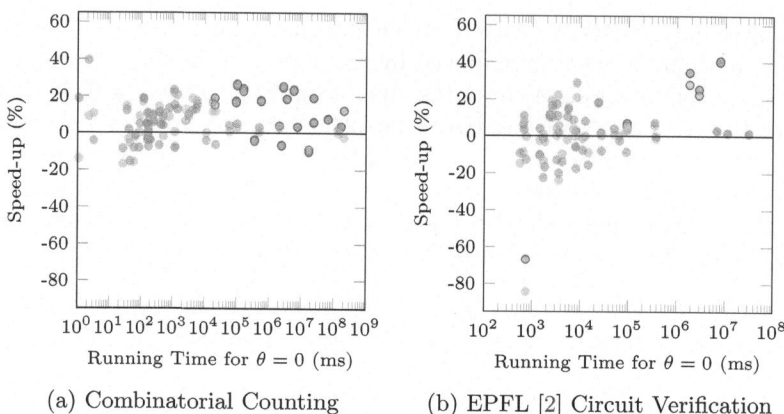

(a) Combinatorial Counting   (b) EPFL [2] Circuit Verification

Fig. 3. Speed-up in running time with • $\theta = B$ and • $\theta = \infty$ relative to $\theta = 0$ (higher is better). Statistically significant measurements have a black border.

a significant improvement of 15.4% on average for $\theta = B$ and 16.3% for $\theta = \infty$. These 14 instances all require 10 s or more to solve with $\theta = 0$. In total, 26 out of the total 64 instances required this amount of time to solve. That is, performance improved for 53.8% of these larger instances.

Similarly to the combinatorial benchmarks, EPFL Verification also gains significant improvements for many of its larger instances. Only one circuit, int2float, requires significantly more time to verify. Yet, while a decrease in performance of 67% seems worrisome, it is only an increase in the computation time from 0.74 s to 1.12 s.

Finally, the optimisation presented in this work further closes the gap between Adiar and conventional BDD packages. For example, CUDD [26] can solve up to the 15-Queens problem with BDDs. In [24] the gap between Adiar and CUDD for this problem's instance was a factor of 1.42. In [21], this was improved to 1.26. With $\theta = B, \infty$, the gap is now further decreased down to 1.07. Simultaneously, this also improves performance for instances not solvable with CUDD. For example, it improves the solving time of 16-Queens by 19.2%.

Table 1. Speed-up ($\triangle$) and slowdown ($\triangledown$) of $\theta = B$ and $\theta = \infty$ for Combinatorial Counting (CC) and EPFL [2] Circuit Verification (EPFL) benchmarks.

			# Significant Instances				Average Relative Difference	
			$\triangle$		$\triangledown$		$\triangle$	$\triangledown$
CC	•	$\theta = B$	14	(21.9%)	3	(4.7%)	15.4%	−6.7%
	•	$\theta = \infty$	14	(21.9%)	3	(4.7%)	16.3%	−6.7%
EPFL	•	$\theta = B$	6	(12.2%)	0	(0.0%)	25.1%	−
	•	$\theta = \infty$	6	(12.2%)	1	(2.0%)	26.9%	−66.8%

5   Conclusion

Let a BDD be *narrow* if its width is smaller than some threshold $\theta < M/2$, where M is the amount of internal memory. Specifically, each individual level of a narrow BDD fits into internal memory. Hence, one can load each of its levels in their entirety and do random access on it. If some input to the Apply algorithm in [24] is narrow then one does not need to synchronise its computation with the sequential order of the narrow one. In this work, we have sketched how this algorithm can be adapted to this case, to save on computation time and I/Os.

For $\theta = B$ (B is the block transfer size), our experiments show a significant improvement in performance for larger instances. In the motivating use case with lots of narrow BDDs, performance increased significantly by 11.8% on average. In the average use case, performance even increased significantly by 25.1%.

Relative to $\theta = B$, our results with an unlimited θ further improves performance significantly for 3 larger combinatorial instances. No instances slowed down significantly. Hence, we have implemented levelised random access as part of Adiar 2.0 and based on the observation above, we use a large θ of $M/8$.

5.1   Future Work

While using levelised cuts [21] improves Adiar's performance for algorithms whose priority queues fit into RAM, it still leaves a gap between the running time of Adiar and conventional BDD packages for the smallest problems.

To efficiently solve the smallest of BDD instances (15 MiB or smaller), the BDD package CAL [20] switches from its breadth-first approach to the conventional depth-first algorithms.

The work presented in this paper can be extended to compute on input BDDs that are stored in an (internal memory) unique node table. Its (unreduced) output, which may exceed main memory, is still stored on the disk. Conversely, one can change Adiar's algorithms to place the final (reduced) BDD back in the node table. Hence, this work provides the basis for an efficient and seamless transition between a conventional depth-first approach and Adiar's external memory algorithms. This will be the final step to make Adiar competitive across the entire spectrum of BDDs.

Furthermore, this work applies to all of Adiar's product construction operations, such as variable quantification. Hence, this work, together with [21], is vital for the design of an I/O-efficient relational product that is usable in practice.

Acknowledgements. Thanks to the Centre for Scientific Computing, Aarhus, (phys.au.dk/forskning/cscaa/) for access to the Grendel cluster.

Data Availability Statement. The data presented in Sect. 4 is available at [25] while the code to run the benchmarks can be found at [27].

References

1. Aggarwal, A., Vitter, Jeffrey, S.: The input/output complexity of sorting and related problems. Commun. ACM **31**(9), 1116–1127 (1988). https://doi.org/10.1145/48529.48535
2. Amarú, L., Gaillardon, P.E., De Micheli, G.: The EPFL combinational benchmark suite. In: 24th International Workshop on Logic and Synthesis (2015)
3. Arge, L.: The buffer tree: a new technique for optimal I/O-algorithms. In: Workshop on Algorithms and Data Structures (WADS). LNCS, vol. 955, pp. 334–345. Springer, Heidelberg (1995).https://doi.org/10.1007/3-540-60220-8_74
4. Arge, L.: The I/O-complexity of ordered binary-decision diagram manipulation. In: 6th International Symposium on Algorithms and Computations (ISAAC). LNCS, vol. 1004, pp. 82–91 (1995). https://doi.org/10.1007/BFb0015411
5. Arge, L.: The I/O-complexity of ordered binary-decision diagram. In: BRICS RS Preprint Series, vol. 29. Department of Computer Science, University of Aarhus (1996). https://doi.org/10.7146/brics.v3i29.20010
6. Ashar, P., Cheong, M.: Efficient breadth-first manipulation of binary decision diagrams. In: IEEE/ACM International Conference on Computer-Aided Design (ICCAD), pp. 622–627. IEEE Computer Society Press (1994). https://doi.org/10.1109/ICCAD.1994.629886
7. Bryant, R.E.: Graph-based algorithms for Boolean function manipulation. IEEE Trans. Comput. **C-35**(8), 677–691 (1986). https://doi.org/10.1109/TC.1986.1676819
8. Chiang, Y.J., Goodrich, M.T., Grove, E.F., Tamassia, R., Vengroff, D.E., Vitter, J.S.: External-memory graph algorithms. In: Proceedings of the Sixth Annual ACM-SIAM Symposium on Discrete Algorithms (SODA 1995), pp. 139—149. Society for Industrial and Applied Mathematics (1995)
9. Elgaard, J., Klarlund, N., Møller, A.: MONA 1.x: new techniques for WS1S and WS2S. In: Proceedings of the 10th International Conference on Computer-Aided Verification, CAV 1998. LNCS, vol. 1427, pp. 516–520. Springer, Heidelberg (1998). https://doi.org/10.1007/3-540-61648-9_56
10. Kant, G., Laarman, A., Meijer, J., Van de Pol, J., Blom, S., Van Dijk, T.: LTSmin: high-performance language-independent model checking. In: Tools and Algorithms for the Construction and Analysis of Systems (TACAS). LNCS, vol. 9035, pp. 692–707. Springer, Heidelberg (2015). https://doi.org/10.1007/978-3-662-46681-0_61
11. Klarlund, N.: Mona & Fido: the logic-automaton connection in practice. In: Computer Science Logic. LNCS, vol. 1414, pp. 311–326. Springer, Cham (1998). https://doi.org/10.1007/BFb0028022
12. Klarlund, N., Rauhe, T.: BDD algorithms and cache misses. In: BRICS Report Series, vol. 26 (1996). https://doi.org/10.7146/brics.v3i26.20007
13. Larsen, C.A., Schmidt, S.M., Steensgaard, J., Jakobsen, A.B., van de Pol, J., Pavlogiannis, A.: A truly symbolic linear-time algorithm for SCC decomposition. In: Tools and Algorithms for the Construction and Analysis of Systems (2). LNCS, vol. 13994, pp. 353–371. Springer, Heidelberg (2023). https://doi.org/10.1007/978-3-031-30820-8_22
14. Long, D.E.: The design of a cache-friendly BDD library. In: Proceedings of the 1998 IEEE/ACM International Conference on Computer-Aided Design (ICCAD), pp. 639–645. Association for Computing Machinery (1998)
15. Meyer, U., Sanders, P., Sibeyn, J.: Algorithms for Memory Hierarchies: Advanced Lectures. Springer, Heidelberg (2003). https://doi.org/10.1007/3-540-36574-5

16. Minato, S.I., Ishiura, N., Yajima, S.: Shared binary decision diagram with attributed edges for efficient Boolean function manipulation. In: 27th Design Automation Conference (DAC), pp. 52–57. Association for Computing Machinery (1990). https://doi.org/10.1145/123186.123225

17. Ochi, H., Yasuoka, K., Yajima, S.: Breadth-first manipulation of very large binary-decision diagrams. In: International Conference on Computer Aided Design (ICCAD), pp. 48–55. IEEE Computer Society Press (1993). https://doi.org/10.1109/ICCAD.1993.580030

18. Pastor, E., Roig, O., Cortadella, J., Badia, R.M.: Petri net analysis using boolean manipulation. In: Valette, R. (ed.) ICATPN 1994. LNCS, vol. 815, pp. 416–435. Springer, Heidelberg (1994). https://doi.org/10.1007/3-540-58152-9_23

19. Pastva, S., Henzinger, T.: Binary decision diagrams on modern hardware. In: Conference on Formal Methods in Computer-Aided Design, pp. 122–131 (2023)

20. Sanghavi, J.V., Ranjan, R.K., Brayton, R.K., Sangiovanni-Vincentelli, A.: High performance BDD package by exploiting memory hierarchy. In: 33rd Design Automation Conference (DAC), pp. 635–640. Association for Computing Machinery (1996). https://doi.org/10.1145/240518.240638

21. Sølvsten, S.C., Van de Pol, J.: Predicting memory demands of BDD operations using maximum graph cuts. In: André, É., Sun, J. (eds.) Automated Technology for Verification and Analysis. LNCS, vol. 14216, pp. 72–92. Springer, Cham (2023). https://doi.org/10.1007/978-3-031-45332-8_4

22. Sølvsten, S.C., Van de Pol, J.: Adiar 1.1: zero-suppressed decision diagrams in external memory. In: NASA Formal Methods Symposium, LNCS, vol. 13903, Springer, Heidelberg (2023). https://doi.org/10.1007/978-3-031-33170-1_28

23. Sølvsten, S.C., Van de Pol, J., Jakobsen, A.B., Thomasen, M.W.B.: Efficient binary decision diagram manipulation in external memory. arXiv preprint arXiv:2104.12101 (2021)

24. Sølvsten, S.C., Van de Pol, J., Jakobsen, A.B., Thomasen, M.W.B.: Adiar: binary decision diagrams in external memory. In: Tools and Algorithms for the Construction and Analysis of Systems. LNCS, vol. 13244, pp. 295–313. Springer, Heidelberg (2022). https://doi.org/10.1007/978-3-030-99527-0_16

25. Sølvsten, S.C., Rysgaard, C.M., van de Pol, J.: Adiar 2.0.0-beta.3 : Experiment Data (2024). https://doi.org/10.5281/zenodo.10493770

26. Somenzi, F.: CUDD: CU decision diagram package, 3.0. Tech. rep., University of Colorado at Boulder (2015)

27. Slvsten, S.C.: BDD Benchmark. Zenodo (2024).https://doi.org/10.5281/zenodo.10803154

Solving Constrained Horn Clauses as C Programs with CHC2C

Levente Bajczi[(✉)] and Vince Molnár

Department of Measurement and Information Systems,
Budapest University of Technology and Economics, Budapest, Hungary
{bajczi,molnarv}@mit.bme.hu

Abstract. Solving Constrained Horn Clauses (CHC) is necessitated by numerous fields in formal methods, from verifying software and smart contracts to modeling systems, yet the competitive scene for academic tools remains fairly sparse, especially compared to more popular fields such as software verification. Comparative evaluation as a competition, such as SV-COMP or CHC-COMP, sparks a more cohesive community around fields in formal methods. Lately, a trend has been emerging with tools such as Btor2C that bridge multiple fields together, thus widening this cohesion. Following that example, we propose and perform an experiment, where we use CHC-to-C transformation to apply software verification tools to linear CHC problems. In the process, we help both fields by diversifying the scene of CHC solvers and providing new and valuable benchmarks to aid the development of software verification tools. Using these benchmarks, we uncovered a previously hidden bug in multiple verification tools that can lead to false positive results. By analysing the results of the experiment, we can confidently make a recommendation for developers of software verifiers to consider supporting CHCs via our pre-verification transformation.

Keywords: CHC · verification · formal methods · software verification

1 Introduction

Formal methods have been gaining significant traction in many new domains in recent years. Facilitating this acceleration of adoption are the many specialized comparative evaluation-based competitions among tools, such as SV-COMP for software verification [2], HWMCC for hardware model checking [6], SMT-COMP for solving queries in satisfiability modulo theories (SMT) [26], or CHC-COMP for solving constrained Horn clauses (CHC) [10]. These competitions boost both academic interest and visibility towards potential users of the competitors, as well as provide a more-or-less standardized benchmark suite for evaluating the tools. However, with certain tools having been developed for multiple decades to work (and compete) in one of these domains, the price of entry into one of the more established fields can be insurmountable to new tools. Thus, instead of developing a brand new tool with specialized algorithms for a new field of study, a more established tool can often be used instead, extended with a pre-transformation layer for adapting to one of the tool's supported formats.

Lately, this trend of adapting the problem to suit the verification tool instead of doing it vice versa has not only been used for solving new problems in new domains but also to close the gap between the many existing domains where formal methods are already being used. A success story in this regard is about Btor2C [4], which brought hardware verification problems in the language of Btor2 to software verification tools by adapting it to standard C. This means that the exact source of the problem can be opaque to the tools themselves, and just by supporting C, they also support Btor2 by extension. Furthermore, the software verification community gained access to valuable new benchmarks in process, available via the SV-COMP benchmark suite[1].

In this paper, we aim to yet again close a gap that exists among use cases of formal methods. Constrained Horn Clauses (CHCs) have long been used as a means to verify software [15,19,23], and lately, conventional software verification tools have also been applied to the reverse problem of representing CHCs via adapting them to a control-flow based representation, akin to imperative software [11,16,25]. However, by adapting the *format* of (linear) CHC problems to a suitable format for software verification, such as plain C, any off-the-shelf C verification tool becomes capable of solving CHC problems. We show that such a transformation is not only possible but beneficial in terms of performance as well.

1.1   Our Motivations and Contributions

The main motivation of our work is the diversification of the field of CHC solving, with the secondary motivation being the addition of valuable benchmark tasks to the software verification community. The importance of the former task is demonstrated well by the multitude of types of problems that necessitate CHCs: various synthesis problems such as syntax- and semantics guided, or functional synthesis [12,17,20]; systems verification [9]; software verification of conventional [15,19], higher-order [21,23], and specialized (e.g., dataflow-description [7]) programs; program equivalence checking [14]; or smart contract verification [27].

While the problem's importance is well understood, the same diverse and competitive scene that characterizes other formal methods fields (e.g., software verification) is yet to form for CHCs. At CHC-COMP, the highest number of competitors in any of its tracks remains well under 10 [10], and this number is not changing by much year-to-year [13,24].

By enabling conventional software verification tools to also support CHC problems, both fields (CHC solving and software verification) will be furthered: CHC solvers will be more diverse, and software verification tools will receive valuable, real-life problems as benchmarks to further tweak and tune their algorithms.

In connection with our motivations outlined above, our contributions regarding this paper are the following:

---

[1] https://gitlab.com/sosy-lab/benchmarking/sv-benchmarks/.

1. We experimentally show that using software verification tools to solve CHC problems can be advantageous
2. We contribute valuable new benchmarks to the software verification community

In furtherance of these contributions, we also *implemented* and *validated* a proof-of-concept transformation tool that takes CHC problems and generates a C-language, error-label reachability-based representation of the problem. Furthermore, we *benchmarked* all ranked participants in this year's SV-COMP'24 on all transformed CHC problems, and *analyzed* their performance.

Note that our contribution is not centered around a fully-fledged and optimized tool for the CHC to C conversion, but rather a demonstration that this theoretical step is *possible*, and even a research prototype is capable of bridging the two domains together in a useful way.

Furthermore, we found and identified some bugs in some of the most widely used software verification tools using these tests, further justifying the point on the value of these new problems.

Novelty. Certain approaches that use a *CHC-to-program* pre-transformation step already exist in the lineup of tools for CHC-COMP [11,16,25]. Details on these approaches are available in Sect. 1.3. The novelty of the approach presented in this paper stems from the genericness of the transformation: by using plain C as the target format, we enable *all* software verification tools to participate in CHC solving, while previous attempts focused on transforming the input into a tool-specific internal formalism. This enables seamless support for CHC problems, even for tools where no development towards supporting CHCs could be justified.

Significance. The significance of our contributions can be demonstrated by the following points:

1. As shown in Sect. 3, some software verification tools perform on par with dedicated CHC solvers on a portion of problems. There are certain tasks that only software verification tools solved, while dedicated CHC solvers all timed out or threw an exception.
2. We uncovered previously hidden faults in well-known software verification tools by testing them on the newly acquired benchmark set of CHC problems.

We believe that these results significantly further the field of formal methods.

1.2 Background and Example

As a slightly simplified definition, let us define a CHC problem in the following way:

- A CHC problem consists of several *deduction rules*, i.e., implications

– A deduction rule may have *uninterpreted functions* (i.e., relations) in both its premise and its consequence
  • The consequence of a deduction rule may have *exactly one* uninterpreted function (besides the queries, which have zero)
  • The premise of a deduction rule may have zero, one, or more uninterpreted functions
       * If zero uninterpreted functions exist in a rule's premise, we call that rule an *atom*
       * If more than one uninterpreted function exists in a rule's premise, the rule is considered to be *non-linear*
  • If any rule is *non-linear*, the CHC is *non-linear*. Otherwise, the CHC is *linear*.
– Any number of deduction rules may deduce the literal *false*, these rules are called *queries*

The domain of variables in a CHC problem is given as SMT theories. In this paper, we mostly concentrate on the *core* theory, with additional support only for *integer* arithmetic. Note that problems requiring support for more theories exist (such as those using *arrays* or *algebraic data types*), but we discount those in the context of this work.

The goal for any CHC problem is to prove whether *false* is deducible (making the system *unsatisfiable*), in which case the query's premise is *true*. In practice, CHC problems are often encoded in the format of SMT-LIBv2 [8], ensuring tools' interoperability. Tasks for CHC-COMP use a strict subset of SMT-LIBv2 [10].

As an example, we can have three simple deduction rules:

$$n = 0 \implies A(n)$$
$$A(n - 2) \implies A(n)$$
$$A(6) \implies false$$

The first rule states that $A(0)$ is *true*, and (given $A(0)$) the second rule makes all positive even numbers n to also evaluate $A(n)$ to *true*. The third rule is a *query*, stating that if $A(6)$ is *true*, then *false* is deduced, and hence the system is unsolvable. In this case, this problem is trivially solved as 6 is an even number, and hence, $A(6)$ must be *true*; thus, the system is *unsatisfiable*.

In software, the same process is easy to implement. We can take the query and start from its premise. This approach, demonstrated in Listing 1.1, is called *top-down*, or *backward*; because it starts the program by evaluating the query. While due to the infinite recursion in line 3 this program may not terminate (depending on the parameter to A), tools that can reason about recursion may find that the exit condition $A(6)$ is reachable.

An alternative, without recursion, is to leverage the support for nondeterminism in software verification tools and rewrite the program as seen in Listing 1.2. In this case, the program constructs all facts by following the deduction rules from the atomic facts; thus, this is called *bottom-up* or *forward* transformation, which is only possible for *linear* CHCs [25]. In lines $2 - 3$ the starting

Listing 1.1. Backward

```
1   int A(int n) {
2       if(n==0) return 1;
3  ↻     else if(A(n−2)) return 1;
4       else return 0;
5   }
6
7   int main() {
8       if(A(6)) return −1;
9       else return 0;
10  }
```

Listing 1.2. Forward

```
1   int main() {
2       int A, n = 0;
3       A = n;
4
5       while(true) {
6           n = nondet();
7           if(A == 6) return −1;
8           else if(A == n − 2) A = n;
9       }
10  }
```

$A(0)$ fact is constructed, then lines $5-9$ construct all further facts by assigning a nondeterministic value to n, then testing whether the previous iteration already deduced $A(6)$ (in which case an error code is returned), followed by testing if the last deduced $A(n)$ value is equal to the current n minus two; in which case the value of A (being the last deduced $A(n)$) is updated. This loop is repeated infinitely, only exiting when $A(6)$ is deduced. Verification tools, however, need only discover that the program is *unsafe*, and the original CHC problem is *unsatisfiable*. Solvers from both SV-COMP'24 and CHC-COMP'23 [10] mostly successfully solved the problem; these results are shown in Table 1.

Table 1. $\sqrt{}$: successful solution; ?: no result; ×: wrong verdict

	Software Verification Tool																	CHC Solver						
	2ls	bubaak	bubaak-split	cpachecker	cpv	emergentheta	esbmc-kind	goblint	infer	mopsa	symbiotic	theta	uautomizer	ukojak	utaipan	veriabs	veriabsl	Eldarica	Golem	LoAT	Theta	U.TreeAut.	U.Unihorn	spacer-22
Forward	√	√	√	√	?	?	√	×	?	?	√	√	√	√	√	√	√	√	√	√	√	√	√	√
Backward	?	√	√	√	?	?	√	×	?	?	√	√	√	√	√	√								

1.3   Related Work

The presented approach of transforming CHC problems into reachability-based software verification tasks can be found in multiple CHC solver tools: Eldarica [16], Ultimate Unihorn [11], and Theta [25]. While the former two are long-time participants in CHC-COMP [10,13,24], their exact pre-transformation steps are not well documented (as even mentioned in the CHC-COMP competition report [10]). The latter, Theta, published an in-depth paper of this process alongside their debut on CHC-COMP [25]. Furthermore, Theta is the only tool out of the three that supports *both* a backward and a forward transformation option, with the other two – to the best of this paper's authors' knowledge – only supporting backward transformation.

As to not re-invent the transformation process from the ground up, in this paper, we take the approach implemented in Theta as the baseline for our experimental evaluation. We build on the existing implementation to create the proof-of-concept tool CHC2C and use the method described in the tool paper of Theta to reason about the theoretical possibilities of the approach.

2   Experiment Proposal and Methodology

To support the claims and goals of this paper, we devise an experiment to answer the following questions.

RQ1 Is a CHC-to-C transformation *sound* and *complete* in terms of support for CHCs?

RQ2 Is a CHC-to-C transformation beneficial for verification tools in terms of *performance*?

2.1   Theoretical Analysis

In short, the answer to RQ1 is *no*. There are several substantial differences between the semantics of CHCs and C programs that will make this transformation impossible, no matter the encoding.

One of the main differences is in handling data types such as integers. C programs are designed to eventually run on actual hardware, which will always have a finite number of bits to represent any value; therefore, the C standard defines different lower and upper bounds on the size of variable types. In contrast, CHCs use infinite integers, akin to conventional SMT solvers, due to their *logical* nature.

However, one familiar with the inner behavior of software model checkers may see a disconnect here: most model-checking software rely on using SMT solvers and will definitely need to pay additional attention to implementing fixed-sized integer or array support when targeting C. Therefore, it is easy to see that this limitation on the expressive power of C over CHC *inside a model checking tool* is entirely artificial and stems from the C standard itself. Therefore, we anticipate that tools supporting a C-syntax input, albeit with a loose definition of semantics (allowing for infinite integers in the model), will fully mirror the semantics of the original CHC problem, even when transformed to C.

As most tools will not support redefining the semantics of C in an easy way, we also devise a safeguard that limits the types of false verdicts to be *false negatives*, akin to a BMC-like behavior, where a bound exists up to which the problem is deemed safe. However, instead of a bound on the steps from an initial state, we use a metric on the *variables* itself. With such safeguarding, the two verdicts a tool may have are the following:

- *Unsafe* output: Unsafe verdict
- *Safe* output: Safe, given all variables are in bound $[-A; +B]$.

Therefore, such tools may iterate with growing bounds until they run out of supported domain size, or find a counterexample and can return *unsafe*.

The problem safeguarding prevents is reaching the bounds of a variable type defined by the standard. When reaching an unsigned integer's upper- or lower bound, the C standard defines the expected behavior of the value to *wrap around*, thus remaining inside the bounds of the type. Therefore, $15 + 2 == 1$ is satisfiable given a 4-bit unsigned integer (bounds $[0; 16)$), but unsatisfiable given mathematical, infinite integers. The same problem should not exist for *signed* integers, as the standard leaves that to be implementation-dependent, and therefore, model checking tools *should* try and be flexible in handling that case. In practice, however, most tools just assume the signed values also wrap around.

The safeguards introduced above make the bounds of the values sufficiently far from the bounds of the variable type so as not to cause the wraparound problem, which would introduce *false positive* results.

Taking the above two problems into account, let us amend RQ1 in the following way:

RQ1a Does safeguarding prevent *false positive* results in verification tools utilizing a CHC-to-C transformation?

RQ1b How often do *false negative* results occur on a realistic problem set?

Note that a similar set of problems arise with the use of *arrays*, as C-like arrays will have some size, outside of which addressing the array leads to undefined behavior, while CHCs use an infinite, mathematical definition of arrays, where every value of the address type's domain must have an associated legal value. However, we discount CHC problems with arrays in the scope of this paper so as not to overcrowd the transformation process with mitigations and safeguards.

2.2   Practical Analysis

To answer RQ1 and RQ2, we need to test the proposed approach on a benchmark suite with diverse and numerous problems (for RQ1) and representative and challenging problems (for RQ2). The GitHub organization associated with CHC-COMP[2] contains a high number of CHCs sourced from a vast selection of sources and each year, a rigorous selection process is applied to the entirety of this collection to select a subset for CHC-COMP [10]. Therefore, we shall use all available problems to answer RQ1 (more specifically, RQ1a and RQ1b), and last year's selection for CHC-COMP'23 [10] to answer RQ2 as to not skew the performance evaluation with the numerous simple tasks some of the sources provide[3].

As mentioned above, we are using the tool Theta to implement the CHC-to-C transformation, as it has a well-documented and versatile pre-transformation step from CHCs to its internal CFA-like representation [25]. We implemented a serialization step, which outputs a C-language program in the format of SV-COMP

---

[2] https://github.com/chc-comp/.

[3] The selection is based on perceived verification difficulty, problem traits, etc.

tasks [2] for compatibility and uses *signed integers* in place of the SMT integers (Boolean values are handled via the _Bool type). We used Theta's default transformation setting, which performs *forward* transformation on linear CHCs. Due to concerns raised about the performance of the *backward* transformation, as well as not to dilute the homogeneous benchmark suite consisting only of linear CHCs with possibly different characteristics over non-linear problems. We also included a new command-line parameter that governs whether to use bounds safeguarding, which uses a limit of $[-1000000000; +1000000000]$ to limit the range of the values. This limit ensures that no single operation may take a value outside the domain's range, which is assumed to be at least 32 bits wide.

The plan for the experimental evaluation is the following:

Getting a Baseline. Having the two benchmark suites (all containing every benchmark, comp containing CHC-COMP'23 benchmarks), we run CHC-COMP'23 participants [10] (and Spacer [22]) on the problems to get the following:

1. a baseline on their performance (in the case of comp), and
2. an expected verdict (for both sets)

The latter is necessary because unlike other benchmark sets such as that of SV-COMP [2], the expected verdicts for CHC-COMP are rarely published together with the input problems. Because verdicts may differ, we take a majority vote among the participants to decide an expected verdict. Note that we use binary classification, and no counterexample or proof validation takes place in these tests.

Note that we ran all participants on the comp benchmark set, but only a subset of the participants on the all benchmark set: Eldarica, Golem, Spacer and Theta.

Generating the Software Benchmark Suits. Using Theta, we generate four benchmark suites: all-range, all-norange, comp-range, comp-norange. Default settings are used for parsing the CHC files. We use the tool indent[4] to pretty-print the results for manual readability, with options -nut -i4.

A step of the transformation process is to generate code for non-deterministic transitions in the control flow automaton (i.e., a location has more than one outgoing edge, and their guards overlap). As in most cases (for CHCs), this means a binary decision, we used the C type _Bool to create a non-deterministic value of either 0 or 1, then used a switch-case statement to choose a transition in the CFA. An example of such a construct can be seen in Fig. 1. Here, reach_error should never be reached because all cases of the switch statement return from $main$[5]. Therefore, the program exits. However, some tools fail to interpret this

---

[4] https://www.gnu.org/software/indent/.
[5] Note that the outcomes do not change when an explicit cast is placed inside the switch statement's head, for which the standard definitely states the value should either be 0 or 1 [18].

correctly and allow a non-existent *default* case to be executed, which may lead to an incorrect verdict. Out of the 17 tools used in the context of this paper, 10 solve the example task correctly (yielding a *safe* verdict), and 4 tools solve it incorrectly (yielding an *unsafe* verdict). Therefore, we included an additional *default* case calling abort(), so that no trace taking the default case may continue after the statement. This modification made all tools previously giving a wrong output to correct their verdict. Therefore, we used this modification throughout the transformation process to get usable results from these 4 tools. We plan to open issues in the offending tools' repositories, where possible, to help developers correct this behavior.

```
extern void reach_error();
extern _Bool __VERIFIER_nondet_Bool();
int main(){
    switch(__VERIFIER_nondet_Bool()) {
        case 0: return 0;
        case 1: return 0;
    }
    reach_error();
}
```

2ls	bubaak	bubaak-split	cpachecker	cpv	emergentheta	esbmc-kind	goblint	infer	mopsa	symbiotic	theta	uautomizer	ukojak	utaipan	veriabs	veriabsl
✓	✓	✓	×	?	✓	✓	✓	?	?	✓	✓	×	×	×	✓	✓

Fig. 1. Verification test case, and verifier outputs ($\checkmark$: safe, $\times$: unsafe, ?: no result)

Executing the Software Benchmarks. We used all non-*hors concours* participants of the ReachSafety category in SV-COMP'24. The full list is shown in Table 1. We used the submitted tool archives archived on Zenodo[6] to run the experiments, with the unreach-call property as an input specification. We recorded all tools' verdicts and CPU time over all benchmarks.

For the *soundness* and *completeness* check (RQ1), we executed all SV-COMP tools [3] on all-range and all-norange. For the *performance* check (RQ2), we executed all SV-COMP tools on comp-range and comp-norange.

Expected Outcomes. We expect that – besides some tool-specific failures – all tools will be able to handle at least a subset of the benchmark tasks, given they only use commonly seen elements of the C language. We expect that tools will report false negative verdicts for tasks where a counterexample cannot be found within the range of the signed integer domain. We also expect that in the case of the *-norange sets, tools that handle signed integer overflow will report false positives where wraparound behavior leads to an infeasible (in terms of feasibility over logical integers) counterexample to be found.

We can discount false positive verdicts, as they can easily be eliminated with a simple feasibility check in an SMT solver after the verification run, which could mark the verdict as *unknown*. We omitted such a check to not group false verdicts together with actual *unknown* verdicts in the experiment's evaluation phase.

---

[6] https://zenodo.org/.

For *false* results, we matched up different tools' outputs and classified the benchmark as *commonly wrong* if multiple tools gave a wrong result for the same task, or *tool-specific fault* if at least one other tool solved the task correctly.

3   Results

We ran the experiments as discussed in Sect. 2.2 [1]. We collected 23958 CHC problems from the CHC-COMP GitHub organization[7]. We removed 8644 tasks containing algebraic data types and a further 8892 tasks containing arrays. Out of the remaining 6422 tasks, Theta could parse (in 60 s) 3076 tasks, out of which 1914 tasks were *linear*. The selected linear CHC problems were then transformed to C, both *with* and *without* the safeguarding technique presented in Sect. 2.1. We also filtered and transformed the CHC-COMP'23 benchmark set[8], yielding us 405 succesfully transformed CHC tasks.

Using the top performing CHC tools from CHC-COMP, we attempted to solve the 1914 tasks and got 1207 *true* (i.e., *safe*); 475 *false* (i.e., *unsafe*); and 232 *unknown* results. For the 405 tasks in CHC-COMP'23, 269 *true*, 71 *false* and 65 *unknown* verdicts were given.

Results of the first experiment regarding RQ1 can be seen in Table 2 and Table 3. Each row corresponds to an SV-COMP participant, and the columns represent the classification of their outputs. The first six number columns show the number of wrong verdicts (*false*) the tool produced. Within the *false* results, there are *common* wrong results (where every software verification tool that solved the task produced the wrong verdict), and *tool-specific* wrong results (where at least one other tool succeeded with the verification). *True* results are correctly classified tasks. In all three groups of columns, *All* denotes the total number of results (*common false, tool-specific false, true*); and + or - denote the output of the tool (i.e., *False/Tool/+* means *tool-specific false positive* results, which are said to be *safe* by the tool yet the expected verdict is *unsafe*, and there is at least one other tool that returned *unsafe* correctly). The last but one column shows the *unconfirmed* verdicts, which are tasks the CHC solvers could not solve. The last column shows the *points* a tool would receive should the scoring system of SV-COMP be applied (1 point per good verdict, -32 points per false negative, -16 points per false positive). Note that false positives are discarded from point calculation due to their inherent easyness in checking: as opposed to software verification, a counterexample to a CHC problem is trivially checked by substituting the values in the SMT formula, and having it solved with a dedicated SMT solver. The table is sorted in descending order by the *points* column.

The number of tasks where a common fault caused all tools to report a wrong verdict upon success is not directly readable from the table. For the safeguarded transformation, 84 tasks were found to cause common wrong verdicts, while for the non-safeguarded transformation, 68 tasks. 16 tasks were found to cause

---

[7] https://github.com/chc-comp/.

[8] https://github.com/chc-comp/chc-comp23-benchmarks.

Table 2. Safeguarded transformation (all tasks)

	False					True			Unconf	Points
	Common			Tool						
	All	+	-	All	+	-	All	+	-		
uautomizer	25	0	25	38	38	0	1054	686	368	4	1054
ukojak	24	0	24	38	38	0	1036	680	356	3	1036
utaipan	25	0	25	35	35	0	952	621	331	4	952
cpachecker	23	0	23	39	39	0	765	444	321	6	765
esbmc-kind	23	0	23	4	1	3	805	427	378	8	709
mopsa	20	0	20	1	0	1	210	210	0	0	178
2ls	0	0	0	0	0	0	101	84	17	0	101
infer	0	0	0	0	0	0	0	0	0	0	0
cpv	0	0	0	0	0	0	0	0	0	0	0
emergentheta	82	0	82	98	0	98	911	832	79	59	-2225
theta	81	0	81	191	0	191	1354	1157	197	55	-4758
bubaak	84	0	84	379	0	379	1183	1171	12	59	-10945
symbiotic	84	0	84	379	0	379	1183	1171	12	59	-10945
bubaak-split	84	0	84	379	0	379	1182	1170	12	59	-10946
veriabs	70	0	70	391	0	391	1117	1117	0	59	-11395
veriabsl	70	0	70	391	0	391	1113	1113	0	59	-11399
goblint	71	0	71	391	0	391	1101	1101	0	59	-11411

Table 3. Non-safeguarded transformation (all tasks)

	False					True			Unconf	Points
	Common			Tool						
	All	+	-	All	+	-	All	+	-		
uautomizer	24	0	24	39	38	1	1142	769	373	5	1110
ukojak	23	0	23	39	38	1	1097	739	358	3	1065
utaipan	24	0	24	36	36	0	1028	690	338	5	1028
cpachecker	22	0	22	78	78	0	792	451	341	5	792
esbmc-kind	22	0	22	60	57	3	827	444	383	8	731
mopsa	22	0	22	1	0	1	212	212	0	0	180
2ls	0	0	0	3	3	0	98	81	17	0	98
infer	0	0	0	0	0	0	0	0	0	0	0
cpv	0	0	0	0	0	0	0	0	0	0	0
theta	67	0	67	208	0	208	1355	1158	197	66	-5301
emergentheta	68	0	68	210	0	210	1344	1147	197	59	-5376
symbiotic	65	0	65	352	1	351	1190	1139	51	173	-10042
bubaak	67	0	67	355	1	354	1229	1176	53	236	-10099
bubaak-split	68	0	68	375	0	375	1201	1170	31	59	-10799
goblint	66	0	66	396	0	396	1120	1120	0	59	-11552
veriabs	65	0	65	407	0	407	1169	1169	0	59	-11855
veriabsl	65	0	65	407	0	407	1169	1169	0	59	-11855

Table 4. Safeguarded transformation (CHC-COMP'23)

	False					True			Unconf	Points
	Common			Tool						
	All	+	-	All	+	-	All	+	-		
utaipan	8	0	8	4	4	0	103	82	21	9	103
uautomizer	8	0	8	6	6	0	94	73	21	7	94
cpachecker	8	0	8	6	6	0	91	71	20	11	91
ukojak	7	0	7	6	6	0	87	68	19	9	87
mopsa	10	0	10	1	0	1	56	56	0	9	56
esbmc-kind	9	0	9	1	0	1	86	67	19	14	54
2ls	0	0	0	0	0	0	17	15	2	0	17
infer	0	0	0	0	0	0	0	0	0	0	0
cpv	0	0	0	0	0	0	0	0	0	0	0
emergentheta	32	0	32	13	0	13	168	163	5	48	-248
theta	37	0	37	14	0	14	198	190	8	48	-250
bubaak	39	0	39	23	0	23	234	234	0	52	-502
symbiotic	37	0	37	23	0	23	208	208	0	51	-528
bubaak-split	34	0	34	23	0	23	198	198	0	50	-538
veriabsl	30	0	30	23	0	23	190	190	0	50	-546
veriabs	31	0	31	23	0	23	189	189	0	50	-547
goblint	28	0	28	23	0	23	182	182	0	50	-554

Table 5. Non-safeguarded transformation (CHC-COMP'23)

	False					True			Unconf	Points
	Common			Tool						
	All	+	-	All	+	-	All	+	-		
utaipan	8	0	8	4	4	0	133	108	25	10	133
cpachecker	12	0	12	15	15	0	117	90	27	10	117
uautomizer	8	0	8	7	6	1	116	93	23	10	84
ukojak	7	0	7	7	6	1	114	91	23	9	82
mopsa	12	0	12	15	14	1	76	76	0	9	76
esbmc-kind	9	0	9	0	0	0	93	72	21	14	61
2ls	0	0	0	0	0	0	17	15	2	0	17
infer	0	0	0	0	0	0	0	0	0	0	0
cpv	0	0	0	0	0	0	0	0	0	0	0
bubaak	34	0	34	20	0	20	258	247	11	70	-382
symbiotic	29	0	29	20	0	20	219	209	10	63	-421
emergentheta	31	0	31	23	0	23	226	218	8	51	-510
theta	29	0	29	23	0	23	212	204	8	50	-524
goblint	28	0	28	25	0	25	190	190	0	50	-610
bubaak-split	28	0	28	28	0	28	207	204	3	50	-689
veriabs	29	0	29	31	0	31	203	203	0	51	-789
veriabsl	29	0	29	31	0	31	202	202	0	51	-790

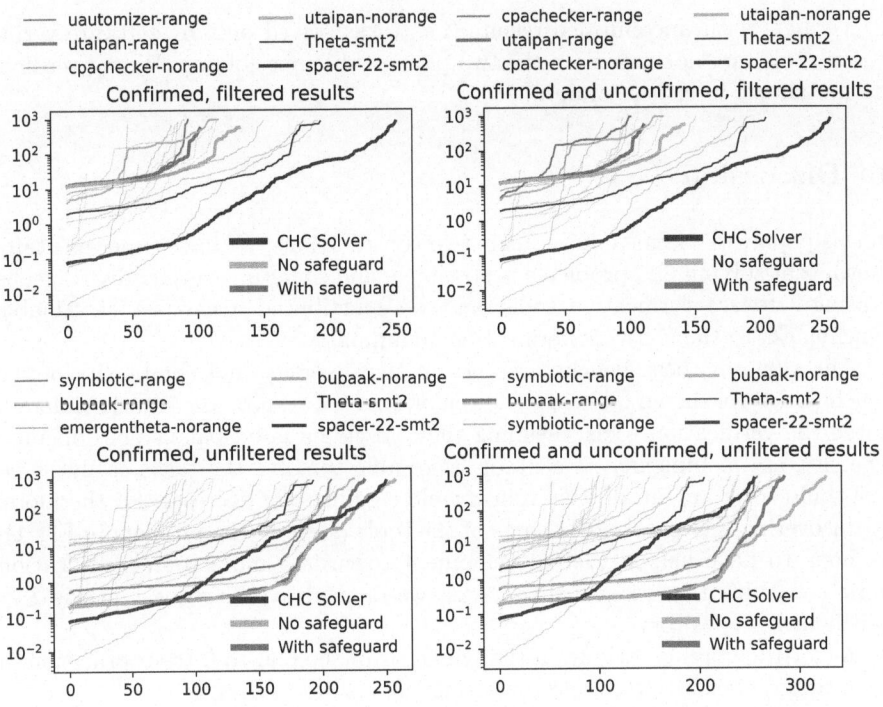

Fig. 2. Software verifiers vs. CHC solvers. Best two tools per category highlighted.

common wrong verdicts only in the safeguarded transformation but not in the non-safeguarded transformation, meaning all 68 tasks from the latter remain commonly wrong for the safeguarded version as well.

Table 4 and 5 show the results for the tasks in CHC-COMP'23 in the same format as described above. Here, 39 and 34 tasks were commonly wrong among the tools for the safeguarded and non-safeguarded transformations, respectively.

The performance of the tools is shown on the quantile plots in Fig. 2. Horizontal axes show the number of solved tasks (out of 405 possible), and vertical axes show verification time, with data points in the series sorted by ascending order. Given any line $y(x) = N$, we can easily determine the tool solving the most tasks under N timelimit by taking the series containing the rightmost point under the said line.

Each quantile plot shows the performance of the native CHC solvers, the software verification tools using safeguarded transformation, and software verification tools with non-safeguarded transformation. The top two participants are highlighted and shown on the legend above each plot in each of the three groups. The top participant is shown in the thickest line.

Tools with negative points in Table 4 and Table 5 are not shown on the two plots in the first row.

Verdicts with unconfirmed results (i.e., tasks solved only by software verification tools) are not shown on the two plots in the first column. Wrong verdicts are not shown in either column.

4 Discussion

In this paper, we claim to have made two contributions: we have experimentally shown that solving CHC problems is possible – and sometimes even advantageous – using software verification tools, and we contributed a new set of valuable benchmarks to the software verification community.

The value of these benchmarks lies in the diversity and complexity of the benchmarks. As shown later in the discussion of the results, these tasks differentiated the verification tools, meaning they provide a good balance of difficulty, diversity, and complexity: if all tools were able to solve the tasks easily or no tool could solve any at all, this value would significantly decrease. Furthermore, by uncovering a latent fault in some of the tools (seen on the problem in Fig. 1), we hope to have helped their development towards a more sound verification workflow. With these new benchmarks, we opened a pull request in the SV-Benchmarks repository[9].

As for the main contribution (i.e., analysis of the CHC-to-C transformation in verification), we show the following using the results in Sect. 3:

Ans1a The transformation can be free from introducing *false positive* results (answering RQ1a)

Ans1b The *false negative* results are significantly less frequent than correct results (answering RQ1b)

Ans2 The performance characteristics of software verification tools on CHC problems make them *competitive* (answering RQ2)

4.1 Analysis of the Results

For **Ans1a** and **Ans1b**, see Table 2 and Table 3: no common *false positive* results exist, meaning no *safe* task was deemed *unsafe* by all tools; but 84 and 68 common *false negative* results exist for the safeguarded and non-safeguarded transformations, respectively. 84 *false negative* results make up around 4% of all tasks (as to answer **Ans1b**). Tool-specific false positive results exist; for most tools, their number is significantly lower in the case of the safeguarded transformation.

Almost half of the tools received negative points (8), and 2 tools produced no results, leaving 7 tools to receive positive points. Out of these 7, the tool 21s is noteworthy: in the case of the safeguarded transformation, it produced *only correct results*, while in the non-safeguarded transformation, it only produced 3 *false*

---

[9] https://gitlab.com/sosy-lab/benchmarking/sv-benchmarks/-/merge_requests/1467.

positive results. These are likely cases where 21s handled signed integer overflow using wraparound. This fairly clean result also provides further evidence for **Ans1a** and RQ1a, showing that safeguarding can prevent false positive results.

The top scoring tool in both cases, `Ultimate Automizer`, experienced 25 and 24 common faults out of the 84 and 68 commonly wrong tasks for the safeguarded and non-safeguarded transformation, respectively. A further 38 and 39 tasks were given a wrong verdict by the tool, where at least one further tool was successful, therefore, we classify these as tool-specific faults. Almost all of these were false *positive* results.

Note that these wrong verdicts might not mean actual faults in the former set of tools, as the handling of signed integer overflow is undefined, and if the 3 tools having more false positive results handle overflow as wraparound, while some other tools just allow values outside of the domain to persist; then tools using wraparound will have false positive results while the others will solve the tasks correctly. This explanation is more likely for such a small number of false positive results. However, for tools on the other end of the scores, having almost a quarter of the results be wrong is more likely to hide a fault in the tool rather than some quirk of such edge cases.

As for the performance of the software verification tools concerning **Ans2**, results are shown in Fig. 2. The best performing native CHC solver throughout the experiments was `Spacer`, with second best being either `Golem` or `Theta` depending on the addition of unconfirmed results[10].

The overwhelming amount of false results in the case of some of the tools makes reasoning about performance difficult. We cannot distinguish between a faulty tool coupled with a lucky guess and a correctly reasoning tool. Therefore, we present multiple views about the participants' performance and the unique flaws these representations may carry.

Filtering. If we exclude all negative scoring tools (as per Table 4 and Table 5) from appearing on the plots, the native CHC solvers cannot be outperformed by software verification tools in terms of number of solved tasks. However, including these tools makes software verification seem on-par (in the case of confirmed results only) or even better (in the case of confirmed and unconfirmed results) than native CHC solvers. Their exclusion favors CHC solvers, while their inclusion favors software verification tools – these tools *did* provide a good verdict for a number of the tasks, but overwhelming false verdicts allow for speculation on hidden faults. A faulty tool may still produce a good result, but it may or may not be correct in its reasoning.

Including Unconfirmed Results. Excluding unconfirmed verdicts skews the results towards CHC solvers, because their majority vote *is* the confirmation – meaning if only a single CHC solver solves a task (e.g., `Spacer`), it is automatically confirmed. On the other hand, including unconfirmed verdicts may skew

---

[10] Note that these results show performance on only a subset of the real problems in CHC-COMP'23, which `Theta` could parse.

the results towards software verification tools because their verdict is accepted without validation from a dedicated (and thoroughly tested) CHC solver.

To summarize, the following problems exist with the four quantile plots representing performance characteristics in Fig. 2:

Confirmed, Filtered Results. Favors CHC solvers because only confirmed results are shown. Favors CHC solvers because the results are filtered only to show positive scoring tools.

Confirmed and Unconfirmed, Filtered Results. Favors software verification tools because unconfirmed results are treated as good verdicts. Favors CHC solvers because the results are filtered only to show positive scoring tools.

Confirmed, Unfiltered Results. Favors CHC solvers because only confirmed results are shown. Favors software verification tools because the results are not filtered only to show positive scoring tools, potentially including tools incorrectly deducing the expected output

Confirmed and Unconfirmed, Unfiltered Results. Favors software verification tools because unconfirmed results are treated as good verdicts. Favors software verification tools because the results are not filtered only to show positive scoring tools, potentially including tools incorrectly deducing the expected output

Even if we ignore the two extremes (*confirmed, filtered*; and *confirmed and unconfirmed, unfiltered*) due to them skewing to one side only, the remaining two plots still do not show a congruent picture. What conclusion we *can* draw is that using software verification tools with the presented approach may be better than *some* native CHC solver tools and worse than or on-par with the leading CHC solvers. Note that these results are produced by a *research prototype* as the transformation step, meaning a theoretical, optimal transformation step may outperform these results. Our contribution concludes by showing that using this transformation, on-par performance is *possible* to native CHC solvers; and should be considered a viable solution in the future.

4.2   Threats to Validity

As the main contribution of this paper is the experiment design and its analysis, the factors that threaten the validity of this experiment are presented in this section.

Internal Validity. Consistency and accuracy of the experiments were ensured by using the BenchExec framework [5]. Memory consumption statistics may deviate between executions due to the managed nature of some languages used in developing the tested tools, therefore, such metrics are not used. CPU time and, therefore, solved tasks may be influenced by external factors such as other processes or environmental temperature fluctuations, therefore, minute differences are disregarded.

External Validity. The results of the experiments are at risk of not being generalizable due to the relatively low number of benchmarks used throughout the experiments. The experiment in this paper is designed to show the characteristics of *one possible* CHC-to-C transformation, and therefore, we can only state observations about the feasibility of the concept, not about the actual performance impact of an optimal transformation. This may skew the results to the detriment of software verification tools because there may be a better, optimized version of the CHC2C tool, which would make them outperform their current behavior.

Construct Validity. To justify the type of metrics used in the evaluation of the experiments, we considered the main use cases these tools would face should they be used in the approach described in this paper. Academic competitions such as CHC-COMP [10] or SV-COMP [2] reflect the performance of tools after careful tuning of them while constantly re-testing on the same benchmark set. Therefore, tools that may produce wrong results when applied first to a problem, are generally fixed before the actual competition. This means that directly comparing the number of successfully solved tasks of dedicated CHC solvers, which have been developed relying on the benchmark set we tested on, to software verification tools that see these problems the first time may skew the results to the benefit of CHC solvers. Small problems and imperfections (such as the one seen in Fig. 1) may introduce false results.

Notice that we have designed our experiment and its analysis to consistently skew *against* our main hypothesis – which is that *software verification tools may be beneficial in solving CHCs* – rather than being *in favor of* it. Therefore, the presented results show a pessimistic view about the value of our contributions.

4.3 Conclusion

We have shown that software verification tools can be useful tools for CHC solving using a CHC2C pre-transformation step before verification. While dedicated CHC solvers produced a better ratio of good results to bad results, just the number of new tools that became capable of solving CHC problems results in a useful diversification of the state-of-the-art.

We plan to use a version of our presented approach in the next CHC-COMP to create a participant that uses a portfolio-based approach and tries to run the optimal software verification tool on any given CHC problem. We hope to show the developers and researchers in both domains, CHC solving and software verification, that it is worth collaborating on tool development and that the exact domain might only matter regarding a pre-verification transformation step, not at the algorithmic level. Thus, we hope to help integrate the knowledge of both fields, resulting in advantages for all parties involved.

References

1. Bajczi, L., Molnár, V.: Solving Constrained Horn Clauses as C Programs with CHC2C (2024). https://doi.org/10.5281/zenodo.10529452

2. Beyer, D.: Competition on software verification and witness validation: SV-COMP 2023. In: Sankaranarayanan, S., Sharygina, N. (eds.) TACAS 2023. LNCS, vol. 13994, pp. 495–522. Springer, Heidelberg (2023). https://doi.org/10.1007/978-3-031-30820-8_29

3. Beyer, D.: Verifiers and Validators of the 12th International Competition on Software Verification (SV-COMP 2023) (2023). https://doi.org/10.5281/ZENODO.7627829

4. Beyer, D., Chien, P., Lee, N.: Bridging hardware and software analysis with Btor2C: a word-level-circuit-to-C translator. In: Sankaranarayanan, S., Sharygina, N. (eds.) TACAS 2023, ETAPS 2022. LNCS, vol. 13994, pp. 152–172. Springer, Heidelberg (2023). https://doi.org/10.1007/978-3-031-30820-8_12

5. Beyer, D., Löwe, S., Wendler, P.: Reliable benchmarking: requirements and solutions. Int. J. Softw. Tools Technol. Transf. **21**(1), 1–29 (2019). https://doi.org/10.1007/s10009-017-0469-y

6. Biere, A., van Dijk, T., Heljanko, K.: Hardware model checking competition 2017. In: Stewart, D., Weissenbacher, G. (eds.) 2017 Formal Methods in Computer Aided Design, FMCAD 2017, Vienna, 2–6 October 2017, p. 9. IEEE (2017). https://doi.org/10.23919/FMCAD.2017.8102233

7. Champion, A., Mebsout, A., Sticksel, C., Tinelli, C.: The kind 2 model checker. In: Chaudhuri, S., Farzan, A. (eds.) CAV 2016. LNCS, vol. 9780, pp. 510–517. Springer, Heidelberg (2016). https://doi.org/10.1007/978-3-319-41540-6_29

8. Cok, D.R.: The SMT-LIBv2 Language and Tools: A Tutorial (2012). https://api.semanticscholar.org/CorpusID:63272811

9. Daniel, J., Cimatti, A., Griggio, A., Tonetta, S., Mover, S.: Infinite-state liveness-to-safety via implicit abstraction and well-founded relations. In: Chaudhuri, S., Farzan, A. (eds.) CAV 2016. LNCS, vol. 9779, pp. 271–291. Springer, Heidelberg (2016). https://doi.org/10.1007/978-3-319-41528-4_15

10. De Angelis, E., K., H.G.V.: CHC-COMP 2022: competition report. In: Hamilton, G.W., Kahsai, T., Proietti, M. (eds.) Proceedings 9th Workshop on Horn Clauses for Verification and Synthesis and 10th International Workshop on Verification and Program Transformation, HCVS/VPT@ETAPS 2022, and 10th International Workshop on Verification and Program Transformation, Munich, 3rd April 2022. EPTCS, vol. 373, pp. 44–62 (2022). https://doi.org/10.4204/EPTCS.373.5

11. Dietsch, D., Heizmann, M., Hoenicke, J., Nutz, A., Podelski, A.: Ultimate TreeAutomizer (CHC-COMP Tool Description). In: Angelis, E.D., Fedyukovich, G., Tzevelekos, N., Ulbrich, M. (eds.) Proceedings of the Sixth Workshop on Horn Clauses for Verification and Synthesis and Third Workshop on Program Equivalence and Relational Reasoning, HCVS/PERR@ETAPS 2019, Prague, 6–7th April 2019. EPTCS, vol. 296, pp. 42–47 (2019).https://doi.org/10.4204/EPTCS.296.7

12. Fedyukovich, G., Gurfinkel, A., Gupta, A.: Lazy but effective functional synthesis. In: Enea, C., Piskac, R. (eds.) VMCAI 2019. LNCS, vol. 11388, pp. 92–113. Springer, Heidelberg (2019). https://doi.org/10.1007/978-3-030-11245-5_5

13. Fedyukovich, G., Rümmer, P.: Competition report: CHC-COMP-21. In: Hojjat, H., Kafle, B. (eds.) Proceedings 8th Workshop on Horn Clauses for Verification and Synthesis, HCVS@ETAPS 2021, Virtual, 28th March 2021. EPTCS, vol. 344, pp. 91–108 (2021). https://doi.org/10.4204/EPTCS.344.7

14. Felsing, D., Grebing, S., Klebanov, V., Rümmer, P., Ulbrich, M.: Automating regression verification. In: Crnkovic, I., Chechik, M., Grünbacher, P. (eds.) ACM/IEEE International Conference on Automated Software Engineering, ASE 2014, Vasteras, 15–19 September 2014, pp. 349–360. ACM (2014). https://doi.org/10.1145/2642937.2642987

15. Gurfinkel, A., Kahsai, T., Komuravelli, A., Navas, J.A.: The SeaHorn verification framework. In: Kroening, D., Pasareanu, C.S. (eds.) CAV 2015. LNCS, vol. 9206, pp. 343–361. Springer, Heidelberg (2015). https://doi.org/10.1007/978-3-319-21690-4_20

16. Hojjat, H., Rümmer, P.: The ELDARICA horn solver. In: Bjørner, N.S., Gurfinkel, A. (eds.) 2018 Formal Methods in Computer Aided Design, FMCAD 2018, Austin, 30 October–2 November 2018, pp. 1–7. IEEE (2018).https://doi.org/10.23919/FMCAD.2018.8603013

17. Hu, Q., Cyphert, J., D'Antoni, L., Reps, T.W.: Exact and Approximate Methods for Proving Unrealizability of Syntax-Guided Synthesis Problems, pp. 1128–1142. ACM (2020). https://doi.org/10.1145/3385412.3385979

18. Information Technology - Programming Languages - C. Standard, International Organization for Standardization (2011)

19. Kahsai, T., Rümmer, P., Sanchez, H., Schäf, M.: JayHorn: a framework for verifying java programs. In: Chaudhuri, S., Farzan, A. (eds.) CAV 2016. LNCS, vol. 9779, pp. 352–358. Springer, Heidelberg (2016). https://doi.org/10.1007/978-3-319-41528-4_19

20. Kim, J., Hu, Q., D'Antoni, L., Reps, T.W.: Semantics-guided synthesis. Proc. ACM Program. Lang. 5(POPL), 1–32 (2021). https://doi.org/10.1145/3434311

21. Kobayashi, N., Sato, R., Unno, H.: Predicate abstraction and CEGAR for higher-order model checking. In: Hall, M.W., Padua, D.A. (eds.) Proceedings of the 32nd ACM SIGPLAN Conference on Programming Language Design and Implementation, PLDI 2011, San Jose, 4–8 June 2011, pp. 222–233. ACM (2011).https://doi.org/10.1145/1993498.1993525

22. Komuravelli, A., Gurfinkel, A., Chaki, S.: SMT-based model checking for recursive programs. Formal Methods Syst. Des. 48(3), 175–205 (2016). https://doi.org/10.1007/s10703-016-0249-4

23. Matsushita, Y., Tsukada, T., Kobayashi, N.: RustHorn: CHC-based verification for rust programs. ACM Trans. Program. Lang. Syst. 43(4), 15:1–15:54 (2021). https://doi.org/10.1145/3462205

24. Rümmer, P.: Competition Report: CHC-COMP-20. In: Fribourg, L., Heizmann, M. (eds.) Proceedings 8th International Workshop on Verification and Program Transformation and 7th Workshop on Horn Clauses for Verification and Synthesis, VPT/HCVS@ETAPS 2020, Dublin, 25–26th April 2020. EPTCS, vol. 320, pp. 197–219 (2020). https://doi.org/10.4204/EPTCS.320.15

25. Somorjai, M., Dobos-Kovács, M., Ádám, Z., Bajczi, L., Vörös, A.: Bottoms up for CHCs: novel transformation of linear constrained horn clauses to software verification. In: 10th Workshop on Horn Clauses for Verification and Synthesis (2023). https://ftsrg.mit.bme.hu/paper-hcvs23-chc/paper.pdf

26. Weber, T., Conchon, S., Déharbe, D., Heizmann, M., Niemetz, A., Reger, G.: The SMT Competition 2015–2018. J. Satisf. Boolean Model. Comput. 11(1), 221–259 (2019). https://doi.org/10.3233/SAT190123

27. Wesley, S., Christakis, M., Navas, J.A., Trefler, R.J., Wüstholz, V., Gurfinkel, A.: Verifying solidity smart contracts via communication abstraction in SmartACE. In: Finkbeiner, B., Wies, T. (eds.) VMCAI 2022. LNCS, vol. 13182, pp. 425–449. Springer, Heidelberg (2022). https://doi.org/10.1007/978-3-030-94583-1_21

Verification Tools

Learning the State Machine Behind a Modal Text Editor: The (Neo)Vim Case Study

Pierre Ganty[(⊠)] [iD]

IMDEA Software Institute, Madrid, Spain
pierre.ganty@imdea.org

Abstract. We use active automata-based learning to extract the state machine underlying the modal text editor Vim. We expose the various challenges to interface an active learning library with the text editor. Furthermore, we report on how we uncovered several issues and how they were dealt with by the (Neo)Vim developers. Finally, we reflect on the possible uses of automatically extracted finite-state machines beyond bug reports.

Keywords: Active Learning · Moore machine · Modal text editor

1 Introduction

Modal text editors such as Vim support multiple editing modes. Depending on the mode, typed characters are interpreted either as sequences of commands or are inserted as text. Figure 1 depicts part of the modes (normal, visual, insert, ...) and how the keystrokes ([esc], [ctrl]+[v], [i], ...) transition between them.

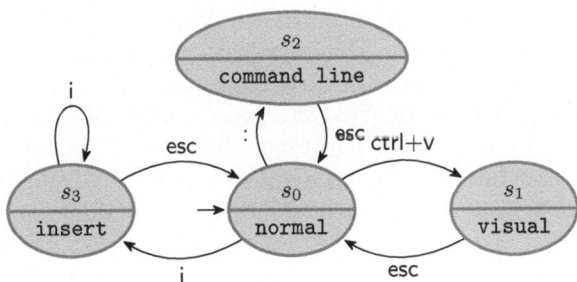

Fig. 1. Four Vim modes and examples of keystrokes to move between them. In insert mode, the text you type is inserted into the buffer (i.e. the in-memory text of a file). visual mode enables the selection of a piece of text via keystrokes. In command line mode, you can enter one line of text, typically commands, at the bottom of the window. In normal mode, the typed keys are interpreted as editor commands to be applied to the text in the buffer.

T. Neele and A. Wijs (Eds.): SPIN 2024, LNCS 14624, pp. 167–175, 2025.
https://doi.org/10.1007/978-3-031-66149-5_9

We report on our experience of using a library for automata-based active learning where the system under learning is Vim. Our main hypothesis is that modal text editors implement a finite-state machine (such as Fig. 1) that can be extracted automatically using active learning automata-based techniques.

Our motivation is twofold: we want to expose (1) active learning automata-based techniques to real-world systems like the Vim text editor and; (2) software developers to automata-like artifacts produced by active learning techniques.

Incidentally, we want to understand how active learning techniques can contribute to software development.

In our endeavor we faced several challenges. A first technical challenge asks which active learning library to interface with Vim and how. There exists several well-maintained publicly available libraries such as learnLib and LibAlf. (See the Automata Wiki [13] for an exhaustive list.) A second challenge is the scalability of the active learning techniques. What to do if the finite-state machine to be computed is too large or even infinite? We took for granted that the state machine underlying the Vim editor had finitely many states, but we quickly realized that in general the finiteness assumption is not true.

In this paper, we solve the interface between an active automata learning library (AALpy) and the Vim text editor (actually, we used the Vim fork called Neovim). We explain how to control the scalability of the active learning process, both via AALpy and the Neovim editor itself. We submitted bugs to the Vim and Neovim developers and report on our interactions, including whether the bugs were acknowledged or not and whether they were fixed.

Based on our experience, we will discuss the lessons learned and possible future work going beyond modal text editors.

2   Preliminaries

(Neo) Vim. We used the Vim[1] text editor for this case study. First and foremost, we used Vim because there are several reports of a state machine underlying its modal capabilities (see, for instance, Vim's Wikipedia page[2] or Darcy Parker's Vim Modes Transition Diagram [19]). Second, Vim is a rather large piece of software, mostly written in C. It implements a highly configurable text editor with a rich set of features. At least a dozen books[3] about Vim are available. Vim has a large user base (see for instance this survey,[4] where Vim came fifth as the most popular Integrated development environment). Vim also has a long history of over 40 years of development [20]. For the above reasons, we claim that it is a battle-tested software. It is also worth pointing that its documentation[5] is well-maintained.

---

[1] https://www.vim.org/.

[2] https://en.wikipedia.org/wiki/Vim_(text_editor)#Modes.

[3] https://iccf-holland.org/vim_books.html.

[4] https://survey.stackoverflow.co/2022/#section-most-popular-technologies-integrated-development-environment.

[5] Available via the :help command of Vim or via dedicated websites on the Internet.

We use a fork of Vim called Neovim[6] because of its remote Python API (pynvim[7]). Neovim shares a lot of its source code with Vim, in particular the core functionality of the editor. Consequently, it is often the case that bugs reported to Neovim are deferred to the Vim developers, who publish patches fixing the issue which are then ported to the Neovim code base by the Neovim developers. The two projects have developers in common.

AALpy. AALpy [17] is a light-weight active automata learning library written in Python. AALpy supports a wide range of modeling formalisms, including Moore machine, which we use as the formal model for the modes transition diagrams of the Vim editor. Indeed, Moore machines are finite-state machines like that of Fig. 1 where transitions between states are labeled by the so-called input symbols while states carry output symbols. In relationship to Vim, we have that the input symbols corresponds to the typed characters (that is, the keystrokes) while the output symbols corresponds to the modes reported by Vim (e.g. INSERT mode in Fig. 2).

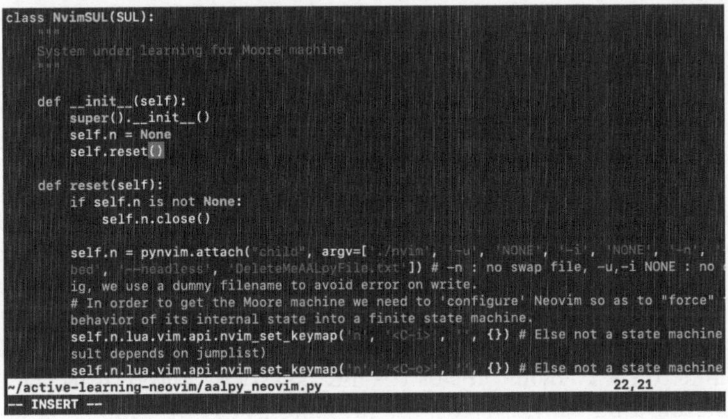

Fig. 2. Screenshot of Neovim running inside a terminal.

A (deterministic) *Moore machine* is a 6-tuple $(S, s_0, \Sigma, O, \delta, G)$, where: S is a finite set of *states* including an *initial state* s_0; Σ is a nonempty finite set called the *input alphabet*, O is a nonempty finite set called the *output alphabet*; $\delta \colon S \times \Sigma \to S$ is a *transition function* mapping a state and the input alphabet to the next state; and $G \colon S \to O$ is an *output function* mapping each state to the output alphabet.

Figure 1 gives an example of Moore machine with states $\{s_0, s_1, s_2, s_3\}$; input alphabet given by $\boxed{:}$, $\boxed{esc}$, $\boxed{ctrl} + \boxed{v}$, $\boxed{i}$; and output alphabet comprising command line, normal, visual and insert.

---

[6] https://neovim.io/.

[7] https://github.com/neovim/pynvim.

In AALpy, active learning of Moore machines is a fully automated process following a paradigm called *minimally adequate teacher* (MAT) initially put forward by Angluin [14]. In our setting, the MAT interacts in rounds with a learner whose task is to compute a Moore machine. The learner asks the teacher Neovim related queries that fall into two categories: (1) *membership queries* asking the teacher to return the output of a sequence of keystrokes applied to a newly spawned Neovim process; and (2) *equivalence queries* asking the teacher whether a Moore machine coincides with the state machine underlying Neovim. In AALpy, equivalence queries can be approximated via *conformance testing strategies*, which performs multiple membership-like queries comparing the output of the Moore machine with that of Neovim when applied the same sequence of keystrokes. For more details, we refer the interested reader to the survey of Vaandrager [24] who also provides an exhaustive list of references on the subject as well as the AALpy website[8] for implementation specific details. In recent years, active learning automata-based techniques have been used successfully on real-world system ranging from virtual private network servers to recurrent neural networks and Bluetooth Low Energy protocols [18, 21–23].

3   Active Learning of Neovim Moore Machine

Interfacing Neovim and AALpy. AALpy interacts with Neovim via its remote API pynvim[9]. One design goal of pynvim is to provide a library for connecting to and scripting Neovim processes, which is the feature we use in this paper. AALpy requires implementing a `step` function as well as a `pre` and a `post` function. The `pre` and `post` functions have to do with initialization/startup of a Neovim process and graceful shutdown of the Neovim process/memory cleanup. For the shutdown, we close the Neovim sub-process and for the initialization we spawn a new child process as shown at the top of Fig. 2. The `step` function submits to the Neovim process the next keystroke and returns the updated mode resulting from the keystroke by invoking `nvim_get_mode()`[10]. The set of keystrokes to be used by AALpy is configurable and, in our latest version of the interface [3], we set it to: { I , ctrl + g , 0 , ctrl + v , c , : , v , g , ctrl + o , r , esc , ↵ , ctrl + c , ctrl + \ + ctrl + n }.

Scaling Up. Generally the larger the set of keystrokes the more time AALpy needs before returning a Moore machine, and therefore we have to select the keystrokes carefully. Indeed, when AALpy adds a state to the Moore machine, it also adds transitions (one per keystroke) leaving that state. Hence, when AALpy returns a Moore machine with 100 states for a set of, say, 15 keystrokes, we know immediately the machine has 1,500 transitions. We settled on the previous set of keystrokes because it allows visiting many modes: : enters the command line

---

[8] https://github.com/DES-Lab/AALpy.
[9] https://github.com/neovim/pynvim.
[10] https://neovim.io/doc/user/api.html#nvim_get_mode().

mode, $\boxed{\mathsf{I}}$ followed by $\boxed{\mathsf{c}}$ the insert mode while also covering operator pending and motions, $\boxed{\mathsf{ctrl}}+\boxed{\mathsf{v}}$ the visual mode,...

Apart from the keystrokes, an independent way to control scalability of the learning process is to customize the Neovim process. Roughly speaking, different settings of Neovim yield different Moore machines. In general, the machine is not even finite, since the result of some keystrokes depends on the previously inserted text inside Neovim. For instance, an editor command in normal mode, like $\boxed{\mathsf{c}}\;\boxed{\mathsf{f}}\;\boxed{\mathsf{z}}$,, deletes the text between the cursor position and the next occurrence of the symbol z on the same line of text, after which it completes by entering the insert mode. (c is for change, f is for forward and z is the symbol looked for). However, if z does not occur between the cursor and the end of the line, then no text is deleted and the editor command completes by staying in normal mode.[11] Because the size of the text being edited is virtually unbounded, the learner is therefore facing the task to compute a Moore machine with infinitely many states. But there is more, assuming we can make the machine finite it can still be very large: the result of the keystrokes $\boxed{\mathsf{ctrl}}+\boxed{\mathsf{i}}$ and $\boxed{\mathsf{ctrl}}+\boxed{\mathsf{o}}$ depends on the so-called jumplist which counts up to 100 entries.

To remedy this problem, we configure Neovim to prevent the behaviors described above to happen. More precisely, we use options controlling Neovim's behavior as well as key mapping which we set after spawning the Neovim subprocess. As shown in Fig. 2, we map the keystrokes $\boxed{\mathsf{ctrl}}+\boxed{\mathsf{i}}$ and $\boxed{\mathsf{ctrl}}+\boxed{\mathsf{o}}$ to do nothing when we are in normal mode to avoid the pitfall of the jumplist mentioned above.

Finally, let us mention that we commented out in the Neovim source code all function calls causing time delays as suggested by developers [6]. These delays are present in the source code to give the user the time needed to read some messages. However, those delays slow down the active learning process.

Current State of the Learning. Using the above set of 14 keystrokes and a customized Neovim as described above AALpy learns a Moore machine of 99 states and 1,386(=14 99) transitions after 58 rounds between the learner and the teacher in less than 10 min by performing 31,479 memberships queries 29,700 of which (or 94%) are carried out in the context of the conformance testing to approximate equivalence queries.

4   Outcomes and Impact

After successfully learning a Moore machine, we set AALpy to produce a file in the dot language of the Graphviz project [16]. The dot language allows for visual inspection of the Moore machine as well as running queries on it using some external tool supporting the dot language format. Our typical workflow has been to first manually inspect the computed Moore machine using a dot file visualization tool (xdot.py[12]) and then formulate queries (e.g. How to reach that state? What

---

[11] See [12] for a more detailed explanation.
[12] https://github.com/jrfonseca/xdot.py.

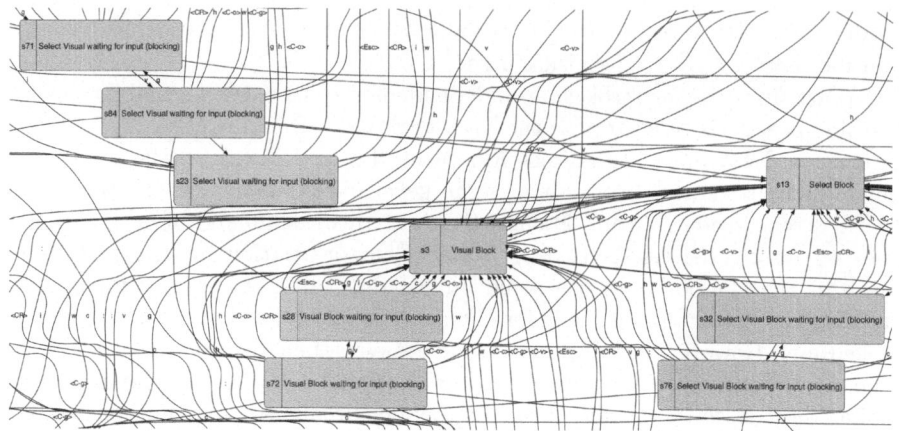

Fig. 3. Detail of a Moore machine learned using our approach.

differentiates these two states?...) to be answered either directly within AALpy (e.g. compare_automata, find_distinguishing_seq, ...) or using an external graph visualization and exploration tool like Gephi[13] [15]. Figure 3 depicts part of such learned Moore machine where states are labelled with modes and transitions with keystrokes.

After inspection and querying we either modify the set of keystrokes and/or Neovim's customization and repeat the learning process; or we find a behavior in the Moore machine requiring further investigation. Such behavior typically consists of one or two short sequences of keystrokes whose resulting modes cannot be explained clearly following the Neovim's documentation. If the unexplained behavior is reproducible in Vim by manually typing the sequences of keystrokes, then we submit an issue to the Vim developers. For instance, according to the documentation, `ctrl`+`v` and `ctrl`+`q` behave the same in insert and replace mode. When learning a Moore machine with a set of keystrokes including `ctrl`+`v` and `ctrl`+`q`, it turned out that `ctrl`+`v` labeled transitions and `ctrl`+`q` labeled transitions leaving the same state ended up in different states. After reproducing and reporting the issue in Vim, the developers fixed it promptly [2].

In general, the reactions to the issue largely vary: convincingly arguing the behavior is not an issue [7,12]; acknowledging a problem in the documentation [8]; acknowledging a problem in Vim's behavior but not fixing it ([9] at first); acknowledging a problem in Vim's behavior and fixing it [1,2,4,9–11]. On one occasion, we fixed the issue ourselves after it was acknowledged [9]. Fixes are typically one-liners and no more than 10 lines of code [11]. A comprehensive list of issues acknowledged and fixed is given on the AALpy discussion thread page New use of AALpy library[14] on their GitHub repository.

---

[13] https://github.com/gephi/gephi.
[14] https://github.com/DES-Lab/AALpy/discussions/13.

5   Discussion

Active Learning and the Software Development Cycle. Both the Vim and Neovim developers have been responsive and helpful when issues were submitted. As we stated in the introduction, the source code of Vim is battle-tested and no issue we reported involved making the text editor unresponsive, mishandle data or let alone crash. Apart from acknowledged issues, the reaction to our reporting ranged from confused[15] to nobody cares[16] and won't fix[17]. We captured the interest of some Neovim developers after showing them the Moore machines. Together with them, we envisioned two potential uses: as a source of **documentation** aimed at end users or as a **formal specification** of behaviors to be tested between (major) releases. More precisely, a conformance testing tool could become part of the software life cycle, where the Moore machine of a release is used as a specification to test subsequent releases. Recall that a conformance testing tool runs a large number of tests arising from a given Moore machine by following a strategy to order and select the tests. The above can be implemented in a few lines of Python using AALpy.

Besides testing, we also claim that a Moore machine model provides valuable information in case of **refactoring**. Case in point is the refactoring Neovim carried out starting from the Vim code base and whose logic[18] [5] leverages an input-driven state machine. We claim that having computed some information about such input-driven state machines helps with the code refactoring.

Enhancing the Model. As we mentioned above the larger the set of keystrokes, the more time AALpy needs before returning a Moore machine. This is because AALpy has to try more sequences of keystrokes the number of which grows exponentially with the length of the sequence: with 3 keystrokes there are 3^n sequences of length n while with 10 keystrokes there are 10^n sequences.

In (Neo)vim, there are some keystrokes that are supposed to have the same effects on the mode. For instance, $\boxed{\text{i}}$ enters insert mode from normal mode and so does $\boxed{\text{I}}$, $\boxed{\text{o}}$ and $\boxed{\text{O}}$ (they differ in other aspects but not from the modal point of view). So if the set of keystrokes already contains $\boxed{\text{i}}$ adding $\boxed{\text{o}}$ should have a predictable effect on the returned Moore machine: for each edge labeled with $\boxed{\text{i}}$ from node s such that s is normal mode to some node t, there is another edge from s to t with label $\boxed{\text{o}}$. Therefore, leveraging (Neo)vim's documentation, one can enrich the Moore machine after the active learning process by adding edges as suggested above. The enriched Moore machine can then be used by a conformance testing tool to test whether the enriched machine still conforms with the implementation. Enriching the Moore machine is as simple as editing its dot file, since the dot format is human-readable.

Disclosure of Interests. The authors have no competing interests to declare that are relevant to the content of this article.

---

[15] https://github.com/vim/vim/issues/13649#issuecomment-1847848471.

[16] https://github.com/vim/vim/pull/13001#issuecomment-1703925253.

[17] https://github.com/vim/vim/issues/12115#issuecomment-1483815493.

[18] https://github.com/neovim/neovim/tree/master/src/nvim#nvim-lifecycle.

Acknowledgments. We thank Xiao Peng Ye, Madeleine Mathieu and for their tremendous help with the interface and uncovering bugs. We also thank the Vim and Neovim developers for their time including Bram Moolenaar, the creator of Vim, who died during the summer 2023. Furthermore, we thank Edi Muskardin for his tremendous help with AALpy. Finally, we thank the SPIN reviewers for their effort and suggestions. This publication is part of the grant PID2022-138072OB-I00, funded by MCIN, FEDER, UE. This work has been partially supported by the PRODIGY Project (TED2021-132464B-I00) funded by MCIN and the European Union NextGeneration.

References

1. Consecutive call of nvim_get_mode() does not return the same mode Issue #15288 neovim/neovim — github.com. https://github.com/neovim/neovim/issues/15288. Accessed 05-01-2024
2. Different behavior between C-v and C-q Issue #12684 vim/vim — github.com. https://github.com/vim/vim/issues/12684. Accessed 05 Jan 2024
3. GitHub - pierreganty/active-learning-neovim: Active automata-based learning of the Moore machine underlying Neovim — github.com. https://github.com/pierreganty/active-learning-neovim. Accessed 10 Jan 2024
4. Non deterministic behavior when querying current mode Issue #8323 vim/vim — github.com. https://github.com/vim/vim/issues/8323. Accessed 05 Jan 2024
5. Nvim lifecycle neovim/neovim — github.com. https://github.com/neovim/neovim/tree/master/src/nvim#nvim-lifecycle. Accessed 05 Jan 2024
6. nvim_get_mode waits 3~4 secs with 'showmode' enabled or when there are error messages Issue #19352 neovim/neovim — github.com. https://github.com/neovim/neovim/issues/19352. Accessed 05 Jan 2024
7. Possible inconsistent behavior Issue #8346 vim/vim — github.com. https://github.com/vim/vim/issues/8346. Accessed 05 Jan 2024
8. Possibly inconsistent behavior Issue #13649 vim/vim — github.com. https://github.com/vim/vim/issues/13649. Accessed 05 Jan 2024
9. Possibly incorrect transitions between modes Issue #12115 vim/vim — github.com. https://github.com/vim/vim/issues/12115. Accessed 05 Jan 2024
10. Possibly undocumented behavior with replace after visual Issue #13091 vim/vim — github.com. https://github.com/vim/vim/issues/13091. Accessed 05 Jan 2024
11. Undocumented (possibly incorrect) behavior for Virtual Replace mode Issue #12045 vim/vim — github.com. https://github.com/vim/vim/issues/12045. Accessed 05 Jan 2024
12. Unexpected behavior? Issue #10693 vim/vim — github.com. https://github.com/vim/vim/issues/10693. Accessed 05 Jan 2024
13. Automata Wiki — automata.cs.ru.nl. https://automata.cs.ru.nl/ (2017). Accessed 05 Jan 2024
14. Angluin, D.: Learning regular sets from queries and counterexamples. Inf. Comput. **75**(2), 87–106 (1987). https://doi.org/10.1016/0890-5401(87)90052-6
15. Bastian, M., Heymann, S., Jacomy, M.: Gephi: an open source software for exploring and manipulating networks. In: Proceedings of the International AAAI Conference on Web and Social Media 3(1), pp. 361–362 (2009). https://doi.org/10.1609/icwsm.v3i1.13937
16. Gansner, E.R., North, S.C.: An open graph visualization system and its applications to software engineering. Softw. Pract. Exper. **30**(11), 1203–1233 (2000)

17. Muškardin, E., Aichernig, B.K., Pill, I., Pferscher, A., Tappler, M.: AALpy: an active automata learning library. Innovations Syst. Softw. Eng. **18**(3), 417–426 (2022). https://doi.org/10.1007/s11334-022-00449-3
18. Muškardin, E., Aichernig, B.K., Pill, I., Tappler, M.: Learning finite state models from recurrent neural networks. In: Integrated Formal Methods, pp. 229–248. LNCS, Springer, Cham (2022). https://doi.org/10.1007/978-3-031-07727-2_13
19. Parker, D.: Vim Modes Transition Diagram in SVG. https://gist.github.com/darcyparker/1886716 (2012). Accessed 05 Jan 2024
20. Pezzi, G.: Understanding the Origins and the Evolution of Vi & Vim. https://pikuma.com/blog/origins-of-vim-text-editor (2023). Accessed 05 Jan 2024
21. Pferscher, A., Aichernig, B.K.: Fingerprinting bluetooth low energy devices via active automata learning. In: FM 2021: Formal Methods, pp. 524–542. LNCS. Springer, Cham (2021). https://doi.org/10.1007/978-3-030-90870-6_28
22. Pferscher, A., Wunderling, B., Aichernig, B.K., Muškardın, E.: Mining digital twins of a VPN server. In: FMDT 2023: Proceedings of the Workshop on Applications of Formal Methods and Digital Twins. CEUR Workshop Proceedings, vol. 3507 (2023)
23. Tappler, M., Aichernig, B.K., Bloem, R.: Model-based testing iot communication via active automata learning. In: ICST 2017: IEEE International Conference on Software Testing, Verification and Validation. IEEE (2017). https://doi.org/10.1109/icst.2017.32
24. Vaandrager, F.: Model Learning. Commun. ACM **60**(2), 86–95 (2017). https://doi.org/10.1145/2967606

Tolerange: Quantifying Fault Masking in Stochastic Systems

Luciano Putruele[1,3]([✉]) [ID], Ramiro Demasi[2,3] [ID], Pablo F. Castro[1,3] [ID],
and Pedro R. D'Argenio[2,3,4] [ID]

[1] Departamento de Computación, FCEFQyN, Universidad Nacional de Río Cuarto,
Río Cuarto, Córdoba, Argentina
{lputruele,pcastro}@dc.exa.unrc.edu.ar
[2] Universidad Nacional de Córdoba, FAMAF, Córdoba, Argentina
{rdemasi,pedro.dargenio}@unc.edu.ar
[3] Consejo Nacional de Investigaciones Científicas y Técnicas (CONICET),
Buenos Aires, Argentina
[4] Saarland University, Saarland Informatics Campus, Saarbrücken, Germany

Abstract. We present Tolerange, an open source tool tailored for measuring the masking fault-tolerance provided by stochastic systems. Tolerange takes as input a nominal model of a system together with the fault-tolerant version of it, both written in a Prism-like notation, and it computes the expected number of faults that the system's fault-tolerant version is able to mask. Our tool supports the analysis of randomized algorithms including the description of faults as probabilistic actions. It combines techniques coming from game theory, linear programming, and probabilistic transition systems. In this paper we describe the tool as well as its use to measure the masking fault-tolerance of some well-known examples.

1 Introduction

Measuring the fault-tolerance provided by systems is crucial for accurately assessing their dependability. An important kind of fault-tolerant systems are those that can mask faults in such a way that users cannot notice their occurrence. This type of fault-tolerance is often referred as *masking fault-tolerance* [7], and it is typically achieved by using some form of redundancy, e.g., replicating memory units, disks, or processes. However, in practice, systems can only mask faults during a finite period of time before exhibiting a failure. Hence, in most cases, designers are interested in quantifying the number of faults that systems are able to mask before failing. This can be thought of as the level of masking tolerance exhibited by systems. Even though critical systems are ubiquitous in modern life, few automated tools are available to perform such measures, which in practice are usually done using ad-hoc methods.

In this paper, we present Tolerange, a tool aimed at measuring the (expected) amount of masking fault-tolerance provided by stochastic systems. This encompasses both, the probabilities of faults, and the possible use of randomized algo-

T. Neele and A. Wijs (Eds.): SPIN 2024, LNCS 14624, pp. 176–183, 2025.
https://doi.org/10.1007/978-3-031-66149-5_10

rithms. To the best of our knowledge, there are presently no other tools available for assessing the masking fault-tolerance of probabilistic systems.

Our tool is based on the notion of *probabilistic bisimulation relations* [11] between *probabilistic transition systems* [13]. The latter are a generalization of Markov chains supporting non-deterministic actions. Tolerange takes as input a specification (a description of the system without faults) and a fault-tolerant implementation of it (a system's version incorporating both faults and fault-tolerance mechanisms) and computes the expected number of *milestones*

```
Process NOMINAL {
  v : INT;
  s : INT;          // 0 = normal,
                    // 1 = refresh

  Initial: v==0 && s==0;

  [write0] !(s==1) -> v=0,s=0;
  [write1] !(s==1) -> v=1,s=0;

  [read0] !(s==1) && v==0 -> v=v;
  [read1] !(s==1) && v==1 -> v=v;

  [tick] <1> s==0 -> 0.05 : s=1
                  ++ 0.95 : v=v;

  [refresh] s==1 && v==0 -> s=0, v=0;
  [refresh] s==1 && v==1 -> s=0, v=1;
}
```

Fig. 1. A Nominal Model for the Memory Example

(events highlighted as important by the users) that the implementation guarantees to preserve under the presence of faults. A simple example of milestone could be the number of packages successfully transmitted by a sender in a communication protocol. The current tool extends the tool MaskD [12], which exclusively focuses on non-stochastic systems and does not support probabilistic models.

Tolerange is meant to be used for measuring the masking fault-tolerance provided by stochastic models of critical software. For instance, it might help the designers to select between different fault-tolerant implementations. We illustrate this via a simple example. Consider the case of a RAM memory that uses redundancy to cope with faults, e.g., bits changing its value because of external noise. We add to the memory an additional fault-tolerant mechanism: a refreshing tick which is performed with certain frequency. In this setting, an important question is what is more convenient: adding redundancy at the hardware level (which can be more expensive) or refreshing with a higher frequency. Our tool helps to find an optimal balance between these two mechanisms. In Sect. 4, we analyse this example with Tolerange, as well as other well-known examples of fault-tolerance.

2   Running Example

In this section we introduce in detail the example mentioned above. Consider a memory cell storing one bit of information that periodically refreshes its value. Figure 1 shows the processes modeling the nominal and a fault-tolerant implementation of this example. Actions read$i$ and write$i$ (for $i = 0, 1$) represent the actions of reading and writing value i, respectively. The bit stored in the memory is saved in variable v. Action tick marks that one time unit has passed and, with probability 0.05, it enables the refresh action (refresh). Variable s indicates whether the system is in write/read mode or producing a refresh.

A potential fault in this scenario occurs when a cell unexpectedly changes its value. In practice, the occurrence of such an error has a certain probability. A typical technique to deal with this situation is using three memory bits instead of one. In such a case, writing operations are performed simultaneously on the three bits, while reading returns the value obtained by majority. Figure 2 shows an augmented version of the nominal model with triple redundancy and the faults mentioned above. Therein, variable v counts the votes for value 1. A tick enables a refreshing as the original model, but also enables the occurrence of a fault with probability 0.1. Now, variable s may get the value 2, representing a state in which a fault may occur.

3   The Tool

Tolerange takes as input a nominal model and a fault-tolerant version of it and produces as output the expected number of milestones that the implementation is able to guarantee under a fault model, which is a value in $\mathbb{R}^+$. To ensure that this value is well-defined we assume that the probability of a system's failure is 1. We also assume that the environment plays in a strongly fair manner (if a fault is infinitely often enabled, then it will occur infinitely often). We call these kinds of systems *almost-surely failing under fairness* [3]. The tool can automatically check whether the input is of this kind. It uses standard algorithms of game theory, together

```
Process FAULTY {
  v : INT;
  s : INT;         // 0 = normal, 1 = refresh
                   // 2 = faulty
  Initial: v==0 && s==0;

  [write0] !(s==1) -> v=0,s=0;
  [write1] !(s==1) -> v=3,s=0;

  [read0] !(s==1) && v<=1 -> v=v;
  [read1] !(s==1) && v>1 -> v=v;

  [tick] <1> s==0 -> 0.05 : s=1
                  ++ 0.1  : s=2
                  ++ 0.85 : v=v;
  [tick] <1> s==2 -> 0.05 : s=1
                  ++ 0.95 : v=v;

  [refresh] s==1 && v<=1 -> s=0, v=0;
  [refresh] s==1 && v>1 -> s=0, v=3;

  [f] faulty s==2 && v<3 -> s=0, v=v+1;
  [f] faulty s==2 && v>=3 -> s=0, v=2;
  [f] faulty s==2 && v>0 -> s=0, v=v-1;
  [f] faulty s==2 && v<=0 -> s=0, v=1;
}
```

Fig. 2. Augmented Model for the Memory Example

with linear programs for reasoning about probabilistic choices, the interested reader is referred to [4] for an in-depth description.

The input models are written in a **Prism**-like language [10]. More precisely, a program is a collection of processes, where each process is composed of a collection of actions of the style: [Label]<reward>Guard->[P]Command++ [Q]Command, where: Guard is a Boolean condition over the actual state of the program; Command is a collection of basic assignments; Label is a name for the action; the positive integer reward is optional, it states that the execution of this action counts as a "milestone" of value reward; and ++ is the probabilistic choice. Here P and Q are the probabilities corresponding to each branch of the choice. There can be many branches, as long as the sum of the probabilities is 1. The language also allows users to label actions as faulty to indicate that they model possible faults.

In order to compute the expected number of milestones guaranteed by the implementation, the tool defines a *two-player stochastic game* [6] using the inputs. The basis of the game is similar to a probabilistic bisimulation game [14], and it is played by two players, named for convenience the Refuter (R) and the Verifier (V). In this game, the Verifier intends to prove that the fault-tolerant version of the system is able to mask faults, while the Refuter intends to disprove that.

Roughly speaking, in each round of the game the Refuter may select a probabilistic transition in any of the models, and then the Verifier has to select a probabilistic transition in the opposite model to match the Refuter's play. The match between the two probabilistic actions consists of a *probabilistic coupling* [9], which shows how the probabilities in one action are redistributed to another action.

Furthermore, the Verifier may play a fault of the augmented model, in such a case the V is obliged to mask the fault and, from the user's point of view, it stays in the same state of the nominal model. This is modeled via a Dirac distribution. For instance, given a state s, the Dirac distribution (denoted Δ_s) states that, with probability 1, the next state is s. In addition, some game states may have associated a value to them, given by a function r, this indicates the reward given to the Verifier for some Refuter's action that she was able to match, for instance, when a fault is masked. The objective of the Verifier is to maximize $\sum_{i=0}^{\infty} r(\rho_i)$, where ρ_i is the ith state of a play, while the objective of the Refuter is to minimize this function. The value of these games are obtained by means of computing the optimal strategies of both players using value iteration and Bellman equations [5], here linear programming is used for coping with the possible (uncountable) set of couplings between distributions. More precisely, a configuration of the game is a tuple: (s, a, s', μ, μ', P) where: s and s' are the current states in the nominal and augmented model,

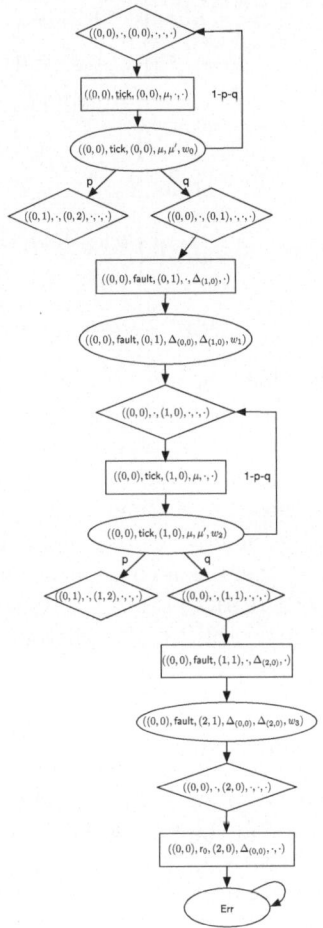

Fig. 3. A fragment of the Memory example game graph.

respectively. a is the last action played by the Refuter. μ and μ' are the probabilistic distribution played by some player, corresponding to the last action played, or to the matching step, a precise definition of the game graph is given in [4].

Consider the graph in Fig. 3. Therein, Verifier's nodes are depicted with boxes, Refuter's nodes with diamonds, and probabilistic nodes with circles. It represents a fragment of the masking game graph between NOMINAL and FAULTY of the running example. The vertices represent the variable values in the following order: $((v, s), \_, (v, s), \_, \_, \_, \_)$. The distributions there are as follows:

$$\mu = p \cdot (0, 1) + (1-p) \cdot (0, 0)$$
$$\mu' = p \cdot (0, 2) + q \cdot (0, 1) + (1-p-q) \cdot (0, 0)$$
$$\mu'' = p \cdot (1, 2) + q \cdot (1, 1) + (1-p-q) \cdot (1, 0)$$
$$w_0 = \begin{cases} p \cdot ((0, 1), (0, 2)) + q \cdot ((0, 0), (0, 1)) + \\ (1-p-q) \cdot ((0, 0), (0, 0)) \end{cases}$$
$$w_1 = \Delta_{((0,0),(1,0))}$$
$$w_2 = \begin{cases} p \cdot ((0, 1), (1, 2)) + q \cdot ((0, 0), (1, 1)) + \\ (1-p-q) \cdot ((0, 0), (1, 0)) \end{cases}$$
$$w_3 = \Delta_{((0,0),(2,0))}$$

Notice that, in the majority of the vertices, many outgoing edges are omitted. In particular, the Verifier's vertex $((0, 0), \texttt{tick}, (0, 0), \mu, \cdot, \cdot, V)$ has infinitely many outgoing edges leading to probabilistic vertices of the form $((0, 0), \texttt{tick}, (0, 0), \mu, \mu', w, P)$, where w is a coupling for (μ, μ'). In the graph, we have chosen to distinguish coupling w_0 which is optimal for the Verifier (similarly later for w_2). We highlighted the path leading to error state v_{err}. Notice that this occurs as a consequence of the Refuter choosing to do a second fault in vertex $((0, 0), \cdot, (1, 1), \cdot, \cdot, \cdot, R)$ steering the game to the bottom part of the graph. Later, the Refuter chooses to read 0 in the Nominal model (at vertex $((0, 0), \cdot, (2, 0), \cdot, \cdot, \cdot, R))$ which the Verifier cannot match.

3.1 Architecture.

Tolerange is an open-source software written in Java and available at [2]. The architecture of the tool is shown in Fig. 4. The key components are:

Parser Module. It performs basic syntactic analysis over the input models, and produces data structures describing the inputs.

PTS Translation. The models obtained from the parser are translated into Probabilistic Transition Systems (PTSs), i.e., graphs whose vertices represent program states and probabilities associated with the transitions that keep information about the actions in the models.

Stochastic Masking Game Generation. A masking stochastic game is generated for the given PTSs where the weighting functions are symbolically captured by means of equation systems (linear programming).

Almost-Sure Failing Under Fairness Check. We provide an algorithm to check whether a game is almost-sure failing under fairness by computing predecessor sets in the symbolic game graph.

Value of the Game Calculation (Value Iteration). The value of the game is computed by solving a collection of functional equations via value iteration. We take such a value as the measure of fault-tolerance.

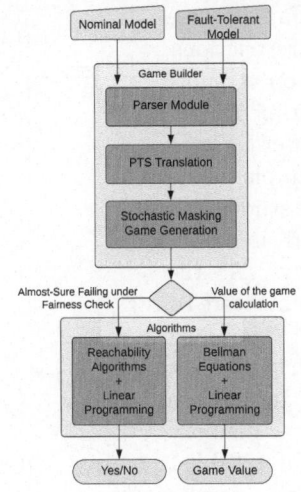

Fig. 4. Architecture of Tolerange.

3.2 Usage

The standard command to execute Tolerange in a Unix operating system is: ./Tolerange <options> <spec_path> <imp_path>. In this case the tool returns the expected accumulated milestones achieved by the implementation. Optional commands are: -f, for checking if the game is almost-sure failing under fairness; -gurobi, which enables Gurobi [8] instead of SCC [1] for linear programming; and b=N to set an upper bound of N for the value of the game. By default, Tolerange computes the value of the game for the given input using SCC and an upper bound of the largest Java double.

4 Experiments

We have performed experiments on several examples. Due to space restrictions, here we only report the experiments performed over the memory example. Further case studies can be found in tool's repository. All the experiments were run on a MacBook Air with processor 1.3 GHz Intel Core i5 and a memory of 4 Gb. Table 1 reports the results obtained for our running example. The top subtable refers to counting the tick action as milestone, the middle subtable takes the refresh as milestone, and the bottom subtable reports the size of the models in number of states and transitions. M_t and M_r are the measurement results for the tick and refresh actions, considered as milestones, respectively.

Table 1. Results on the redundant memory cell with time and size information.

Fault Prob.	Refr. Prob.	3 bits redundancy		5 bits redundancy		7 bits redundancy	
		M_t	Time	M_t	Time	M_t	Time
0.5	0.5	6	30s	14	1m43s	30	4m48s
	0.1	4.44	13s	7.28	32s	10.74	1m4s
	0.05	4.2	11s	6.62	26s	9.28	43s
0.1	0.5	70	4m21s	430	29m24s	2590.1	162m27s
	0.1	30	1m22s	70	4m23s	150	13m58s
	0.05	25	1m3s	47.5	2m41s	81.25	6m43s
0.05	0.5	240.02	12m10s	2660.1	128m37s	29281	68m49s
	0.1	80	30m31s	260.01	13m29s	800.03	89m20s
	0.05	60	2m39s	140	9m29s	300	23m26s

Fault Prob.	Refr. Prob.	3 bits redundancy		5 bits redundancy		7 bits redundancy	
		M_r	Time	M_r	Time	M_r	Time
0.5	0.5	3	35s	7	2m13s	15	4m54s
	0.1	0.44	14s	0.72	35s	1.07	1m8s
	0.05	0.21	12s	0.33	27s	0.46	58s
0.1	0.5	35	4m19s	215.02	37m55s	1295.05	199m34s
	0.1	3	1m33s	7	5m58s	15	14m24s
	0.05	1.25	1m19s	2.38	3m15s	4.06	8m6s
0.05	0.5	120.01	15m11s	1330.06	163m3s	14640.5	713m25s
	0.1	8	3m40s	26.01	16m54s	80.03	63m2s
	0.05	3	2m58s	7	8m58s	15	25m

size (states/transitions)		3 bits	5 bits	7 bits
	spec.	4/12	4/12	4/12
	impl.	12/56	18/84	24/112
	game	505/2304	757/3688	1009/5012

From the results in the table we can conclude that either, increasing the redundancy or augmenting the frequency of refreshing, has positive effects on the measures. In practice, these values can be taken into account when designing a fault-tolerant component that provides an optimal balance between efficiency and hardware costs. For example, assuming a fault probability of 0.05, one might prefer 3 bits and more frequent refreshing, over 5 bits with a less often refreshing, despite the software overhead.

As a final remark, we observe that small probabilities have a great impact in running times (in many cases even more than the model sizes), this is mainly due to the linear programming procedures that are used to solve the stochastic games. We plan to improve the tool by explicitly using the solutions of the linear programs (vertices of a polytope) instead of embedding the linear programs into Bellman equations.

References

1. SCC tool. https://www.ssclab.org/
2. Tolerange tool. https://github.com/cl-unrc-lab/Tolerange
3. Castro, P.F., D'Argenio, P.R., Demasi, R., Putruele, L.: Playing against fair adversaries in stochastic games with total rewards. In: CAV 2022, Haifa, Israel (2022)
4. Castro, P.F., D'Argenio, P.R., Demasi, R., Putruele, L.: Quantifying masking fault-tolerance via fair stochastic games. In: Proceedings of the Combined 30th International Workshop on Expressiveness in Concurrency and 20th Workshop on Structural Operational Semantics (2022)
5. Chatterjee, K., Henzinger, T.A.: Value iteration. In: Grumberg, O., Veith, H. (eds.) 25 Years of Model Checking. LNCS, vol. 5000, pp. 107–138. Springer, Heidelberg (2008). https://doi.org/10.1007/978-3-540-69850-0_7
6. Filar, J., Vrieze, K.: Competitive Markov Decision Processes, 1st edn. Springer, New York (1997)
7. Gärtner, F.C.: Fundamentals of fault-tolerant distributed computing in asynchronous environments. ACM Comput. Surv. **31** (1999)
8. Gurobi Optimization, LLC: Gurobi Optimizer Reference Manual (2022). https://www.gurobi.com
9. Jonsson, B., Larsen, K.G.: Specification and refinement of probabilistic processes. In: Proceedings of the Sixth Annual Symposium on Logic in Computer Science (LICS '91), Amsterdam, The Netherlands, July 15-18, 1991, pp. 266–277. IEEE Computer Society (1991)
10. Kwiatkowska, M., Norman, G., Parker, D.: PRISM 4.0: verification of probabilistic real-time systems. In: Gopalakrishnan, G., Qadeer, S. (eds.) CAV 2011. LNCS, vol. 6806, pp. 585–591. Springer, Heidelberg (2011). https://doi.org/10.1007/978-3-642-22110-1_47
11. Larsen, K., Skou, A.: Bisimulation through probabilistic testing. Inf. Comput. **94**(1), 1–28 (1991)
12. Putruele, L., Demasi, R., Castro, P.F., D'Argenio, P.R.: MaskD: A tool for measuring masking fault-tolerance. In: Fisman, D., Rosu, G. (eds.) TACAS 2022. LNCS, vol. 13243, pp. 396–403. Springer, Cham (2022)
13. Segala, R.: Modelling and Verification of Randomized Distributed Real Time Systems. Ph.D. thesis, MIT (1995)
14. Stirling, C.: The joys of bisimulation. In: Brim, L., Gruska, J., Zlatuska, J. (eds.) MFCS'98, vol. 1450, pp. 142–151. Springer, Cham (1998). https://doi.org/10.1007/BFb0055763

Software Verification Witnesses 2.0

Paulína Ayaziová [1], Dirk Beyer [2]✉, Marian Lingsch-Rosenfeld [2],
Martin Spiessl [2], and Jan Strejček [1]

[1] Masaryk University, Brno, Czech Republic
{xayaziov,strejcek}@fi.muni.cz
[2] LMU Munich, Munich, Germany
{dirk.beyer,marian.lingsch,spiessl}@sosy.ifi.lmu.de

Abstract. Verification witnesses are now widely accepted objects used
not only to confirm or refute verification results, but also for general
exchange of information among various tools for program verification. The
original format for witnesses is based on GraphML, and it has some known
issues including a semantics based on control-flow automata, limited tool
support of some format features, and a large size of witness files. This
paper presents version 2.0 of the witness format, which is based on YAML
and overcomes the above-mentioned issues. We describe the new format,
provide an experimental comparison of various aspects of the original
and the new witness format showing that both witness formats perform
similarly, and report on its adoption in the community.

Keywords: Verification Witness · Software Verification · Validation ·
Exchange Format · Invariant · Counterexample

1 Introduction

Software verification is a process that detects bugs in computer programs or
proves their absence. Unfortunately, software verifiers can also contain bugs and
their verdicts can thus be incorrect. To increase the reliability of the verification process, starting eight years ago, software verifiers have accompanied their
verification results with witnesses that justify the verdict and can be indepen-
dently analyzed by witness validators developed by various teams and based on
different techniques.

The first generic format for witnesses of verification results [1] was introduced
in 2015. It supported only *violation witnesses* (also called *counterexamples*)
produced when a verifier reports that a given program violates a considered
safety specification. In 2016, the format was extended to accommodate also
witnesses for the cases when a verifier decides that a given program satisfies a
given specification [2]. Such witnesses are called *correctness witnesses*, and they
should contain invariants that help to prove that the program is correct. The
format was soon adopted by the verification community and by the *Competition
on Software Verification (SV-COMP)* [3], which led to fast adoption of the format
by many verification tools and to the development of numerous witness validators.

© The Author(s) 2025
T. Neele and A. Wijs (Eds.): SPIN 2024, LNCS 14624, pp. 184–203, 2025.
https://doi.org/10.1007/978-3-031-66149-5_11

The overview of existing validators can be found in a recent survey [4]. Since 2023, SV-COMP has a new track on witness validation [5].

While the format was originally intended for validation of verification results, some witness validators can also refute a witness [4, 5]. The format soon found also some applications that were not intended at the time of its development. In particular, it is used to exchange information between different verifiers in the context of cooperative verification [6, 7], as a way to provide feedback to a software developer [8, 9], or as a way to combine automatic and interactive verifiers [10]. In 2022, the authors of the format published a paper [11] with its detailed description and with an extensive experimental study on its applications.

Despite the indisputable success of the format, it has also some weaknesses. The format is based on GraphML [12] and witnesses have the form of automata, which makes them easy to visualize, but also lengthy and unsuitable for reading in their textual form. More importantly, the semantics of the format is formally defined over programs represented by *control-flow automata (CFA)*. Unfortunately, there is no standardized translation of programs written in common programming languages like C or Java to CFA. As a result, the semantics of the format over programs in standard languages has some ambiguities. The SV-COMP community even found a part of the semantics related to implicit loop edges as inappropriate and decided to change it. Another issue of the original witness format is connected to the high number of features it provides. For example, if an invariant or an assumption uses variables that appear in different functions or scopes, the format allows to specify the scope for their interpretation. Another example is that the location of some witness event can be specified very loosely by an interval of lines. Practical experience shows that some of these features are not used in any witness generated by verifiers and, what is more alarming, unsupported or even ignored by witness validators. In fact, there is probably no witness validator fully implementing the format. This can lead to the situation in which a valid verification witness employing some less frequent feature is not confirmed or even refuted, or an invalid witness is confirmed.

This paper presents a new generation of the witness format that avoids the mentioned weaknesses. In particular, we use a concise format that is based on YAML, which makes the witnesses shorter in general. Further, the format provides only features that were really used by verification witnesses in the original format. As the format is significantly simpler, it is easy to fully support it by validators. Finally, the semantics of the format is formulated over programming languages using terms and concepts from their standards.

The format itself is described in Sect. 2 and referred to as version 2.0. In its current state, the format supports only sequential programs written in C and basic program properties, namely unreachability of a given error function, unreachability of signed integer overflow, and unreachability of invalid pointer dereference and deallocation. We have adopted two verification tools, namely CPACHECKER [13] and SYMBIOTIC [14], to produce verification witnesses in the new format. We have also developed two witness validators, namely CPACHECKER [11] (as an extension of the exiting tool) and WITCH3 (new validator based on the concept of WITCH3 [15]) to validate witnesses in this format. Using these tools and other tools from the competition, Sect. 3 evaluates the impact of the new format on the witnesses and their validation. Sect. 4 summarizes the differences between the original and the new format and shows the current adoption of the new format by verification and witness validation tools.

Contributions. In this paper, we contribute:

- a new generation of the format for verification witnesses that solves most problems that were present in the previous format,
- a preliminary evaluation of the impact of the new format on the effectivity and performance of witness validation, and
- an overview of a few measures that characterize the new witnesses.

Related Work. Our work certainly stands on the shoulders of the original format for verification witnesses [1, 2], but we claim to provide a substantial improvement over the original format by addressing its weaknesses (see Sect. 4). Witnesses are used ubiquitously in areas where algorithms have a high computational complexity. For example, witnesses are used for certifying graph algorithms [16]. Turing used assertions [17] already and argued that one should justify the correctness of programs. In the area of logic solvers, witnesses for the results are of essence for competitions, and important competitions require witnesses and their validation. For example, the *Termination Competition (termCOMP)* [18] uses the format CPF [19], the competition of SAT solvers [20] uses the DRAT format [21] together with the validator DRAT-trim [22], and the competition on SMT solving verifies models with DOLMEN [23].

Witnesses are not only important to certify the correctness of a solver's answer, it is also important for the goal of explainability: The *true / false* answers alone are not as valuable compared to also providing the reasons to understand the answer. For example, witnesses can be used to derive test cases [9] and to aid debugging with visualizations [8]. Execution reports [24] help organize the analysis results, and the format SARIF [25] is used by static analyzers to represent results.

2   Witness Format 2.0

The witness format 2.0 is an extension of the YAML format, version 1.2. Individual verification witnesses are represented by *entries*. Each entry has three key-value pairs. The key `entry_type` has the value `invariant_set` or `violation_sequence` corresponding to the type of the witness: a correctness witness is represented by one or more entries of type `invariant_set`, while a violation witness is represented by a single entry of type `violation_sequence`. Further, the key `metadata` refers to a mapping that describes mainly the context of the witness: the format version used by the entry, the unique identifier of the entry, the creation time of the entry, the tool that produced the entry, and the verification task the witness relates to. Finally, the value of the key `content` represents the semantical content

Table 1: Structure of entries common for violation and correctness witnesses; some nodes are nested; optional items are marked with *; the term *scalar* in YAML refers also to strings

Key	Value	Description
`entry_type`	`invariant_set`	the entry type of a correctness witness
	`violation_sequence`	the entry type of a violation witness
`metadata`	mapping	the context of the witness; see below
`content`	sequence	the witness content; see Tables 2 and 3
———————— content of `metadata` ————————		
`format_version`	`2.0`	the used version of the format
`uuid`	scalar	a unique identifier of the entry; it uses the UUID format defined in RFC4122
`creation_time`	scalar	the date and time of the entry creation; it uses the format given by ISO 8601
`producer`	mapping	the tool that produced the entry; see below
`task`	mapping	the verification task to which the entry is related; see below
———————— content of `producer` ————————		
`name`	scalar	the name of the tool
`version`	scalar	the version of the tool
`configuration`*	scalar	the configuration in which the tool ran
`command_line`*	scalar	the command line with which the tool ran; it should be a bash-compliant command
`description`*	scalar	any additional information
———————— content of `task` ————————		
`input_files`	sequence	the list of files given as input to the verifier, e.g. `["path/f1.c", "path/f2.c"]`
`input_file_hashes`	mapping	SHA-256 hashes of all files in `input_files`, e.g. `{"path/f1.c": 511..., "path/f2.c": f70...}`
`specification`	scalar	the property considered by the verifier; it uses the SV-COMP format given at https://sv-comp.sosy-lab.org/2024/rules.php
`data_model`	`ILP32` or `LP64`	the data model considered for the task
`language`	`C`	the programming language of the input files; the format currently supports only C

of the entry. The key-value pairs are presented in a structured way in Table 1. The table also presents the key-value pairs of the nested mapping `metadata` and its nested mappings `producer` and `task`. We describe the possible values of the

```
1 void reach_error () {}
2 extern unsigned char __VERIFIER_nondet_uchar (void);
3 int main () {
4    unsigned char n = __VERIFIER_nondet_uchar ();
5    if (n == 0) {
6       return 0;
7    }
8    unsigned char v = 0;
9    unsigned int  s = 0;
10   unsigned int  i = 0;
11   while (i < n) {
12      v = __VERIFIER_nondet_uchar ();
13      s += v;
14      ++i;
15   }
16   if (s < v) {
17      reach_error ();
18      return 1;
19   }
20   if (s > 65025) {
21      reach_error ();
22      return 1;
23   }
24   return 0;
25 }
```

Specification:

G ! call(reach_error()),

i.e., all calls of reach_error()
are unreachable

```
1 - entry_type: invariant_set
2   metadata: <...>
3   content:
4   - invariant:
5       type: loop_invariant
6       location:
7         file_name: "inv-a.c"
8         line: 11
9         column: 1
10        function: main
11      value: "s <= i*255 && 0 <=
        ↪ i && i <= 255 && n <= 255"
12      format: c_expression
```

Fig. 1: Example C program inv-a.c taken from [11] (top left) satisfying the given specification (bottom left) and equivalent correctness witnesses in format 1.0 (top right, visualized as automaton) and format 2.0 (bottom right), with a single nontrivial invariant

key content in the following subsections separately for correctness witnesses and violation witnesses as they are conceptually different.

2.1   Correctness Witnesses

Correctness witnesses provide invariants that should help to prove the program correct. In the old format (1.0), invariants are tied to automata nodes and these nodes can correspond to multiple program locations and various moments of program executions. The new format (2.0) simply assigns invariants to program locations. Figure 1 provides an example of a correctness witness in the old format and in the new format.

Syntax. In entries of type invariant_set which represent a correctness witness, the key content contains a sequence of zero or more *invariants*. An invariant is a mapping with the following four keys.

type has the value loop_invariant if the invariant is assigned to a loop head and the value location_invariant if it is assigned to another location.

Table 2: Structure of the `content` part of entries representing correctness witnesses; optional items are marked with *

Key	Value	Description
`content`	sequence	a sequence of one or more `invariant` elements
		description of `invariant`
`invariant`	mapping	a basic building block of correctness witnesses; see below
		content of `invariant`
`type`	`loop_invariant`	the invariant type for iteration statements
	`location_invariant`	the invariant type for arbitrary statements
`location`	mapping	the location of the invariant; see below
`format`	`c_expression`	the invariant is a C expression
`value`	scalar	the actual invariant
		content of `location`
`file_name`	scalar	the file of the location
`line`	scalar	the line number of the location
`column`*	scalar	the column of the location
`function`*	scalar	the name of the function containing the location

`location` of a `loop_invariant` must point to the first character of a keyword at the beginning of a loop (i.e., `for`, `while`, or `do`). The `location` of a `location_invariant` must point to the first character of a statement or a declaration that is within a compound statement.

`format` has the value `c_expression` as the format currently supports only invariants that are C expressions.

`value` holds the actual invariant string (e.g., `"s <= i*255 && i > 0"`), which is a side-effect-free C expression over variables in the scope where the invariant is placed.

The `location` is a mapping with mandatory keys `file_name` that holds the name of the file and `line` representing the line number (the first line has the number 1). Additionally, there are two optional keys called `column` and `function`. The key `column` specifies the column number of the location (value 1 is the position of the first character on the `line`). If the column is not given, then it is interpreted as the leftmost suitable position on the line, where suitability is given by `type` and the restrictions given above. The key `function` provides the name of the function containing the location. Technically, this information is superfluous as it is determined by the `file_name`, `line`, and `column`. It is therefore not intended for any algorithmic processing of the witness, but only to improve human readability of the witness.

The structure of `content` and its nested items are summarized in Table 2.

Semantics. The correctness witness is *valid* if it fulfills the following requirements.

- Each `loop_invariant` must always hold immediately before evaluating the condition of the corresponding loop.
- Each `location_invariant` must always hold immediately before evaluating the corresponding statement or declaration.
- The specification must be satisfied for all program executions.
- No invariant evaluation causes undefined behavior and no undefined behavior occurs during any execution of the program.

Note that the order of invariants in an `invariant_set` or their division into several entries of type `invariant_set` is not important. The semantics also reveals the difference between the two types of invariants: if we replace `loop_invariant` with `location_invariant`, then the invariant has to hold only before the loop is executed, but not after each loop iteration.

2.2   Violation Witnesses

A violation witness should describe a program execution violating the considered property. For brevity, the violating execution is described loosely and the witness thus represents a set of such executions. In the old format (1.0), a violation witness is an automaton with edges prescribing consecutive restrictions on program executions. The automaton can contain various branches and loops. In the new format (2.0), a violation witness is a sequence of *waypoints* that have to be passed by the executions. To make the witness validation more efficient, the format also allows specifying waypoints that have to be avoided. Figure 2 provides an example of a violation witness in the old format and in the new format.

Syntax. The basic building blocks of violation witnesses in the new format are waypoints. Technically, a `waypoint` is a mapping with four keys, namely `type`, `location`, `constraint`, and `action`. The values of the first three keys specify a requirement on a program execution to pass a waypoint: `type` describes the type of the requirement, `location` ties the requirement to some program location, and `constraint` gives the requirement itself. The key `action` then states whether the executions represented by the witness should pass through the waypoint (value `follow`) or avoid it (value `avoid`). The format currently supports five possible values of `type` with the following meanings:

`assumption` The `location` has to point to the first character of a statement or a declaration within a compound statement. A requirement of this type says that a given constraint holds before evaluating the pointed statement or declaration. The `constraint` is a mapping with two keys: `format` specifies the language of the assumption and `value` contains a side-effect-free assumption over variables in the current scope. The value of `format` is `c_expression` as C expressions are the only assumptions currently supported. In the future, we plan to support also assumptions in ACSL [26].

`branching` A requirement of this type says that a given branching is evaluated in a given way. The `location` points to the first character of a branching keyword like `if`, `while`, `switch`, or to the character `?` in the ternary operator (`?:`).

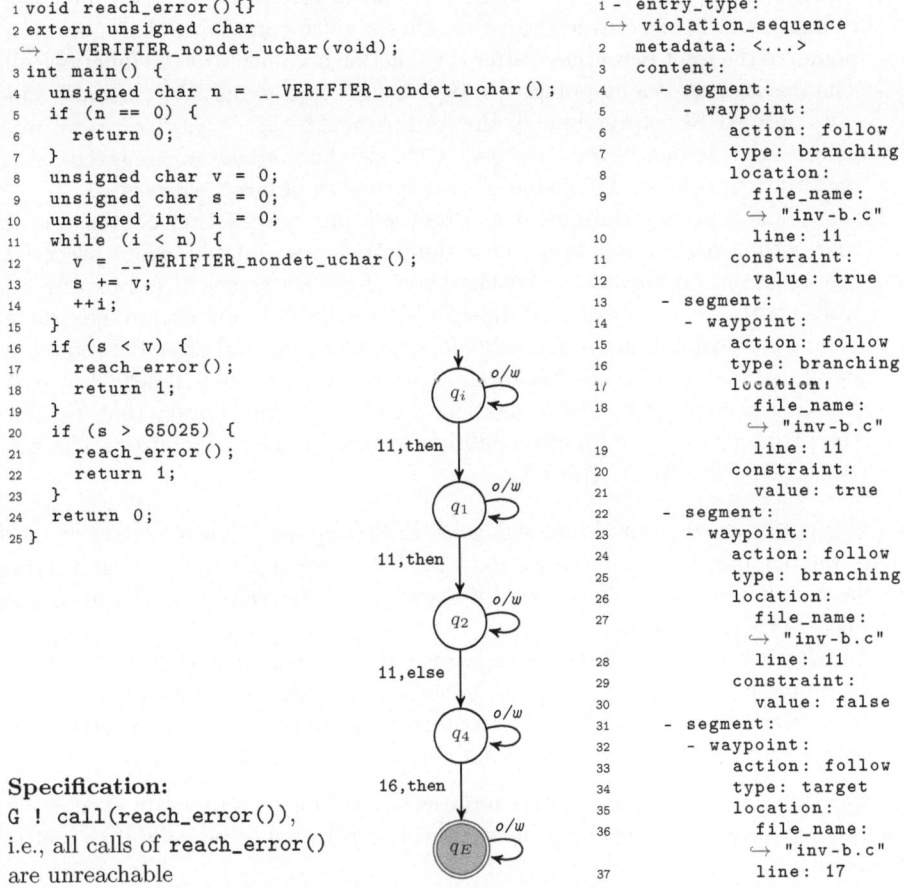

```
1 void reach_error(){}
2 extern unsigned char
   __VERIFIER_nondet_uchar(void);
3 int main() {
4    unsigned char n = __VERIFIER_nondet_uchar();
5    if (n == 0) {
6       return 0;
7    }
8    unsigned char v = 0;
9    unsigned char s = 0;
10   unsigned int  i = 0;
11   while (i < n) {
12      v = __VERIFIER_nondet_uchar();
13      s += v;
14      ++i;
15   }
16   if (s < v) {
17      reach_error();
18      return 1;
19   }
20   if (s > 65025) {
21      reach_error();
22      return 1;
23   }
24   return 0;
25 }
```

```
1 - entry_type:
      violation_sequence
2   metadata: <...>
3   content:
4     - segment:
5       - waypoint:
6           action: follow
7           type: branching
8           location:
9             file_name:
                "inv-b.c"
10            line: 11
11          constraint:
12            value: true
13    - segment:
14      - waypoint:
15          action: follow
16          type: branching
17          location:
18            file_name:
                "inv-b.c"
19            line: 11
20          constraint:
21            value: true
22    - segment:
23      - waypoint:
24          action: follow
25          type: branching
26          location:
27            file_name:
                "inv-b.c"
28            line: 11
29          constraint:
30            value: false
31    - segment:
32      - waypoint:
33          action: follow
34          type: target
35          location:
36            file_name:
                "inv-b.c"
37            line: 17
```

Specification:
G ! call(reach_error()),
i.e., all calls of reach_error()
are unreachable

Fig. 2: Example C program inv-b.c taken from [11] (top left) violating the given specification (bottom left) and similar violation witnesses in format 1.0 (middle, visualized as automaton) and format 2.0 (right)

The `constraint` is then a mapping with only one key `value`. For binary branchings, `value` can be either `true` or `false` saying whether the true branch is used or not. For the keyword `switch`, `value` can be an integer constant or `default`. The integer constant specifies the value of the controlling expression of the `switch` statement. The value `default` says that the value of this expression does not match any case of the `switch` with the exception of the `default` case (if it is present).

`function_enter` The `location` points to the right parenthesis after the function arguments of a function call. The requirement says that the called function is entered. The key `constraint` has to be omitted in this case.

`function_return` Such a requirement says that a given function call has been evaluated and the returned value satisfies a given constraint. The `location` points to the right parenthesis after the function arguments at the function call. The `constraint` is a mapping with keys `format` and `value`. We currently support only ACSL expressions of the form `\result <op> <const_expression>`, where `<op>` is one of `==`, `!=`, `<=`, `<`, `>`, `>=` and `<const_expression>` is a constant expression. The value of `format` has to be `acsl_expression`.

`target` This type of requirement can be used only with action `follow` and it marks the program location where the property is violated. More precisely, the `location` points at the first character of the statement or full expression whose evaluation is sequenced directly before the violation occurs, i.e., there is no other evaluation sequenced before the violation and after the sequence point associated with the `location`. This also implies that it can point to a function call only if it calls a function of the C standard library that violates the property or if the function call itself is the property violation. The key `constraint` has to be omitted.

Waypoints are organized into *segments*. Each `segment` is a sequence of zero or more waypoints with action `avoid` and exactly one waypoint with action `follow` at the end. A segment is called *final* if it ends with the `target` waypoint and it is called *normal* otherwise.

Finally, we can describe the `content` part of `violation_sequence` entries which represent violation witnesses. The value of `content` is a sequence of zero or more normal segments and exactly one final segment at the end. The structure of `content` and its nested items are summarized in Table 3.

Semantics. Each violation witness represents a set of some program executions violating the specified property. The witness is considered to be *valid* if the set is nonempty.

Let us consider a violation witness with $n \geq 1$ segments. An execution is represented by this witness, if the execution can be divided into n parts such that, for each $1 \leq i \leq n$, the i-th part matches the corresponding segment of the witness. An execution part matches a normal segment if

- it does not pass any avoid waypoint of the segment,
- it ends in the moment when the sequence point corresponding to the follow waypoint of the segment is entered for the first time in the execution part, and
- the follow waypoint is passed in this moment.

The final execution part matches the final segment if

- it does not pass any avoid waypoint of the segment and
- it violates the considered property during execution of the statement identified by the target waypoint.

Moreover, the execution must not contain any instruction that causes undefined behavior. An exception to this are witnesses of undefined behavior, in

Table 3: Structure of the `content` part of entries representing violation witnesses

Key	Value	Description
`content`	sequence	a sequence of zero or more normal `segment` elements and one final `segment` at the end
description of `segment`		
`segment`	sequence	a sequence of zero or more `waypoint` elements with `action: avoid` and one `waypoint` with `action: follow` at the end; the final segment ends by `waypoint` with `type: target`
description of `waypoint`		
`waypoint`	mapping	a basic building block of violation witnesses
content of `waypoint`		
`action`	`follow`	the waypoint should be passed through
	`avoid`	the waypoint should be avoided
`type`	`assumption`	restriction on variable values given by an expression
	`branching`	restriction specifying the result of a branching
	`function_enter`	restriction saying that a function is entered
	`function_return`	restriction on the result of a function call
	`target`	identification of a location of the property violation
`location`	mapping	the location of the waypoint; see Table 2
`constraint`	mapping	the constraint of the waypoint; not allowed with `type: function_enter` and `type: target`
content of `constraint`		
`format`	`c_expression`	for `type: assumption`, constraints are C expressions
	`acsl_expressions`	for `type: function_return`, constraints are specific ACSL expressions; not allowed for other `type` values
`value`	scalar	the actual constraint

which case the only instruction that causes undefined behavior must be the one represented by the target waypoint.

In each execution part, only the waypoints of the corresponding segment are evaluated. An `assumption` waypoint is evaluated at the sequence point immediately before the waypoint location. The evaluation must not lead to undefined behavior; otherwise the witness is incorrect. The waypoint is passed if the given constraint evaluates to true. A `branching` waypoint is evaluated at the sequence point immediately after evaluation of the controlling expression of the corresponding branching statement. The waypoint is passed if the resulting value of the controlling expression corresponds to the given constraint. A `function_enter` waypoint is evaluated at the sequence point immediately after evaluation of all arguments of the function call. The waypoint is passed without any additional constraint. A `function_return` waypoint is evaluated immediately after evaluation of the corresponding function call. The waypoint is passed if the returned value satisfies the given constraint.

3   Evaluation

This section presents experiments with validation of verification results using both formats (1.0 and 2.0) to answer the following research questions:

- **RQ 1:** How does the performance of the validation of the new witness format compare to the old witness format?
- **RQ 2:** Does the new format improve attributes related to readability when compared to the old format?

In the experiments, the following tools were used.

- CPACHECKER [13, 27] is a verifier and witness validator that can produce and validate both correctness and violation witnesses in both formats. The experiments are based on version 0af0e41240.
- SYMBIOTIC [28] is a verifier that can produce violation witnesses in both formats. We use version 9c278f9.
- SYMBIOTIC-WITCH2 [15] is a witness validator for violation witnesses in the old format (1.0). The experiments are based on version svcomp24.
- WITCH3 [29] is a new witness validator based on similar principles as SYMBIOTIC-WITCH2, but designed for violation witnesses in format 2.0. The tool is made of SYMBIOTIC in version b011ec9 and WITCH-KLEE in version 6dabb94.
- UAUTOMIZER [30] is a verifier and witness validator that can produce both correctness and violation witnesses in both formats and validate correctness witnesses in both formats and violation witnesses only in format 1.0. We use version 0.2.4-?-8430d5a-m and version 0.2.4-dev-0e0057c for validation of YAML and GraphML correctness witnesses respectively.

Note that the support of the new witness format in all mentioned tools except UAUTOMIZER has been implemented by authors of this paper.

For the experiments, we used all verification tasks of SV-COMP 2024 where the property to be verified is *unreachabilty of error function*, i.e., the specification used in Figs. 1 and 2. We did not use the witnesses produced during the competition [31], but rather based our experiments on fresh witnesses produced by the latest versions of SYMBIOTIC and CPACHECKER; the results are much better compared to the results from SV-COMP 2024 [32].

Benchmark Environment. For conducting our evaluation, we use BENCHEXEC to ensure reliable benchmarking [33]. All benchmarks are performed on machines with an Intel Xeon E5-1230 CPU (4 physical cores with 2 processing units each), 33 GB of RAM, and running Ubuntu 22.04 as operating system. Each verification and witness validation task is executed with resource limits used in SV-COMP, i.e., 900 s of CPU time[3], 15 GB of memory, and 1 physical core (2 processing units).

---

[3] Except violation witness validation, where the convention is to use 90 s of CPU time.

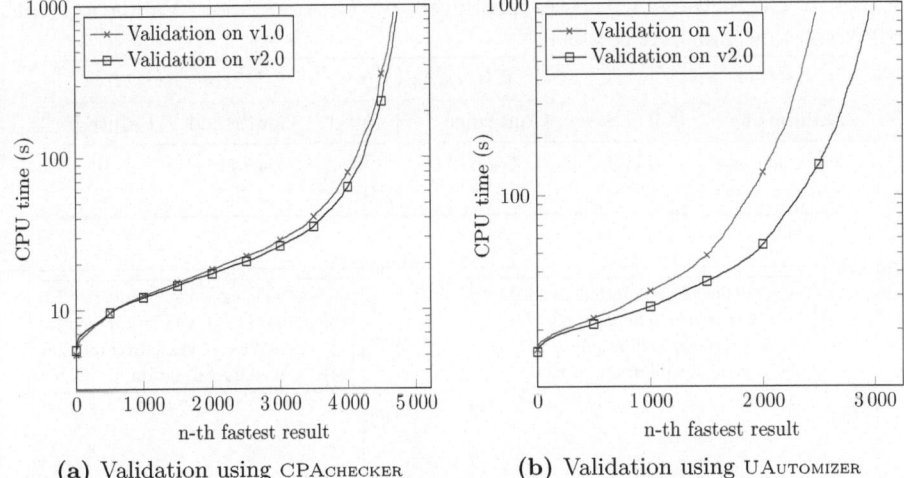

(a) Validation using CPACHECKER   (b) Validation using UAUTOMIZER

Fig. 3: Correctness witnesses produced by CPACHECKER: Quantile plots for the time taken for validation of the old and new witnesses for two different validators

3.1 Evaluation Results for RQ 1 (Validation Performance)

One of the most important questions is whether the validation using the new format is as effective and efficient as with the old witness format.

Correctness Witnesses. CPACHECKER can generate correctness witnesses in both formats. The witnesses from CPACHECKER were then validated by CPACHECKER and UAUTOMIZER. This allows for a direct comparison as shown in Fig. 3. We can observe in Fig. 3a that the validation performance of CPACHECKER is largely identical when comparing both formats. This is to be expected, as the only thing that CPACHECKER extracts from the GraphML witnesses are the invariants and their locations, and this is also the information that is present in and extracted from the witnesses in format 2.0.

For UAUTOMIZER, the new format 2.0 substantially improves both the speed of validation and the number of witnesses that can be validated. Besides aiding in verification of the original property during validation, a witness can also add additional obligations for the validator to validate. This is the case here, where the extensive automaton that is embedded into CPACHECKER's witnesses in format 1.0 is harder to prove correct for UAUTOMIZER than the much simpler set of invariants that is present in the witnesses in format 2.0. Table 4 shows numbers of confirmed and refuted witnesses.

Violation Witnesses. We present the results of our evaluation regarding RQ 1 for violation witnesses in Fig. 4, which is complemented by Table 5 with the concrete number of validated and refuted witnesses. For witnesses generated by CPACHECKER (cf. Fig. 4a), WITCH3 is able to confirm significantly more witnesses in the new format than SYMBIOTIC-WITCH2 is able to confirm in the old

Table 4: Correctness witnesses produced by CPAchecker: Validation with CPAchecker and UAutomizer

		Witnesses v1.0		Witnesses v2.0	
Validator	Witnesses	Confirmed	Refuted	Confirmed	Refuted
CPAchecker	6 729	4 685	0	4 741	0
UAutomizer	6 729	2 478	109	2 959	2

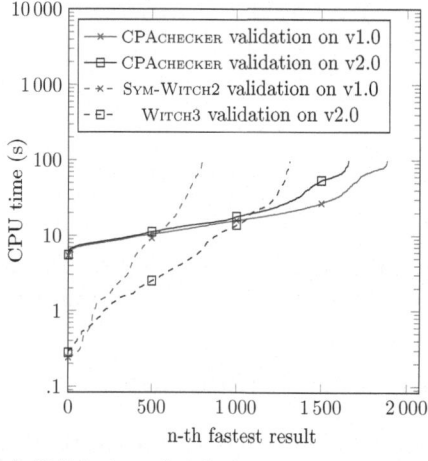

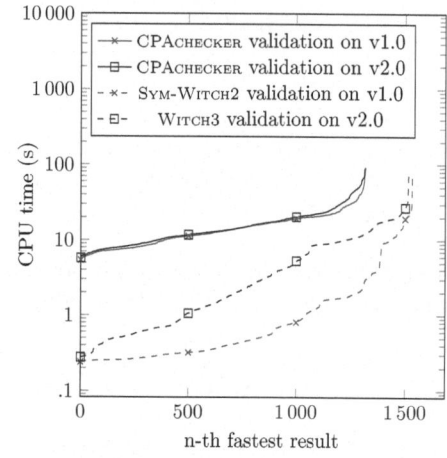

(a) Validation of violation witnesses generated by CPAchecker

(b) Validation of violation witnesses generated by Symbiotic

Fig. 4: Violation witnesses: Quantile plots for the time taken for validation of the old and the new witnesses generated by two different verifiers for two different validators

format. Due to the large number of features and underspecified semantics of the GraphML format, Symbiotic-Witch2 does not support all the attributes used in the GraphML witnesses. Ignoring these features leads to a larger state-space that needs to be explored during validation, which results in more timeouts, or misinterpreting information in the witness and missing the described error. This is not the case for Witch3 as it supports the full set of features of the new format and makes use of all the information provided in the witness. For CPAchecker there is still a relatively small performance gap between validation with the new and old format. This is not surprising, as the GraphML-based format is inspired by the specification-automata language that CPAchecker uses internally, so achieving a similar performance requires still some more engineering.

For witnesses generated by Symbiotic (cf. Fig. 4b), we can observe that the number of new witnesses confirmed by Witch3 is almost the same as the number of old witnesses confirmed by Symbiotic-Witch2. Thus, both validation approaches are very close when it comes to effectiveness. This is also the case for

Table 5: Results of validating CPAchecker's and Symbiotic's **violation** witnesses with different validators; Witch stands for Symbiotic-Witch2 when validating old witnesses, and for Witch3 when validating new witnesses.

Verifier	Validator	Produced	Witnesses v1.0		Witnesses v2.0	
			Confirmed	Refuted	Confirmed	Refuted
CPAchecker	CPAchecker	2 011	1 880	35	1 657	9
CPAchecker	Witch	2 011	798	0	1 312	17
Symbiotic	Witch	1 556	1 533	5	1 516	0
Symbiotic	CPAchecker	1 556	1 319	29	1 315	27

Table 6: Different attributes of correctness witnesses in version 1.0 and 2.0 generated by CPAchecker

Attribute	Witnesses v1.0			Witnesses v2.0		
	Min	Median	Max	Min	Median	Max
Length in Lines of Code	53	1 536	1 014 533	18	28	1 058
Size in kB	3	52	35 573	1	1	965
Number of Nodes	3	114	26 899	-	-	-
Number of Edges	2	198	142 016	-	-	-
Number of Invariants	0	1	162	0	1	104

CPAchecker, which manages to confirm almost the same number of witnesses in both formats.

> The new format for correctness witnesses does not reduce validation performance, for UAutomizer it shows a significant advantage over the old format. For violation witnesses, Witch3 handling the new format performs better than Symbiotic-Witch2 on the old format, while there is still room for improvement of CPAchecker as it performs slightly better on the old format.

3.2   Evaluation results for RQ 2 (Witness Readability)

Another important question is concerned with the attributes corresponding to the readability of the files encoding the witnesses. In particular, we are interested in the size and length of the witnesses, since this has a large effect on how easy they are to be read and understood by humans and machines.

Table 6 provides an overview of different attributes of the two versions of witnesses produced by CPAchecker for correctness. Table 7 does the same for violation witnesses produced by CPAchecker and Symbiotic. Some attributes are only applicable to one of the two versions of witnesses.

For correctness witnesses we can see that the new witnesses are usually very small in comparison to the old witnesses. This is because the new witnesses encode only the invariants, while the old witnesses encode information about the control-flow of the program. One explanation for the difference is that witnesses in version 1.0 roughly scale with the size of the program. While witnesses in version 2.0 scale

Table 7: Different attributes of violation witnesses in version 1.0 and 2.0 generated by CPACHECKER and SYMBIOTIC

Attribute	Witnesses v1.0			Witnesses v2.0		
	Min	Median	Max	Min	Median	Max
Length in Lines of Code	12	372	258 730	27	171	114 460
Size in kB	2	14	9 098	1	6	3 071
Number of Nodes	1	38	28 304	-	-	-
Number of Edges	0	42	28 793	-	-	-
Number of Waypoints	-	-	-	1	13	9 537

only with respect to the amount of invariants, which for CPACHECKER is roughly correlated to the amount of function calls and loops. As we saw in Sect. 3.1, this extra information is not necessarily relevant for validation.

For violation witnesses, we see that, apart from a small factor due to over-head in describing the automaton, both formats are similar in all metrics. This is not surprising, as both formats encode similar information about an error path. Therefore, they both roughly scale with the amount of assumption for nondeterministic variables and the amount of branching decisions in the error path.

The tables show that the new witnesses are usually much shorter than the old witnesses. As we have seen in Sect. 3.1, this does not have a negative impact on the validation performance, since the information most relevant for validation is retained. Having less information makes it much easier for a verification engineer to understand the witness and use it in some further processing steps.

> In summary, witnesses in version 2.0 are generally much smaller and easier to read than witnesses in version 1.0, while retaining all important data.

3.3   Threats to Validity

Internal Validity. We used the benchmarking framework BENCHEXEC [33] to run the experiments, which uses the most modern Linux features for reliable benchmarking. This tool also makes sure to never run two different executions on the same physical core, in order to avoid interference of shared computing resources. Our validation tools might contain bugs, which could lead to wrong conclusions, however, our claim is that the new format works already sufficiently well to serve as an alternative format.

External Validity. The conclusions about the validators might not hold for other validators that will be developed in the future, also, witnesses generated by other verifiers might have different characteristics. However, other tools are not expected to deviate much from the presented witnesses, because they would serve the same purpose of testifying the bug or proof. Our experiments were done on a large benchmark set, which is also used in competitions, but it could still be the case that there are witnesses and programs for which the results presented are not applicable. Since extending and improving the witness format is an ongoing process, we expect that if this is the case, it will be adequately addressed in the future.

4   Version 1.0 vs. Version 2.0

The witness format 2.0 is closely tied to the actual program syntax. While the format 1.0 uses an automaton largely independent of the program syntax and closely tied to the program representation as control-flow automata internally used by some verifiers. Due to this, the format 2.0 is more succinct, has well-defined semantics, and is easier to understand by humans. On the other hand, format 1.0 is more expressive, for example it can define different loop invariants for the same loop, when two different paths are taken to reach the loop.

Currently, the format 2.0 has the same practical limitations as the format 1.0. In the case of correctness witnesses, they have not yet been defined for concurrency safety, memory safety and for termination. Violation witnesses have not yet been defined for concurrency safety. There are also features which are not yet supported by the new format but which are straightforward extensions, such as support for Java and violation termination witnesses. Extending witnesses to be able to validate more programs and specifications is ongoing work, we expect that the simplification of the syntax and clarification of the semantics with format version 2.0 will make it easier to extend the format in the future.

In order to validate our concept of the new format, we reported our initiative to the SV-COMP community, and the jury made a decision to immediately include the new format as an alternative to the existing format, in order to quickly adopt it and improve the state of the art. This was seen in SV-COMP 2024 [32], where 8 verifiers and 4 validators supported the correctness witnesses v2.0 and 2 verifiers and 2 validators supported the violation witnesses v2.0. Table 8 shows all these tools and their support of witnesses formats in detail.

This also shows the large interest the software verification community has in the new format, since the first mention of the format for correctness witnesses was only in September 2021[4] and the work on the violation witnesses part of the new format started only in April 2023.

5   Conclusion

Verification witnesses are an important part of the software-verification ecosystem. Just like verification tools, specification formats, and witness validators, there is also a need to improve the *format* for verification witnesses. This paper introduces the witness format version 2.0, which changes the container format from GraphML to YAML, has more concise data representation, and has a clearly

---

[4] https://gitlab.com/sosy-lab/benchmarking/sv-witnesses/-/merge_requests/44

Table 8: Tools with some support of witnesses format 2.0 and their abilities to generate/verify correctness/violation witnesses in format 1.0/2.0 in SV-COMP 2024; tools where the support of witness format 2.0 was implemented by the authors of this paper are typeset in bold

| | Witness Generation | | | | Witness Validation | | | |
| | Correctness | | Violation | | Correctness | | Violation | |
Tool	v1.0	v2.0	v1.0	v2.0	v1.0	v2.0	v1.0	v2.0
CPAchecker	•	•	•	•	•	•	•	•
Symbiotic	•		•	•				
Symbiotic-Witch2							•	
Witch3								•
UAutomizer	•	•	•		•	•	•	
UKojak	•	•	•					
UTaipan	•	•	•					
UGemCutter	•	•	•					
Mopsa	•	•					•	
CPV	•	•	•					
Goblint	•	•					•	

defined semantics independent from control-flow automata. Besides describing the syntax and semantics of the new format, we also evaluated the effectiveness and efficiency induced by the new format. In sum, the new witnesses are much smaller and the experimental results show a significantly improved confirmation rate for some validators: using the new format, UAutomizer can confirm 481 more correctness witnesses (Table 4) and Witch3 can confirm 514 more violation witnesses (Table 5). Furthermore, shortly after we proposed this new format, already seven other tools support the format, which is an indicator that the developers value the new format.

Data-Availability Statement. A reproduction package (that includes all software and data that we used for our experiments) is available on Zenodo [34].

Funding Statement. P. Ayaziová and J. Strejček were supported by the Czech Science Foundation grant GA23-06506S. D. Beyer, M. Lingsch-Rosenfeld, and M. Spiessl were supported by the Deutsche Forschungsgemeinschaft (DFG) – 378803395 (ConVeY) and 496588242 (IdeFix).

References

1. Beyer, D., Dangl, M., Dietsch, D., Heizmann, M., Stahlbauer, A.: Witness validation and stepwise testification across software verifiers. In: Proc. FSE. pp. 721–733. ACM (2015). https://doi.org/10.1145/2786805.2786867
2. Beyer, D., Dangl, M., Dietsch, D., Heizmann, M.: Correctness witnesses: Exchanging verification results between verifiers. In: Proc. FSE. pp. 326–337. ACM (2016). https://doi.org/10.1145/2950290.2950351

3. Beyer, D.: Software verification and verifiable witnesses (Report on SV-COMP 2015). In: Proc. TACAS. pp. 401–416. LNCS 9035, Springer (2015). https://doi.org/10.1007/978-3-662-46681-0_31

4. Beyer, D., Strejček, J.: Case study on verification-witness validators: Where we are and where we go. In: Proc. SAS. pp. 160–174. LNCS 13790, Springer (2022). https://doi.org/10.1007/978-3-031-22308-2_8

5. Beyer, D.: Competition on software verification and witness validation: SV-COMP 2023. In: Proc. TACAS (2). pp. 495–522. LNCS 13994, Springer (2023). https://doi.org/10.1007/978-3-031-30820-8_29

6. Beyer, D., Wehrheim, H.: Verification artifacts in cooperative verification: Survey and unifying component framework. In: Proc. ISoLA (1). pp. 143–167. LNCS 12476, Springer (2020). https://doi.org/10.1007/978-3-030-61362-4_8

7. Beyer, D., Haltermann, J., Lemberger, T., Wehrheim, H.: Decomposing Software Verification into Off-the-Shelf Components: An Application to CEGAR. In: Proc. ICSE. pp. 536–548. ACM (2022). https://doi.org/10.1145/3510003.3510064

8. Beyer, D., Dangl, M.: Verification-aided debugging: An interactive web-service for exploring error witnesses. In: Proc. CAV (2). pp. 502–509. LNCS 9780, Springer (2016). https://doi.org/10.1007/978-3-319-41540-6_28

9. Beyer, D., Dangl, M., Lemberger, T., Tautschnig, M.: Tests from witnesses: Execution-based validation of verification results. In: Proc. TAP. pp. 3–23. LNCS 10889, Springer (2018). https://doi.org/10.1007/978-3-319-92994-1_1

10. Beyer, D., Spiessl, M., Umbricht, S.: Cooperation between automatic and interactive software verifiers. In: Proc. SEFM. p. 111–128. LNCS 13550, Springer (2022). https://doi.org/10.1007/978-3-031-17108-6_7

11. Beyer, D., Dangl, M., Dietsch, D., Heizmann, M., Lemberger, T., Tautschnig, M.: Verification witnesses. ACM Trans. Softw. Eng. Methodol. **31**(4), 57:1–57:69 (2022). https://doi.org/10.1145/3477579

12. Brandes, U., Eiglsperger, M., Herman, I., Himsolt, M., Marshall, M.S.: GraphML progress report. In: Graph Drawing. pp. 501–512. LNCS 2265, Springer (2001). https://doi.org/10.1007/3-540-45848-4_59

13. Beyer, D., Keremoglu, M.E.: CPACHECKER: A tool for configurable software verification. In: Proc. CAV. pp. 184–190. LNCS 6806, Springer (2011). https://doi.org/10.1007/978-3-642-22110-1_16

14. Chalupa, M., Rechtáčková, A., Mihalkovič, V., Zaoral, L., Strejček, J.: SYMBIOTIC 9: String analysis and backward symbolic execution with loop folding (competition contribution). In: Proc. TACAS (2). pp. 462–467. LNCS 13244, Springer (2022). https://doi.org/10.1007/978-3-030-99527-0_32

15. Ayaziová, P., Strejček, J.: SYMBIOTIC-WITCH 2: More efficient algorithm and witness refutation (competition contribution). In: Proc. TACAS (2). pp. 523–528. LNCS 13994, Springer (2023). https://doi.org/10.1007/978-3-031-30820-8_30

16. McConnell, R.M., Mehlhorn, K., Näher, S., Schweitzer, P.: Certifying algorithms. Computer Science Review **5**(2), 119–161 (2011). https://doi.org/10.1016/j.cosrev.2010.09.009

17. Turing, A.: Checking a large routine. In: Report on a Conference on High Speed Automatic Calculating Machines. pp. 67–69. Cambridge Univ. Math. Lab. (1949), https://turingarchive.kings.cam.ac.uk/publications-lectures-and-talks-amtb/amt-b-8

18. Giesl, J., Mesnard, F., Rubio, A., Thiemann, R., Waldmann, J.: Termination competition (termCOMP 2015). In: Proc. CADE. pp. 105–108. LNCS 9195, Springer (2015). https://doi.org/10.1007/978-3-319-21401-6_6

19. Sternagel, C., Thiemann, R.: The certification problem format. In: Proc. UITP. pp. 61–72. EPTCS 167, EPTCS (2014). https://doi.org/10.4204/EPTCS.167.8

20. Järvisalo, M., Berre, D.L., Roussel, O., Simon, L.: The international SAT solver competitions. AI Magazine **33**(1) (2012)
21. Heule, M.J.H.: The DRAT format and drat-trim checker. CoRR **1610**(06229) (October 2016)
22. Wetzler, N., Heule, M.J.H., Jr., W.A.H.: DRAT-TRIM: Efficient checking and trimming using expressive clausal proofs. In: Proc. SAT. pp. 422–429. LNCS 8561, Springer (2014). https://doi.org/10.1007/978-3-319-09284-3_31
23. Bury, G., Bobot, F.: Verifying models with DOLMEN. In: Proc. SMT Workshop. CEUR Workshop Proceedings, CEUR (2023)
24. Castaño, R., Braberman, V.A., Garbervetsky, D., Uchitel, S.: Model checker execution reports. In: Proc. ASE. pp. 200–205. IEEE (2017). https://doi.org/10.1109/ASE.2017.8115633
25. OASIS: Static analysis results interchange format (sarif) version 2.0 (2019)
26. Baudin, P., Cuoq, P., Filliâtre, J.C., Marché, C., Monate, B., Moy, Y., Prevosto, V.: ACSL: ANSI/ISO C specification language version 1.17 (2021), available at https://frama-c.com/download/acsl-1.17.pdf
27. Baier, D., Beyer, D., Chien, P.C., Jankola, M., Kettl, M., Lee, N.Z., Lemberger, T., Lingsch-Rosenfeld, M., Spiessl, M., Wachowitz, H., Wendler, P.: CPACHECKER 2.3 with strategy selection (competition contribution). In: Proc. TACAS. LNCS , Springer (2024)
28. Jonáš, M., Kumor, K., Novák, J., Sedláček, J., Trtík, M., Zaoral, L., Ayaziová, P., Strejček, J.: SYMBIOTIC 10: Lazy memory initialization and compact symbolic execution (competition contribution). In: Proc. TACAS. LNCS , Springer (2024)
29. Ayaziová, P., Strejček, J.: WITCH 3: Validation of violation witnesses in the witness format 2.0 (competition contribution). In: Proc. TACAS. LNCS , Springer (2024)
30. Heizmann, M., Bentele, M., Dietsch, D., Jiang, X., Klumpp, D., Schüssele, F., Podelski, A.: Ultimate automizer and the abstraction of bitwise operations (competition contribution). In: Proc. TACAS. LNCS , Springer (2024)
31. Beyer, D.: Verification witnesses from verification tools (SV-COMP 2024). Zenodo (2024). https://doi.org/10.5281/zenodo.10669737
32. Beyer, D.: State of the art in software verification and witness validation: SV-COMP 2024. In: Proc. TACAS. LNCS , Springer (2024)
33. Beyer, D., Löwe, S., Wendler, P.: Reliable benchmarking: Requirements and solutions. Int. J. Softw. Tools Technol. Transfer **21**(1), 1–29 (2019). https://doi.org/10.1007/s10009-017-0469-y
34. Ayaziová, P., Beyer, D., Lingsch-Rosenfeld, M., Spiessl, M., Strejček, J.: Reproduction package for SPIN 2024 article 'Software verification witnesses 2.0'. Zenodo (2024). https://doi.org/10.5281/zenodo.10826204

Fault Localization on Verification Witnesses

Dirk Beyer[iD], Matthias Kettl[iD], and Thomas Lemberger[iD]

LMU Munich, Germany

Abstract. When verifiers report an alarm, they export a violation witness (exchangeable counterexample) that helps validate the reachability of that alarm. Conventional wisdom says that this violation witness should be very precise: the ideal witness describes a single error path for the validator to check. But we claim that verifiers overshoot and produce large witnesses with information that makes validation unnecessarily difficult. To check our hypothesis, we reduce violation witnesses to that information that automated fault-localization approaches deem relevant for triggering the reported alarm in the program. We perform a large experimental evaluation on the witnesses produced in the International Competition on Software Verification (SV-COMP 2023). It shows that our reduction shrinks the witnesses considerably and enables the confirmation of verification results that were not confirmable before.

1 Introduction

Nowadays, a large body of automated formal-verification tools is available [4]. But when these tools present a verification alarm to the user without any additional information, it requires expensive reasoning about the program. So when a modern verifier reports an alarm, it also produces a *violation witness* [7] that describes a set of paths through the program of which at least one provokes the reported alarm. This allows to validate the claimed alarm with a separate, independent validator.[1] Developers may also use witnesses to debug. The description of paths through the program can convey more information than a single test input. For example, a violation witness may describe a full range of input values that trigger an alarm, or describe that the found program paths that trigger the alarm all pass through a specific code location.

A trivial violation witness that describes all program paths is valid, but it is not helpful. Conventional wisdom in the community assumes that a precise witness that only describes a single program path is the most helpful and the easiest to check for a validator. We observe that this makes some verifiers produce

---

A poster version of this paper is published in Proc. ICSE 2024 [15].

[1] All participants of the International Competition on Software Verification (SV-COMP) [4] produce violation witnesses. SV-COMP validates them and expects that this is significantly faster than the original verification: participating verifiers get 900 s of CPU time to analyze a program, but the verifier only receives points if a validator can confirm the produced violation witness within 90 s.

© The Author(s) 2025
T. Neele and A. Wijs (Eds.): SPIN 2024, LNCS 14624, pp. 204–223, 2025.
https://doi.org/10.1007/978-3-031-66149-5_12

large witnesses that describe details about a program path that are not relevant with regards to the reported alarm. For example, an overly-detailed witness may describe a concise sequence of loop iterations in the program, even though the program reaches the alarm independent of that loop. Such details may slow down the validator when reasoning about the violation.

As a countermeasure, we propose to reduce witnesses to a selection of relevant information with fault localization. Given an error path, fault localization reports the statements that it suspects responsible for the reachability of that path. This makes witnesses more comprehensive for both machines and humans. We use three different fault-localization techniques based on SMT formulas: MAXSAT [29], its derivative MINUNSAT, and the baseline UNSAT where we ask for an arbitrary set of statements that are relevant for the reachability of an error path.

We perform large-scale experiments on the data of SV-COMP 2023. These show that our reduction does not influence the quality of the witnesses, but reduces their size by 25 %. In addition, most validators confirm more alarms after our reduction.

Contributions. We make the following contributions:

- We design the first formal approach to apply fault localization to software verification witnesses (violation witnesses) [8].
- Our implementation is available open source. Fault localization is implemented as part of the verification framework CPACHECKER [10], and witness reduction is implemented in the new tool FLOW [14].
- We perform a large experimental evaluation that shows that the approach can effectively reduce the size of the witnesses (by 25 %) and that the confirmation rate of validators is improved through this (by up to 34 %). We use 21 356 original witnesses that were produced by 14 verifiers on 3 225 verification tasks during SV-COMP 2023.
- Our code and all experimental data are available in an evaluated reproduction artifact [11].

Example. Figure 1 shows an example program that computes the prime factors for a number num that is provided through function nondet(). If a factor i is found for the number (num % i == 0) (line 9), it is checked whether i is a prime number (lines 12 to 18). If it is (isPrime = 1), the number num is divided by i and the process is repeated until num is 1 (lines 19 to 22). This way, all prime factors for num could be found. But the program has a fault in the computation of the division (line 20): Instead of computing num = num / i, there is an off-by-one error: num = num / (i + 1). Therefore, the program calls reach_error() for input num = 2: The first iteration of the for-loop in line 8 assigns num = 0 instead of the correct num = 1.

Assume we do not know that this error exists and we want to know whether a call to reach_error() is reachable. When we run the formal verifier UAu-TOMIZER [24, 25] on the program with property "reach_error is never called", it reports an alarm. It also provides a violation witness that represents at least one claimed counterexample to the property. This violation witness is presented

```
1  int nondet();
2  void reach_error();
3
4  void main() {
5    int num = nondet();
6    if (num < 1) return;
7
8    for (int i=2; i <= num; i++) {
9      if (!(num % i == 0)) {
10       continue;
11     }
12     int isPrime = 1;
13     for (int j=2; j <= i/2; j++) {
14       if (i % j == 0) {
15         isPrime = 0;
16         break;
17       }
18     }
19     if (isPrime) {
20       num = num / (i + 1);
21       i--;
22     }
23   }
24
25   if (num != 1) {
26     reach_error();
27   }
28 }
```

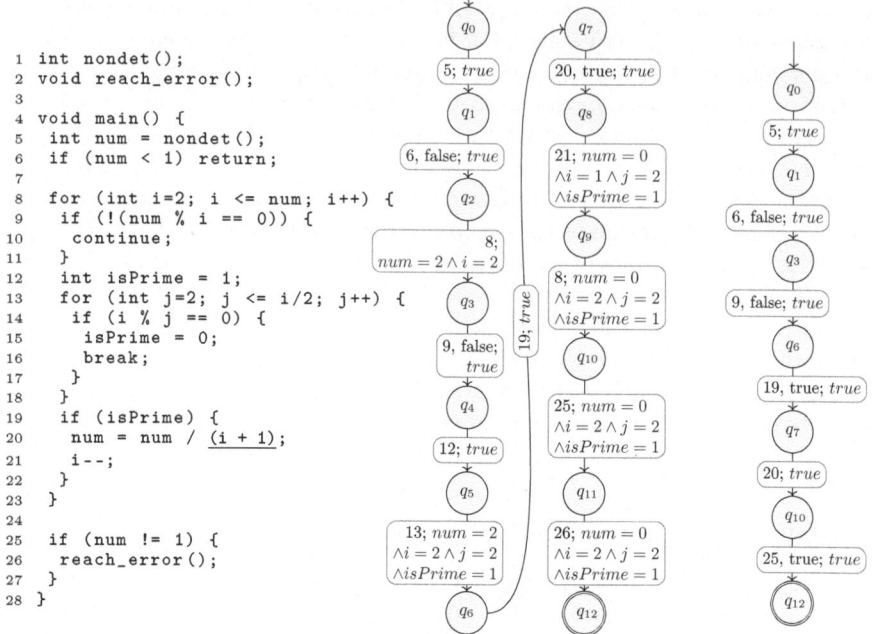

Fig. 1: Program that computes factorization, with an off-by-one error in line 20 (underlined)

Fig. 2: Detailed violation witness that describes the program path to the violation in line 26

Fig. 3: Less detailed violation witness, reduced with our approach

as automaton in Fig. 2. The witness describes a detailed sequence of source-code lines and variable values that it claims an execution must pass to reach a specification violation. Transition labels are denoted as pairs of a syntax guard and state-space guard, divided by a semicolon. Syntax guards name a line number and, optionally, after a comma, a branching condition true or false. State-space guards describe the required program state space at the specific transition.

All states ($q_0, \ldots, q_{12}$) have an implicit self-transition that is matched only if no other outgoing transition is valid for an execution step. To transition from q_2 to q_3, for example, the values of num and i have to be equal to 2. Otherwise, execution stays in q_2. The witness in Fig. 2 is very detailed and describes more information than necessary. This may limit a witness validator that tries to reason about the property violation, because the witness steers the validator to a precise path.

The strong contrast to this is a violation witness that enters an accepting state on every transition. It does not contain any information and does not restrict the program executions; it just says that there is a property violation *somewhere* in the program. It fits any violation that a validator finds.

We aim for the middle of these two extremes: We reduce the witness to only include information about the program that do not unnecessarily restrict the state-space, but restrict it to (a subset of) executions that actually reach the error. To do this, we run a fault-localization approach on the counterexample described by the witness and produce a new, reduced witness (Fig. 3). This reduced witness only contains those transitions of the original witness that the fault localization deems relevant for the feasibility of the counterexample. In our example, this is the following lines: line 5, which initializes num, line 6, which restricts the value of num, line 9, which instructs analysis to continue with line 12, line 19, which instructs analysis to continue with line 20, line 20, which is the erroneous statement, and line 25, which guards the call to reach_error(). The reduced witness is significantly smaller and still contains relevant information for reasoning about the witness.

Related Work. Execution-based witness validation [9] eases the use of a violation witness by turning it into an executable test. A developer can then debug this test in the used manner to reason about the reported violation. Other approaches also exist for the execution of counterexamples [6, 22], but these do not use a common exchange format. Other approaches [20, 33, 36] provide simulators for counterexamples that allow to directly "execute" a counterexample step-by-step, similar to prominent debuggers for program execution. Test execution and execution-simulation are orthogonal to our approach: with an executable test, the developer has the full program path to debug, and is not hinted towards statements of potentially high interest. Additionally, error paths that are not described by the test case are lost.

Multiple fault-localization approaches exist, based on code-coverage statistics [1, 27, 38], slicing-based approaches [26], and logic reasoning [19, 23, 28, 29].

There exists a preliminary study [3] on the structure of verification witnesses, but it focuses on witnesses for correctness proofs and does not examine the indication of violation-witness detail on their confirmation rate. We close this gap with our new approach.

Performing fault localization on witnesses has been studied before [30] in the context of timed automata. Given a counterexample, the approach computes the maximum satisfiable core of the trace and tries to fix the automaton.

2   Background

Program Representation. For the sake of presentation, we consider an imperative, sequential programming language with two types of operations: assign operation (x = x + 1) and assume operation ([x <= 0]). We use assign operation x = nondet() to signal introduction of a new non-deterministic value, and special statement reach_error() to signal an error.[2] All program variables are of type integer.[3] Set X is the set of all program variables in a program and set Ops is the set of all possible program operations over X.

---

[2] All safety properties on programs can be reduced to this property.
[3] Our implementation supports the GNU C programming language.

We represent programs as control-flow automata (CFAs). A CFA $P = (L, l_0, G)$ consists of the set of program locations L, the initial program location $l_0 \in L$, and a set of control-flow edges $G \subseteq L \times Ops \times L$.

We use formulas in first-order logic. Given a symbol s, we define $s^{\langle i \rangle}$ as the instantiation of s with i primes. For example, $s^{\langle 0 \rangle} = s$ and $s^{\langle 2 \rangle} = s''$.

The set $\mathcal{C}$ contains all possible concrete program states. A concrete program state $c : X \mapsto \mathbb{Z}$ consists of one value assignment for each program variable. A program path $pp = l_0 \xrightarrow{op_0} \ldots \xrightarrow{op_n} l_{n+1}$ is a run through the CFA so that $(l_i, op_i, l_{i+1}) \in G$. A program execution $ex(pp, c_0) = (l_0, c_0) \xrightarrow{op_0} \ldots \xrightarrow{op_n} (l_{n+1}, c_{n+1})$ for program path pp and initial state $c_0 \in \mathcal{C}$ is feasible iff each c_{i+1} is a valid product of evaluating op_i on c_i. Trace formula $\mathrm{TF}(pp)$ is a sequence of formulas that exactly describes the valid program states in all possible program executions for pp. For example, $pp = l_5 \xrightarrow{\texttt{num = nondet()}} l_6 \xrightarrow{\texttt{[!(num < 1)]}} l_{8_0} \xrightarrow{\texttt{i = 2}} l_{8_1} \xrightarrow{\texttt{[i <= num]}} \ldots \xrightarrow{\texttt{i = i + 1}} l_{8_1} \xrightarrow{\texttt{[!(i <= num)]}} l_{25} \xrightarrow{\texttt{[num != 1]}} l_{26}$ is represented by $\mathrm{TF}(pp) = \langle num^{\langle 0 \rangle} = n\dot{o}ndet^{\langle 0 \rangle}, \neg(num^{\langle 0 \rangle} < 1), i^{\langle 0 \rangle} = 2, i^{\langle 0 \rangle} \leq num^{\langle 0 \rangle}, \ldots, i^{\langle 2 \rangle} = i^{\langle 1 \rangle} + 1, \neg(i^{\langle 2 \rangle} \leq num^{\langle 1 \rangle}), num^{\langle 1 \rangle} \neq 1 \rangle$. We define $\mathrm{TF}(pp)\{i\}$ as the i-th formula in $\mathrm{TF}(pp)$, starting with index 0. We expect that, for each operation op_i in pp, there is exactly one corresponding formula $\mathrm{TF}(pp)\{i\}$. We define subset $\mathrm{TF}(pp)\{:j\} = \{\mathrm{TF}(pp)\{0\}, \ldots, \mathrm{TF}(pp)\{j\}\}$ as the prefix of $\mathrm{TF}(pp)$ up to index j.

A counterexample is a finite program path $l_0 \xrightarrow{op_0} \ldots \xrightarrow{op_{n-1}} l_n \xrightarrow{\texttt{reach\_error()}} l_{n+1}$ that ends with an operation $\texttt{reach\_error()}$. We say that a counterexample cex is feasible if there exists a $c_0 \in \mathcal{C}$ so that $ex(cex, c_0)$ is feasible. For the sake of our presentation, we assume that a $\texttt{reach\_error()}$ is always directly preceded by a single assume operation op_{n-1}.[4]

We represent program properties as *observer automata*: A program property φ for a program $P = (L, l_0, G)$ is a finite-state automaton $\varphi = (\Omega, \Sigma_\varphi, \delta, \omega_0, F)$ with states Ω, alphabet $\Sigma_\varphi = 2^G$, transitions $\delta \subseteq \Omega \times \Sigma_\varphi \times \Omega$, initial state $\omega_0 \in \Omega$ and accepting states $F \subseteq \Omega$, where $\omega_0 \notin F$. The accepting states F represent a violation to the property. The observer automaton gets as input a program path.

Fault Localization. A *suspect* is a subset $f \subseteq G$ of CFA edges whose operations may be responsible for a feasible counterexample.

Given a feasible counterexample cex, fault localization has the goal to determine a set $\mathcal{F} = \{f_0, f_1, \ldots\}$ of suspects.

Because cex is feasible, we know that the conjunction $\bigwedge \mathrm{TF}(cex)$ of the trace formula's elements is satisfiable. We define the *precondition* ψ, the *faulty trace* π, and the *post condition* ϕ. The precondition $\psi = (nondet^{\langle 0 \rangle} = v_0 \wedge \ldots \wedge nondet^{\langle k \rangle} = v_k)$ describes one satisfying variable assignment of $\bigwedge \mathrm{TF}(cex)$ for all non-deterministic values (variables $nondet^{\langle i \rangle}$) that occur in $\bigwedge \mathrm{TF}(cex)$. It can be generated by extracting the relevant variable assignments from a model of $\bigwedge \mathrm{TF}(cex)$. For example, one precondition for the counterexample described by Fig. 2 is $\psi = (nondet^{\langle 0 \rangle} = 2)$. It initializes $num = 2$.

---

[4] In our implementation, we find the last assume operation before the error location and of these we use all assume operations that originate from the same code line.

Algorithm 1 MaxSat(ψ, π, ϕ)

Input: Precondition ψ, faulty trace π, postcondition ϕ
Output: Set $\mathcal{F}$ of candidate faults
1: $\mathcal{F} = \{\}$
2: $susp = 2^\pi$
3: **for** $size$ in $1 \ldots |\pi|$ **do**
4:     $\mathcal{F} = \mathcal{F} \cup \{f \in susp \mid size = |f| \text{ and } \psi \wedge \bigwedge(\pi \setminus f) \wedge \phi \text{ sat}\}$
5:     $susp = \{t \in susp \mid \forall f \in \mathcal{F}. \, f \not\subseteq t\}$
6: **return** $\mathcal{F}$

Algorithm 2 MinUnsat(ψ, π, ϕ)

Input: Precondition ψ, faulty trace π, postcondition ϕ
Output: Set $\mathcal{F}$ of candidate faults
1: $susp = 2^\pi$
2: $\mathcal{F} = \{\}$
3: **for** $size$ in $1 \ldots |\pi|$ **do**
4:     $\mathcal{F} = \mathcal{F} \cup \{f \in susp \mid size = |f| \text{ and } \psi \wedge \bigwedge f \wedge \phi \text{ unsat}\}$
5:     $susp = \{t \in susp \mid \forall f \in \mathcal{F}. \, f \not\subseteq t\}$
6: **return** $\mathcal{F}$

The faulty trace $\pi = \text{TF}(cex)\{: n - 2\}$ is the prefix of $\text{TF}(cex)$ that includes all possible program faults. It excludes the formula of the last assume operation op_{n-1}. The post condition $\phi = \neg \text{TF}(cex)\{n - 1\}$ represents the final assume operation that guards the `reach_error()`. The `reach_error()` is reachable when $\neg \phi = \text{TF}(cex)\{n - 1\}$ is fulfilled. For example, the postcondition for the counterexample that is described by Fig. 2 is $\phi = \neg(num \neq 1)$. If $\neg \phi = num \neq 1$ is fulfilled, the error location is reached. When formula $\psi \wedge \bigwedge \pi \wedge \neg \phi$ represents a feasible counterexample, we can assume that it is satisfiable. Because ψ defines a definite initial assignment for all variables in ψ and π, we can then also assume that the formula $\psi \wedge \bigwedge \pi \wedge \phi$ with a fulfilled postcondition ϕ is unsatisfiable.

MaxSAT. MaxSat [29] (Alg. 1) computes a set of candidate faults for a counterexample by finding all minimal combinations of program operations that together contribute to the feasibility of the counterexample.

MaxSat initializes the set $susp$ with all subsets of π. This are all suspects for a program fault. Then, MaxSat starts with the smallest possible candidate subsets that only consist of a single clause ($size = 1$) and increasingly considers candidate subsets of larger $size$. For each $size$, MaxSat selects all suspects $f \subseteq \pi$ that consist of $size$ clauses and for which $\psi \wedge \bigwedge(\pi \setminus f) \wedge \phi$ is satisfiable. The set $susp$ of possible candidate subsets is then updated to only include subsets of π that are not supersets of any previously selected subset. After all relevant combinations of program operations have been considered, the set $\mathcal{F}$ of selected subsets is returned as the set of possible suspects.

MinUnsat. Analogous to MaxSat, algorithm MinUnsat (Alg. 2) computes a set of candidate faults for a counterexample by computing all subsets $f \subseteq \pi$

that suffice so that $\psi \wedge \bigwedge f \wedge \phi$ is unsatisfiable, and for which no smaller subset $f' \subset f$ fulfills the same criterion.

Unsat. Last, we use algorithm UNSAT as a baseline for fault localization: UNSAT asks an SMT solver for an arbitrary subset $f \subseteq \pi$ that makes $\psi \wedge \bigwedge f \wedge \phi$ unsatisfiable. This may even be π itself.

Example. To illustrate MAXSAT and MINUNSAT, assume the precondition $\psi = (nondet^{\langle 0 \rangle} = 2)$, the trace π with

$$\pi = \{num^{\langle 0 \rangle} = nondet^{\langle 0 \rangle}, \ \neg(num^{\langle 0 \rangle} < 1), \ i^{\langle 0 \rangle} = 2, \ i^{\langle 0 \rangle} \leq num^{\langle 0 \rangle},$$
$$num^{\langle 0 \rangle} \% i^{\langle 0 \rangle} = 0, \ isPrime^{\langle 0 \rangle} = 1, \ j^{\langle 0 \rangle} = 2, \ \neg(j^{\langle 0 \rangle} \leq num^{\langle 0 \rangle}/2),$$
$$isPrime^{\langle 0 \rangle} \neq 0, \ num^{\langle 1 \rangle} = num^{\langle 0 \rangle}/(i^{\langle 0 \rangle} + 1), \ i^{\langle 1 \rangle} = i^{\langle 0 \rangle} - 1,$$
$$i^{\langle 2 \rangle} = i^{\langle 1 \rangle} + 1, \ \neg(i^{\langle 2 \rangle} \leq num^{\langle 1 \rangle})\}$$

and the postcondition $\phi = (num^{\langle 1 \rangle} = 1)$. Given these three, MAXSAT produces the set $\mathcal{F} = \{\{num^{\langle 1 \rangle} = num^{\langle 0 \rangle}/(i^{\langle 0 \rangle}+1)\}, \{num^{\langle 0 \rangle} = nondet^{\langle 0 \rangle}\}, \{i^{\langle 0 \rangle} = 2\}\}$ of three suspects, because it is enough to remove any of the three assignments f from the trace to make the formula $\psi \wedge \bigwedge(\pi \setminus f) \wedge \phi$ satisfiable. MINUNSAT produces the set $\mathcal{F} = \{\{num^{\langle 1 \rangle} = num^{\langle 0 \rangle}/(i^{\langle 0 \rangle} + 1), num^{\langle 0 \rangle} = nondet^{\langle 0 \rangle}, i^{\langle 0 \rangle} = 2\}\}$ of a single suspect f, because all three assignments are necessary to make the formula $\psi \wedge \bigwedge f \wedge \phi$ unsatisfiable.

To make fault localization terminate faster, we introduce an option to immediately return the first found suspect, instead of collecting all.

Violation Witness. Given a CFA (L, l_0, G), a *source-code guard* $S \subseteq G$ is a set of control-flow edges. A *state-space guard* p is a Boolean assumption over program variables X. The type of state-space guards is Φ. A violation-witness automaton [8] is a non-deterministic finite-state automaton $W = (Q, \Sigma_W, \delta, q_0, F)$ with the following components: States Q, alphabet $\Sigma_W = 2^G \times \Phi$, transitions $\delta \subseteq Q \times 2^G \times \Phi \times Q$, initial state $q_0 \in Q$, and accepting states $F \subseteq Q$. The violation-witness automaton gets as input a program execution. For program-execution step $(l, c) \xrightarrow{op_i} (l', c')$, a transition $(q_i, (S, p), q_j)$ is possible if $(l, op_i, l') \in S$ and if c' is a satisfying assignment for p; i.e., formula $p \wedge \bigwedge_{x \in X} x = c'(x)$ is satisfiable. Each state also includes an implicit 'otherwise' self-transition, which only activates if there is no other matching transition from the current state. A violation-witness automaton describes a set of counterexamples by restricting the set of all possible program executions through source-code guards and state-space guards: Source-code guards restrict the control flow, and state-space guards restrict the potential program states. In SV-COMP, a violation witness[5] is described in the GraphML format [16], which is based on XML.

Witness Validation. Given a program P, a specification φ, and a violation witness W, a witness validator [7] checks whether W accepts any program execution on P that violates φ. If it does, W is *confirmed*. Otherwise it is *rejected*.

---

[5] https://gitlab.com/sosy-lab/benchmarking/sv-witnesses/-/blob/0ca5dbf/doc/README-GraphML.md

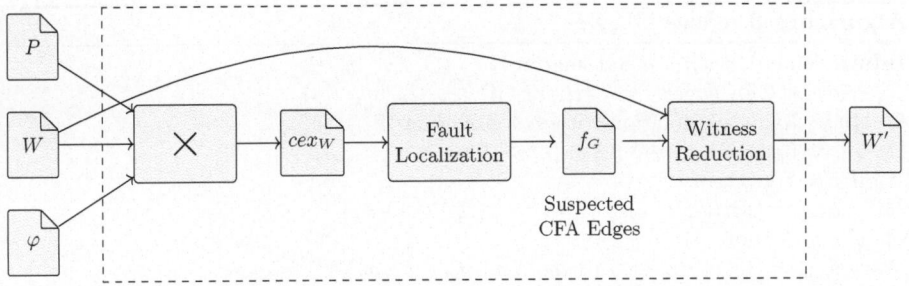

Fig. 4: Workflow of fault localization on violation witnesses. By creating the product automaton of program P, witness W, and specification φ, we obtain counterexample cex_W and apply fault localization to find suspects f_G. With these, witness W is reduced to W'.

3   Fault Localization on Violation Witnesses

Figure 4 illustrates our approach. Given a CFA $P = (L, l_0, G)$, a violation-witness automaton $W = (Q, \Sigma_W, \delta_W, q_0, F_W)$, and a program property $\varphi = (\Omega, \Sigma_\varphi, \delta_\varphi, \omega_0, F_\varphi)$, we (1) reconstruct a counterexample cex_W from W, (2) compute suspects f_G through fault localization on cex_W, and (3) produce a new violation-witness automaton W' that only contains those states and transitions of W that match the program operations contained in f_G.

Reconstruct Counterexample. To reconstruct a counterexample from the witness, we first build the product automaton $P \times W \times \varphi = (Q_\times, \Sigma_\times, \delta_\times, q_{\times 0}, F_\times)$, with states $Q_\times = L \times Q \times \Omega$, alphabet $\Sigma_\times = Ops \times \Sigma_W \times \Sigma_\varphi$, transition relation $\delta_\times$, and initial state $q_{\times 0} = (l_0, q_0, \omega_0)$. The transition relation $\delta_\times$ contains transition $\left(q_\times, (op, (S_W, p), S_\varphi), q_\times'\right)$ with $q_\times = (l, q, \omega)$ and $q_\times' = (l', q', \omega')$ if the following three conditions hold: $(l, op, l') \in G$, $(q, (S_W, p), q') \in \delta_W$, $(\omega, S_\varphi, \omega') \in \delta_\varphi$. The product automaton accepts all states for which the violation-witness automaton W and observer automaton φ agree on a violation. Formally, the set of accepting states is $F_\times = \{(l, q, \omega) \mid q \in F_W, \omega \in F_\varphi\}$.

The product automaton gets as input a program execution. For the program-execution step $(l, c) \xrightarrow{op_i} (l', c')$, a transition $\left(q_\times, (op, (S_W, p), S_\varphi), q_\times'\right) \in \delta_\times$ from state $q_\times = (1, q, \omega)$ to state $q_\times' = (1', q', \omega')$ is possible if all of the following conditions hold: $op_i = op$, $l = 1$, $l' = 1'$, $(l, op_i, l') \in S_W \cap S_\varphi$, and c' fulfills p.

The product automaton's accepting runs describe all counterexamples that are described by the violation-witness automaton, that exist in the program under analysis, and that violate the specification. From these, we select one arbitrary run and use the program-path information of that run as counterexample cex_W.

Fault Localization. We apply fault localization to counterexample cex_W. This returns the set $\mathcal{F}$ of suspects. From that set, we consider the suspect $f = \{\mathrm{TF}(cex_W)\{i\}, \ldots, \mathrm{TF}(cex_W)\{k\}\}$ with the smallest size and map it to the set $f_G = \{(l_i, op_i, l_i'), \ldots, (l_k, op_k, l_k')\}$ of corresponding, suspected CFA edges. If

---

Algorithm 3 $\text{reduce}_c(W, f_G)$

Input: Violation-witness automaton $W = (Q, \Sigma, \delta, q_0, F)$,
    relevant CFA edges $f_G = \{(l_i, op_i, l_i'), \ldots, (l_k, op_k, l_k')\}$
Output: Reduced violation-witness automaton W'
1: $\delta_S = \{(q, (S, p), q') \in \delta \mid S \cap f_G \neq \emptyset\}$
2: **if** $c = full$ **then**
3:    $\delta_{nop} = \{(q, (G, true), q') \mid (q, (S, p), q') \in \delta \setminus \delta_S\}$
4: **if** $c = \Phi$ **then**
5:    $\delta_{nop} = \{(q, (S, true), q') \mid (q, (S, p), q') \in \delta \setminus \delta_S\}$
6: $\delta_{|f_G} = \delta_S \cup \delta_{nop}$
7: $W' = (Q, \Sigma, \delta_{|f_G}, q_0, F)$
8: **return** W'

Algorithm 4 $\text{collapse}(W)$

Input: Violation-witness automaton $W = (Q, \Sigma, \delta, q_0, F)$
Output: Collapsed violation-witness automaton W_c
1: $Q_c = Q$
2: $\delta_c = \delta$
3: $F_c = F$
4: $waitlist = \{q' \in Q \mid (q_0, (\cdot, \cdot), q') \in \delta\}$
5: **while** $waitlist \neq \emptyset$ **do**
6:    choose $q' \in waitlist$
7:    $waitlist = waitlist \setminus \{q'\}$
8:    **if** $\{(q, (S, p), q') \in \delta \mid S \neq G \vee p \neq true\} \neq \emptyset$ **then**
9:       **continue**
10:   $\delta_{q'|nop} = \{(q, (G, true), q') \in \delta_{nop}\}$
11:   **if** $|\delta_{q'|nop}| > 1 \vee q' = q_0$ **then**
12:      **continue**
13:   $Q_c = Q_c \setminus \{q'\}$
14:   $\delta_{del} = \{(q', (S, p), q'') \in \delta_c\}$
15:   $\delta_{adj} = \{(q, (S, p), q'') \mid (q', (S, p), q'') \in \delta_{del}\}$
16:   $\delta_c = \delta_c \cup \delta_{adj} \setminus (\delta_{q'|nop} \cup \delta_{del})$
17:   **if** $q' \in F$ **then**
18:      $F_c = F_c \cup \{q\} \setminus \{q'\}$
19:   $waitlist = waitlist \cup \{q'' \mid (q, (\cdot, \cdot), q'') \in \delta_c\}$
20: $W_c = (Q_c, \Sigma, \delta_c, q_0, F_c)$
21: **return** W_c

---

multiple suspects of the same size exist, an arbitrary suspect is selected. The suspected CFA edges f_G are used to reduce the violation witness.

Witness Reduction. We define three variants of witness reduction. Reduction $r_{\text{state}}(W, f_G) = \text{reduce}_\Phi(W, f_G)$ deletes all state-space guards that are deemed irrelevant by fault localization, but keeps all source-code guards. Reduction $r_{\text{match}}(W, f_G) = \text{reduce}_{full}(W, f_G)$ deletes all state-space guards and source-code guards that are deemed irrelevant by fault localization. Finally, reduction

$r_{all}(W, f_G) = \text{collapse}(\text{reduce}_{full}(W, f_G))$ reduces the witness like r_{match}, but also collapses the resulting witness to produce a smaller violation witness.

Algorithm 3 describes methods reduce$_\Phi$ and reduce$_{full}$: The algorithm first selects those transitions $\delta_S \subseteq \delta$ for which at least one edge in f_G matches the transition's source-code guard S. It then selects all remaining transitions $\delta \setminus \delta_S$. Reduction reduce$_{full}$ then replaces the source-code guards with G (we forget the source-code guard and match everything) and state-space guards with $true$ (line 3). Reduction reduce$_\Phi$ (line 5) only replaces the state-space guards with $true$, but keeps the source-code guards to steer the witness validation. In both cases, we call the resulting set δ_{nop}.

After the initial reduction, reduction r_{all} collapses the witness (Alg. 4). The intuition behind this is that transitions with the trivial guards $(G, true)$ still restrict the set of program paths because they define a minimum number of program operations that must occur between two transitions with non-trivial guards. We remove this implicit restriction on a best-effort basis. We do not perform a precise elimination of δ_{nop} because this may greatly alter the structure of the automaton. Instead, we only collapse sequential sequences of transitions: Algorithm 4 first initializes the states Q_c after collapse, the transitions δ_c after collapse, and the accepting states F_c after collapse, with the original values of W. It then starts traversal of the violation-witness automaton W at all direct successors of q_0. For each visited automaton state q' (lines 6-7), Alg. 4 checks whether q' has any ingoing transition with a non-trivial guard (line 9), has more than one ingoing transition, or is the entry state (line 11). In both cases, Alg. 4 continues with the next state. Otherwise, q' only has a single ingoing edge with only trivial guards, and it can be collapsed into its successors q''. First, q' is removed from Q_c (line 13). Next, Alg. 4 deletes the ingoing and outgoing transitions ($\delta_{q'|nop}$ and δ_{del}) of q' from δ_c, and instead adds new transitions that directly go from q to q'' (lines 14–16). If q' is removed from Q_c and it is an accepting state (line 17), then its predecessor q becomes an accepting state instead (line 18). Algorithm 4 then continues with all successor states (line 19). When no more states can be removed, Alg. 4 exhausts the waitlist by running into line 9 or line 12 for all states. Finally, the violation-witness automaton W_c is defined and returned.

To ensure that the reduced violation witness describes the same set of program paths as the original witness, we do not remove source-code guards that restrict the control flow. In particular, we always keep source-code guards in the GraphML witness that contain one of the following keys: control, enterLoopHead, enterFunction, or returnFromFunction. Our replication package [11] provides experimental results for a witness reduction that removes all source-code guards. The data shows that this reduction is not beneficial since the produced witnesses have a low confirmation rate.

Soundness. Algorithm 3 produces a sound overapproximation W' of W. Both the removal of source-code guards and state-space guards can only increase the set of program paths that are described by a violation-witness automaton, due to the implicit otherwise-self loops at each automaton state. In consequence,

Table 1: Data on the reduction with r_{all} and the fault-localization techniques

Producer	Total	Number of witnesses where fault localization was successful				Average number of edges with		
		MaxSat	MinUnsat	Unsat	Union	MaxSat	MinUnsat	Unsat
2LS	490	134	112	93	138	409	256	728
Bubaak	1 464	1 048	870	267	1 124	17	16	102
CBMC	2 256	605	505	105	630	438	220	466
CPAchecker	1 932	1 276	1 201	340	1 386	762	697	3 269
ESBMC-kind	2 164	587	565	243	716	36	26	4 810
Graves	2 065	1 266	1 097	360	1 405	974	612	4 365
PeSCo	2 147	855	683	269	944	809	670	2 837
Symbiotic	1 418	1 076	897	190	1 088	15	15	25
UAutomizer	1 060	380	343	205	390	272	167	641
UKojak	379	184	179	96	184	85	72	47
UTaipan	869	331	302	177	336	293	189	619
VeriAbs	1 588	404	363	128	438	701	595	2 321
VeriAbsL	1 777	447	410	134	485	686	625	2 300
VeriFuzz	1 747	801	673	144	863	18	24	42
Overall	21 356	9 394	8 200	2 801	10 127	4 379	3 232	18 513
Mean	1 525	671	586	200	723	394	299	1 612
Median	1 668	596	535	184	673	351	205	685

the reduced witness W' is an overapproximation: it describes a superset of the program paths that are described by the original witness W. If the original witness describes a feasible counterexample, then the reduced witness describes the same counterexample (and maybe more). Analogous, Alg. 4 produces a sound overapproximation because the removal of violation-witness-automaton edges can only increase the set of program paths that are described by the violation-witness automaton.

4 Evaluation

We answer the following research questions:

RQ 1 Witness Reduction. Can fault localization reduce the number of transitions in violation-witness automata?
RQ 2 Confirmation Rate. Does the level of detail influence the confirmation rate of violation witnesses?
RQ 3 Reduction Variants. How do the different reduction variants r_{match}, r_{state}, and r_{all} influence the confirmation rate?

4.1 Experiment Setup

We run all experiments on machines with an Intel Xeon E3-1230 v5 @ 3.40 GHz 8-core processor and 33 GB RAM. We limit the machines to only use 2 cores and

7 GB of RAM. We execute all our experiments with BenchExec v3.11[6], a reliable tool for benchmarking [12]. The above setup is equal to the setup of SV-COMP 2023 [4] to ensure comparability between the validation results. The timeout for witness validation is set to 90 s (SV-COMP standard).

We use all 3 225 non-recursive, unsafe verification tasks violating the **unreach-call** property (`reach_error` is reachable) in the sv-benchmarks[7] version of SV-COMP 2023. We use 14 non-hors concours participants in the *Reach-Safety* category. We exclude the verifiers Mopsa [32] and Goblint [37] since they did not produce any violation witnesses. We also exclude the verifier Theta [39] because Theta produces violation-witness automata with invalid source-code guards. We use all 21 356 violation witnesses [5] that the considered participants produce on the verification tasks.

We use the three fault-localization approaches MaxSat, MinUnsat, and Unsat. All three are implemented in CPAchecker and run with a time limit of 900 s. We use CPAchecker in revision 44191[8]. In CPAchecker, we use the SMT solver MathSAT5 [18]. CPAchecker exports a JSON format that describes all located suspects. We match these entries against the source-code guards of the witnesses. We implement the witness reduction algorithms in our tool Flow[9]. All tools, scripts, and evaluation data are available in our replication package [11]. To validate our witnesses, we use the SV-COMP 2023 versions of CPAchecker [10], UAutomizer [25], MetaVal [13], and Symbiotic-Witch [2].

The following evaluation only considers those witnesses for which at least one fault-localization approach works. The left half of Table 1 shows the success rate of the three fault-localization approaches (columns) on each producer's witnesses (rows). Looking at the mean and median values on the bottom of the table, we find that Unsat only reduces approximately 15 % (2 801 of 21 356) of the witnesses. Asking MathSAT5 for an arbitrary UNSAT core fails more often than repeatedly checking for satisfiability of smaller subsets. However, we will later see that for large programs Unsat works better as checking many subsets becomes expensive. MaxSat and MinUnsat, on average, successfully reduce a third of the witnesses. Fault localization is not guaranteed to succeed in some cases, e.g., if we have to deal with pointers, or if the SMT solver times out when querying for (un)satisfiability. Column 'union' displays the number of witnesses where at least one fault localization technique was successful. In total, we successfully apply fault localization to 10 127 different witnesses and consider $3 \cdot (9\,394 + 8\,200 + 2\,801) = 61\,185$ witnesses that were produced across the three reduction approaches.

---

[6] https://github.com/sosy-lab/benchexec/tree/637de81c0d

[7] https://gitlab.com/sosy-lab/benchmarking/sv-benchmarks/-/tree/svcomp22

[8] https://svn.sosy-lab.org/software/cpachecker/trunk/?p=44191

[9] https://gitlab.com/sosy-lab/software/fault-localization-on-witnesses

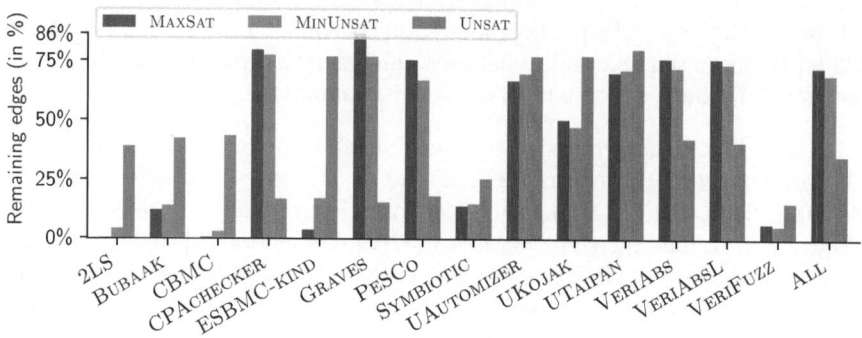

Fig. 5: Number of source-code guards in violation witnesses after applying reduction r_{all} with the respective fault-localization approach; the plot shows the producers of the original witnesses on the x-axis; the lower the bars, the greater the reduction

4.2 Experiment Results

RQ 1: Witness Reduction. Our approach significantly reduces the number of transitions of the original violation-witness automata. To show this, we use reduction variant r_{all}. We also observe a similar picture for variant r_{match}.

Figure 5 shows the averaged reduction of transitions per verifier on successfully reduced tasks per producer. The x-axis names the tools from which we take the original witnesses. For every of the 14 producers we plot 3 bars corresponding to the three fault-localization approaches MaxSat, MinUnsat, and Unsat. A value of 25 % means that the original witnesses are 4 times as big as the reduced witnesses. The bars for one producer do not necessarily talk about the same witnesses because one technique may fail for a task where the others succeed. The lower the bar, the higher the reduction. The last column ALL indicates that Unsat has the highest reduction rate, followed by MinUnsat, and finally MaxSat. ALL shows the ratio of the sum of remaining edges after the reduction and the sum of edges in the original witnesses over *all* producers. Despite Unsat often keeping more edges for individual producers (e.g., 2Ls and Bubaak), it significantly reduces larger witnesses (e.g., CPAchecker and PeSCo) causing less edges to survive on average. For the comparison, we exclusively consider witnesses where fault localization succeeded. Overall, fault localization has a significant impact on the reduction of violation witnesses. On average, MinUnsat and MaxSat remove approximately 30 % of the transitions for the vast majority of witnesses. Unsat reduces the witnesses even to one third of the original size. The right half of Table 1 shows the average number of edges per fault-localization approach (columns) for each verifier (rows). Columns MaxSat, MinUnsat, and Unsat show the average number of transitions in witnesses where the respective fault localization approach worked. We also see why Unsat has, in

```
1  int x = 0;
2  while (x < 20) {
3    x = x + 1;
4  }
5  if (x == 20) {
6    assert(x % 5 != 0);
7  }
```

Fig. 6: Fault localization creates two minimal unsat cores: (1) the assignment x = x + 1 and (2) the condition x == 20 of the then-branch in line 5

general, less remaining edges in the witnesses: the average number of transitions of witnesses where the approach works is higher. One reason is that it does not need to enumerate all possible subsets and can just return an arbitrary unsat core. Depending on which operations are marked relevant by fault localization reduced witnesses have different sizes although the unsat core marks equally many operations as relevant. Consider the program in Figure 6. We initialize variable x with 0 and enter a while loop where we increment x 20 times. Eventually, we exit the loop with x = 20 causing the program to reach an error location since 20 is divisible by 5. A detailed witness describes the complete path with 20 guards at line 3, the assumptions x >= 20 and x < 20 at line 2, and the assumption x == 20 at line 5. Fault localization gives us 2 unsat cores: (1) x = x + 1 at line 3, and (2) x == 20 at line 5. Although all unsat cores are of same size, the reduction of the detailed witness varies. Unsat core (1) leaves 20 transitions while unsat core (2) removes all but one transition on line 5. This property of our reduction techniques causes the varying reduction percentage across different fault-localization approaches since a smaller set of suspects does not automatically lead to a higher reduction.

> Yes, fault localization can significantly reduce the size of violation witnesses.

RQ 2: Confirmation Rate. Our approaches to witness reduction can have both positive and negative effects on the success of witness validation. For fairness reasons, we only evaluate the confirmation rate on witnesses that were deemed syntactically correct by the respective validator before and after reduction. To examine the effect, we consider, for each fault-localization approach, all original violation witnesses from 14 participants of SV-COMP 2023 for which the respective fault localization was successful. We use reduction variant r_{all}. From our data, we showcase the validators UAUTOMIZER and METAVAL because they show the strongest positive and negative effect, respectively: METAVAL benefits the most from the reduction while UAUTOMIZER performs significantly worse after reduction[10].

---

[10] The analysis of the other validators and all reduction approaches is available in our replication artifact [11].

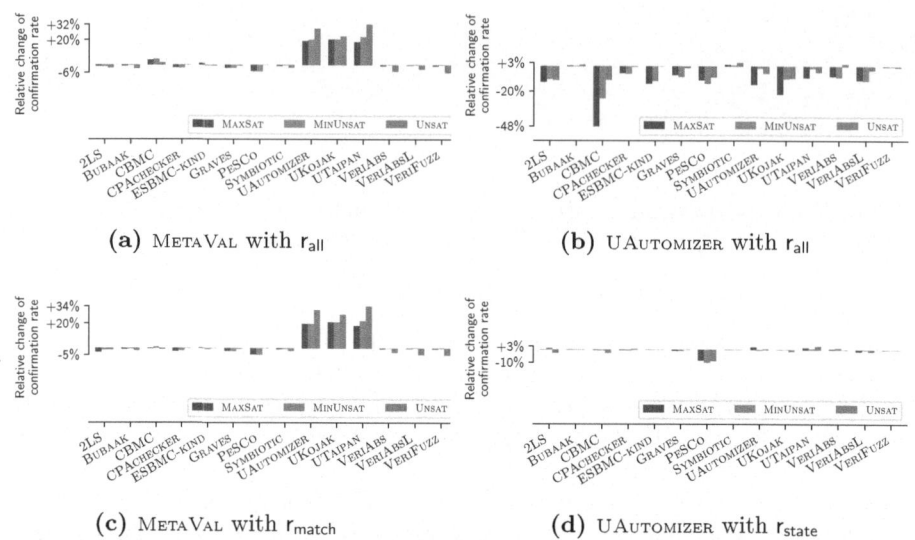

(a) METAVAL with r_{all} (b) UAUTOMIZER with r_{all}

(c) METAVAL with r_{match} (d) UAUTOMIZER with r_{state}

Fig. 7: Relative change of the confirmation rate of METAVAL and UAUTOMIZER per fault-localization approach with different reduction variants

Figure 7a and Fig. 7b show the relative change of the confirmation rate after witness reduction with r_{all} for the validators METAVAL and UAUTOMIZER. Bars that reach positive values indicate a relative increase in the confirmation rate compared to the confirmation rate on the original witnesses. A value of 20 % means that after reduction, 20 % more of the considered witnesses can be confirmed by the respective validator. A value of −20 % means that after reduction, −20 % less of the considered witnesses can be confirmed by the respective validator.

Figure 7a shows that our witness reduction has small negative effects (worst −6 %) on the confirmation rate of METAVAL for some witness producers, but also tremendous positive effects (best 32 %) on the confirmation rate of METAVAL for some other witness producers. METAVAL uses verification engines as backend and annotates information in witnesses directly in the program. Therefore, giving the verifiers more freedom in analysis through more abstract violation witnesses benefits the METAVAL approach. In contrast, UAUTOMIZER performs significantly worse for many fault-localization approaches when reduction variant r_{all} is used. The confirmation rate decreases up to −48 % after reduction. This decreases the strongest for fault-localization with MAXSAT.

The confirmation rate of validators can also decrease, for two reasons: First, the reduction might remove all information and transform the new witness to the trivial witness. With trivial witnesses, validators have no other choice than performing a complete re-verification of the original program. We observe this for producers BUBAAK [17], VERIFUZZ [31], and ESBMC-KIND [21] (cf. Table 1 and Fig. 5). Second, fault localization has no knowledge about the structure of the program and might therefore remove information that would exclude certain branches from being explored. This becomes especially visible when looking at the different reduction techniques. Technique r_{all} has less negative effects on UAUTOMIZER.

> Yes. Depending on the reduction strategy, the confirmation rate of witness validators increases significantly on reduced witnesses.

RQ 3: Reduction Variants. In RQ 1 and RQ 2 we mainly focused on reduction variant r_{all}. In the following, we pick two examples to illustrate the effect that the reduction variant can have on the confirmation rate of reduced witnesses.

Figure 7c shows the confirmation rate of METAVAL with r_{match} across the three considered fault-localization approaches and across all witness producers. It shows that METAVAL performs slightly better with r_{match} compared to r_{all} (Fig. 7a). The highest increase in confirmation rate is 34 %, compared to 32 % with r_{all}.

Figure 7d shows the confirmation rate of UAUTOMIZER with r_{state} across the three considered fault-localization approaches and across all witness producers. It shows that UAUTOMIZER performs better on witnesses that are reduced with r_{state} than with r_{all}: With r_{state}, UAUTOMIZER improves its confirmation rate for all producers. It now only experiences a decrease of -10% compared to the decrease of -48% with r_{all} (Fig. 7b). Additionally, the confirmation rate is now similar to the confirmation rate of the original witnesses for all producers but PESCO [34, 35].

> Our three reduction strategies have different advantages: r_{all} shrinks violation witnesses significantly, while r_{state} keeps the structure of the witness but removes superfluous assumptions. Strategy r_{match} combines the advantages and disadvantages of the other two strategies.

4.3   Threats to Validity

Internal Validity. CPACHECKER exports fault candidates in a JSON format and describes their location by providing the line number and a character offset in the input program. Verifiers export witnesses in a `GraphML` format and also describe file locations of source-code guards with four values, namely `startline`, `endline`, `startoffset`, and `endoffset`. These values should help to uniquely identify statements in the program. However, most witnesses only contain the key `startline` in their guards Therefore, matching the suspects to source-code guards can cause problems as it might be ambiguous.

To prevent the introduction of a bias to the existing witnesses, we apply the reduction directly to the original witnesses with our tool FLOW instead of using the witness export of CPACHECKER. Additionally, we evaluate our approach on 14 verifiers and 4 validators of SV-COMP 23 since all tools in SV-COMP support witnesses. A high number of verifiers and validators minimizes potential biases in our approach towards specific tools.

We re-implement MAXSAT, an established fault localization technique in CPACHECKER. The soundness of MAXSAT was evaluated in the original publication [29]. We are confident that our implementation is sound as well. MINUNSAT and UNSAT are heavily inspired by MAXSAT as seen in the similarities of the algorithms sketched in Algs. 1 and 2.

External Validity. We use the SV-COMP benchmark set for evaluating our approach. It is the largest available benchmarks set for verification tasks in the programming language C. Naturally, every benchmark set has a bias and we cannot be sure if the approach performs as well on other benchmark sets. However, this set is community maintained and everyone can contribute. It consists of a diverse task set.

To run our experiments, we use BENCHEXEC, a tool for reliable benchmarking. Nonetheless, there is always the risk of imprecise measurements. Tasks near our timelimit of 90 s may be flaky. We implement the fault localization and reduction techniques strictly deterministic.

5   Conclusion

We presented three reduction strategies to reduce violation witnesses based on three different fault-localization approaches. An extensive evaluation on 21 356 original violation witnesses of the *Reach-Safety* category of SV-COMP 2023 showed that our reductions can significantly reduce the size of witnesses and increase their confirmation rate. This makes it easier to store and read them, as well as easier to handle them automatically.

Data-Availability Statement. The experiment setup and all experimental data are archived and available at Zenodo [11]. Our supplementary webpage at https://www.sosy-lab.org/research/fl-witnesses provides easy access to the data and additional information.

Funding Statement. This project was funded in part by the Deutsche Forschungsgemeinschaft (DFG) – 378803395 (ConVeY), 418257054 (Coop), and 496588242 (IdeFix).

References

1. Abreu, R., Zoeteweij, P., Golsteijn, R., van Gemund, A.J.C.: A practical evaluation of spectrum-based fault localization. J. Syst. Softw. **82**(11), 1780–1792 (2009). https://doi.org/10.1016/j.jss.2009.06.035
2. Ayaziová, P., Strejček, J.: SYMBIOTIC-WITCH 2: More efficient algorithm and witness refutation (competition contribution). In: Proc. TACAS (2). pp. 523–528. LNCS 13994, Springer (2023). https://doi.org/10.1007/978-3-031-30820-8_30
3. Beyer, D.: A data set of program invariants and error paths. In: Proc. MSR. pp. 111–115. IEEE (2019). https://doi.org/10.1109/MSR.2019.00026

4. Beyer, D.: Competition on software verification and witness validation: SV-COMP 2023. In: Proc. TACAS (2). pp. 495–522. LNCS 13994, Springer (2023). https://doi.org/10.1007/978-3-031-30820-8_29

5. Beyer, D.: Verification witnesses from verification tools (SV-COMP 2023). Zenodo (2023). https://doi.org/10.5281/zenodo.7627791

6. Beyer, D., Chlipala, A.J., Henzinger, T.A., Jhala, R., Majumdar, R.: Generating tests from counterexamples. In: Proc. ICSE. pp. 326–335. IEEE (2004). https://doi.org/10.1109/ICSE.2004.1317455

7. Beyer, D., Dangl, M., Dietsch, D., Heizmann, M., Lemberger, T., Tautschnig, M.: Verification witnesses. ACM Trans. Softw. Eng. Methodol. **31**(4), 57:1–57:69 (2022). https://doi.org/10.1145/3477579

8. Beyer, D., Dangl, M., Dietsch, D., Heizmann, M., Stahlbauer, A.: Witness validation and stepwise testification across software verifiers. In: Proc. FSE. pp. 721–733. ACM (2015). https://doi.org/10.1145/2786805.2786867

9. Beyer, D., Dangl, M., Lemberger, T., Tautschnig, M.: Tests from witnesses: Execution-based validation of verification results. In: Proc. TAP. pp. 3–23. LNCS 10889, Springer (2018). https://doi.org/10.1007/978-3-319-92994-1_1

10. Beyer, D., Keremoglu, M.E.: CPAchecker: A tool for configurable software verification. In: Proc. CAV. pp. 184–190. LNCS 6806, Springer (2011). https://doi.org/10.1007/978-3-642-22110-1_16

11. Beyer, D., Kettl, M., Lemberger, T.: Reproduction package for article 'Fault localization on witnesses'. Zenodo (2024). https://doi.org/10.5281/zenodo.10952383

12. Beyer, D., Löwe, S., Wendler, P.: Reliable benchmarking: Requirements and solutions. Int. J. Softw. Tools Technol. Transfer **21**(1), 1–29 (2019). https://doi.org/10.1007/s10009-017-0469-y

13. Beyer, D., Spiessl, M.: MetaVal: Witness validation via verification. In: Proc. CAV. pp. 165–177. LNCS 12225, Springer (2020). https://doi.org/10.1007/978-3-030-53291-8_10

14. Beyer, D., Kettl, M., Lemberger, T.: Flow: Fault localization on witnesses. https://gitlab.com/sosy-lab/software/fault-localization-on-witnesses (2023), [Online; accessed 22-January-2024]

15. Beyer, D., Kettl, M., Lemberger, T.: Fault localization on verification witnesses (poster paper). In: Proc. ICSE. ACM (2024). https://doi.org/10.1145/3639478.3643099

16. Brandes, U., Eiglsperger, M., Herman, I., Himsolt, M., Marshall, M.S.: GraphML progress report. In: Graph Drawing. pp. 501–512. LNCS 2265, Springer (2001). https://doi.org/10.1007/3-540-45848-4_59

17. Chalupa, M., Henzinger, T.: Bubaak: Runtime monitoring of program verifiers (competition contribution). In: Proc. TACAS (2). pp. 535–540. LNCS 13994, Springer (2023). https://doi.org/10.1007/978-3-031-30820-8_32

18. Cimatti, A., Griggio, A., Schaafsma, B.J., Sebastiani, R.: The MathSAT5 SMT solver. In: Proc. TACAS. pp. 93–107. LNCS 7795, Springer (2013). https://doi.org/10.1007/978-3-642-36742-7_7

19. Ermis, E., Schäf, M., Wies, T.: Error invariants. In: Proc. FM. pp. 187–201. LNCS 7436, Springer (2012). https://doi.org/10.1007/978-3-642-32759-9_17

20. Ernst, G., Blau, J., Murray, T.: Deductive verification via the debug adapter protocol. In: Proença, J., Paskevich, A. (eds.) Proceedings of the 6th Workshop on Formal Integrated Development Environment, F-IDE@NFM 2021, Held online, 24-25th May 2021. EPTCS, vol. 338, pp. 89–96 (2021). https://doi.org/10.4204/EPTCS.338.11

21. Gadelha, M.Y., Ismail, H.I., Cordeiro, L.C.: Handling loops in bounded model checking of C programs via k-induction. Int. J. Softw. Tools Technol. Transf. **19**(1), 97–114 (February 2017). https://doi.org/10.1007/s10009-015-0407-9

22. Gennari, J., Gurfinkel, A., Kahsai, T., Navas, J.A., Schwartz, E.J.: Executable counterexamples in software model checking. In: Proc. VSTTE. pp. 17–37. LNCS 11294, Springer (2018). https://doi.org/10.1007/978-3-030-03592-1_2

23. Groce, A., Visser, W.: What went wrong: Explaining counterexamples. In: Proc. SPIN. pp. 121–135. LNCS 2648, Springer (2003). https://doi.org/10.1007/3-540-44829-2_8

24. Heizmann, M., Barth, M., Dietsch, D., Fichtner, L., Hoenicke, J., Klumpp, D., Naouar, M., Schindler, T., Schüssele, F., Podelski, A.: ULTIMATE AUTOMIZER 2023 (competition contribution). In: Proc. TACAS (2). pp. 577–581. LNCS 13994, Springer (2023). https://doi.org/10.1007/978-3-031-30820-8_39

25. Heizmann, M., Hoenicke, J., Podelski, A.: Software model checking for people who love automata. In: Proc. CAV. pp. 36–52. LNCS 8044, Springer (2013). https://doi.org/10.1007/978-3-642-39799-8_2

26. Jhala, R., Majumdar, R.: Path slicing. In: Proc. PLDI. pp. 38–47. ACM (2005). https://doi.org/10.1145/1065010.1065016

27. Jones, J.A., Harrold, M.J.: Empirical evaluation of the Tarantula automatic fault-localization technique. In: Proc. ASE. pp. 273–282. ACM (2005). https://doi.org/10.1145/1101908.1101949

28. Jose, M., Majumdar, R.: Bug-assist: Assisting fault localization in ANSI-C programs. In: Proc. CAV. pp. 504–509. LNCS 6806, Springer (2011). https://doi.org/10.1007/978-3-642-22110-1_40

29. Jose, M., Majumdar, R.: Cause clue clauses: Error localization using maximum satisfiability. In: Proc. PLDI. pp. 437–446. ACM (2011). https://doi.org/10.1145/1993498.1993550

30. Kölbl, M., Leue, S., Wies, T.: Tartar: A timed automata repair tool. In: Lahiri, S.K., Wang, C. (eds.) Computer Aided Verification. pp. 529–540. Springer International Publishing, Cham (2020). https://doi.org/10.1007/978-3-030-53288-8_25

31. Metta, R., Yeduru, P., Karmarkar, H., Medicherla, R.K.: VERIFUZZ 1.4: Checking for (non-)termination (competition contribution). In: Proc. TACAS (2). pp. 594–599. LNCS 13994, Springer (2023). https://doi.org/10.1007/978-3-031-30820-8_42

32. Monat, R., Ouadjaout, A., Miné, A.: MOPSA-C: Modular domains and relational abstract interpretation for C programs (competition contribution). In: Proc. TACAS (2). pp. 565–570. LNCS 13994, Springer (2023). https://doi.org/10.1007/978-3-031-30820-8_37

33. Müller, P., Ruskiewicz, J.N.: Using debuggers to understand failed verification attempts. In: Proc. FM. pp. 73–87. LNCS 6664, Springer (2011). https://doi.org/10.1007/978-3-642-21437-0_8

34. Richter, C., Hüllermeier, E., Jakobs, M.C., Wehrheim, H.: Algorithm selection for software validation based on graph kernels. Autom. Softw. Eng. **27**(1), 153–186 (2020). https://doi.org/10.1007/s10515-020-00270-x

35. Richter, C., Wehrheim, H.: PESCO: Predicting sequential combinations of verifiers (competition contribution). In: Proc. TACAS (3). pp. 229–233. LNCS 11429, Springer (2019). https://doi.org/10.1007/978-3-030-17502-3_19

36. Rockai, P., Barnat, J.: DivSIM, an interactive simulator for LLVM bitcode. STTT **24**(3), 493–510 (2022). https://doi.org/10.1007/s10009-022-00659-x

37. Saan, S., Schwarz, M., Erhard, J., Pietsch, M., Seidl, H., Tilscher, S., Vojdani, V.: GOBLINT: Autotuning thread-modular abstract interpretation (competition

contribution). In: Proc. TACAS (2). pp. 547–552. LNCS 13994, Springer (2023). https://doi.org/10.1007/978-3-031-30820-8_34

38. Wong, W.E., Debroy, V., Gao, R., Li, Y.: The DStar method for effective software fault localization. IEEE Trans. Reliab. **63**(1), 290–308 (2014). https://doi.org/10.1109/TR.2013.2285319

39. Ádám, Z., Bajczi, L., Dobos-Kovács, M., Hajdu, A., Molnár, V.: THETA: Portfolio of cegar-based analyses with dynamic algorithm selection (competition contribution). In: Proc. TACAS (2). pp. 474–478. LNCS 13244, Springer (2022). https://doi.org/10.1007/978-3-030-99527-0_34

Software Verification

Augmenting Interpolation-Based Model Checking with Auxiliary Invariants

Dirk Beyer⬤, Po-Chun Chien⬤, and Nian-Ze Lee⬤

LMU Munich, Munich, Germany

Abstract. Software model checking is a challenging problem, and generating relevant invariants is a key factor in proving the safety properties of a program. Program invariants can be obtained by various approaches, including lightweight procedures based on data-flow analysis and intensive techniques using Craig interpolation. Although data-flow analysis runs efficiently, it often produces invariants that are too weak to prove the properties. By contrast, interpolation-based approaches build strong invariants from interpolants, but they might not scale well due to expensive interpolation procedures. Invariants can also be injected into model-checking algorithms to assist the analysis. Invariant injection has been studied for many well-known approaches, including k-induction, predicate abstraction, and symbolic execution. We propose an augmented interpolation-based verification algorithm that injects external invariants into *interpolation-based model checking* (McMillan, 2003), a hardware model-checking algorithm recently adopted for software verification. The auxiliary invariants help prune unreachable states in Craig interpolants and confine the analysis to the reachable parts of a program. We implemented the proposed technique in the verification framework CPACHECKER and evaluated it against mature SMT-based methods in CPACHECKER as well as other state-of-the-art software verifiers. We found that injecting invariants reduces the number of interpolation queries needed to prove safety properties and improves the run-time efficiency. Consequently, the proposed invariant-injection approach verified difficult tasks that none of its plain version (i.e., without invariants), the invariant generator, or any compared tools could solve.

Keywords: Software model checking · Program invariants · Invariant injection · Craig interpolation · Data-flow analysis · SMT · SAT

1 Introduction

Assuring that programs execute correctly with respect to their specifications is fundamental for deploying them in our daily lives. In the software industry, testing [2] is the most popular methodology to validate the quality of programs. However, software testing can only spot the presence of bugs in programs but not guarantee their absence. To prove the correctness of programs with mathematical

An extended version of this article is available on arXiv [1].

© The Author(s) 2025
T. Neele and A. Wijs (Eds.): SPIN 2024, LNCS 14624, pp. 227–247, 2025.
https://doi.org/10.1007/978-3-031-66149-5_13

rigorousness, automatic software model checking [3] has been extensively studied. One of the greatest challenges to formally verify a program is to deduce suitable *program invariants* that can be used to prove the program's safety properties. A program invariant is a logical formula over some program variables that must hold at a certain program location for all feasible program paths.

There are numerous approaches to invariant generation, with different performance characteristics and strengths of the produced invariants. Data-flow analysis [4, 5, 6, 7, 8, 9, 10] is a category of methods that sacrifice path sensitivity for scalability. To reduce the size of the abstract reachability graph, data-flow analysis merges abstract program states arising from different execution paths when the control flow meets. Since merging abstract states usually loses information, the resulting invariants might not be strong enough to prove the safety properties. In contrast to data-flow analysis, techniques [11, 12, 13, 14, 15, 16] based on Craig interpolation [17] iteratively derive interpolants[1] to construct strong invariants. Although these approaches are capable of generating useful invariants, they might take too many refinement iterations and fail to converge.

To leverage the information carried by invariants, researchers also use invariant generators as an auxiliary component to aid various model-checking techniques. Algorithms that have been augmented by *invariant injection* include bounded model checking (BMC) [18, 19, 20], k-induction [21, 22, 23, 24, 25], predicate abstraction [26, 27], symbolic execution [28], and IC3/PDR [29]. The idea is to confine the scope of a verification algorithm to the reachable state space of a program, as safety properties often do not hold in the unreachable parts [22].

1.1   Our Research Goal

Despite the extensive study of injecting invariants to various software-verification methods, the possibilities of leveraging auxiliary invariants to assist *interpolation-based model checking* [11], a hardware model-checking algorithm published by McMillan in 2003 and adopted to software recently [30], have not been explored yet. Although this algorithm was invented two decades ago, it still remains state-of-the-art for safety verification. In this paper, our goal is to find out *whether invariant injection can enhance McMillan's interpolation-based model-checking algorithm from 2003 for software verification*. In particular, we aim to reduce the number of interpolation queries required to prove the safety property of a program, since interpolation is usually the most time-consuming step in interpolation-based algorithms. To avoid confusion between McMillan's algorithm from 2003 [11] and other interpolation-based verification approaches, we refer to McMillan's approach from 2003 as IMC from now on. IMC has been combined with expensive SAT-based invariant generators for hardware model checking [31, 32], but its characteristics when assisted by lightweight invariant generators [33] based on data-flow analysis remain unknown for software verification.

To motivate how auxiliary invariants could assist IMC, we briefly introduce the algorithm first. IMC extends BMC via constructing inductive invariants by

---

[1] An interpolant τ for an implication $A \Rightarrow B$ is a formula that uses only the common variables of formulas A and B such that $A \Rightarrow \tau$ and $\tau \Rightarrow B$ are valid.

Craig interpolants [17] arising from unsatisfiable BMC queries. Whenever a BMC query starting from initial states is unsatisfiable, IMC derives an interpolant from the query and replaces the initial states in the query with the interpolant. If the new query is unsatisfiable again, another interpolant can be obtained. The process is repeated until either (1) the newest interpolant is contained in the union of initial states and previous interpolants, or (2) the query starting from the newest interpolant becomes satisfiable. In case (1), an inductive invariant (i.e., a fixed point) is found and the property is proven. In case (2), the satisfied query might be a spurious counterexample, so IMC will continue to unfold the program.

Interpolants tend to abstract away irrelevant information and may intersect with the unreachable state space. Therefore, conjoining them with auxiliary invariants prunes away some unreachable states. In the IMC algorithm outlined above, if the newest interpolant is strengthened by an auxiliary invariant, it is more likely to be contained in the union of initial states and previous interpolants, and the query starting from a strengthened interpolant is less likely to be satisfiable. In other words, a fixed point might be found with fewer interpolation queries in case (1), and a spurious alarm plus extra program unrollings could be avoided in case (2). An example will be shown in Sect. 1.3 to illustrate these benefits.

1.2   Our Contributions

Augmenting IMC with Auxiliary Invariants. We devise two methods to augment IMC with auxiliary invariants. The first method confines the containment check between the newest interpolant and the union of initial states and previous interpolants with auxiliary invariants. The strengthened check is more likely to succeed, so a fixed point can be found with fewer unrollings and interpolation calls. The second method conjoins derived interpolants with auxiliary invariants. The BMC query starting from a strengthened interpolant is more likely to be unsatisfiable, and more interpolants can be derived to form a fixed point before IMC further unrolls the program. We rigorously prove the correctness of the proposed techniques. The proposed augmenting approaches are **novel** because they are the first invariant-injection techniques for IMC applied to program analysis. Moreover, the theoretical results in this paper are applicable to IMC for hardware verification even though we focus on software verification.

Open-Source Implementation and Extensive Evaluation. We implemented the proposed approaches in the open-source framework CPACHECKER [34] and conducted an extensive evaluation on more than 1 600 difficult verification tasks of C programs from the 2022 Intl. Competition on Software Verification [35].[2] In our experiments, the *plain* IMC algorithm (i.e., without invariant injection), three other mature SMT-based algorithms in CPACHECKER, and two other software verifiers were used as references to evaluate our implementation. Our experimental results show that invariant injection (1) effectively reduces the numbers of program unrollings and interpolation queries needed by plain IMC to prove safety

---

[2] Due to space limitation, the pseudocode of the proposed approaches and some experimental results are not included in this article, but in the extended version [1].

properties, (2) reduces the wall-time usage, (3) proves 16 tasks that neither plain IMC nor the data-flow analyzer used to generate invariants could solve, and (4) outperformed other well-established software verifiers. These observations are **significant** because they enhance the knowledge about the effect of auxiliary invariants on IMC. Furthermore, our open-source implementation of the proposed approaches helps other researchers to understand the details of the algorithms better and provides a solid baseline for future studies.

1.3   Motivating Example

We use the example C program in Fig. 1 to explain why invariant injection can reduce the numbers of program unrollings and interpolation queries for IMC. In the program, variables x and i are both initialized to 0. The loop will be executed nondeterministically many times depending on the values returned by function nondet(). The values of x and i are modified in each loop iteration: x will be incremented by 2; if i equals 3, x will be further incremented by 1. Variable i will be incremented by 1 and reset to 0 if its value equals 2 after being incremented. An error occurs if x is odd when the control flow exits the loop.

```
1   extern int nondet();
2   int main(void) {
3       unsigned int x = 0;
4       unsigned int i = 0;
5       while (nondet()) {
6           x += 2;
7           if (i == 3) x++;
8           i++;
9           if (i == 2) i = 0;
10      }
11      if (x % 2) {
12          ERROR: return 1;
13      }
14      return 0;
15  }
```

Fig. 1: An example C program to motivate how auxiliary invariants help IMC

IMC first checks if the error location line 12 is reachable by skipping the loop (i.e., assuming nondet() returns 0). As the conjunction of the initial states $x = 0 \wedge i = 0$ and the guard to the error location $x \% 2 \neq 0$ is unsatisfiable, namely, line 12 is unreachable if the loop is not entered, IMC will try to build an inductive invariant at the loop head line 5 to prove that all paths entering the loop cannot reach the error location. IMC unrolls the program and builds a BMC query[3] starting from the initial location line 3, going through the loop once (i.e., visiting line 5 twice), and ending at the error location line 12. This BMC query is unsatisfiable, so an interpolant at the loop head line 5 can be derived to overapproximate the program states reachable after one loop iteration. Assume the interpolant is $x \% 2 = 0$ (instantiated with variables of the initial states). IMC will replace the initial states $x = 0 \wedge i = 0$ in the BMC query by $x \% 2 = 0$ and pose another query to derive the next interpolant. Unfortunately, the new query is satisfiable, indicating that there exists a feasible path from line 5 to the error location, which assumes x to be even at the beginning of the path and goes through the loop body once. Indeed, a solution to the new query is $(x, i) = (0, 3)$, which leads to $x = 3$ after exiting the

---

[3] The query is $(x = 0 \wedge i = 0) \wedge (i = 3 \Rightarrow x' = x + 3) \wedge (i \neq 3 \Rightarrow x' = x + 2) \wedge (i + 1 = 2 \Rightarrow i' = 0) \wedge (i + 1 \neq 2 \Rightarrow i' = i + 1) \wedge (x' \% 2 \neq 0)$, where the prime symbols indicate variables after a loop iteration, and the returned values of function nondet() are omitted for simplicity.

loop. To decide whether this solution is a spurious counterexample or not, IMC has to further unroll the program and fails to converge to a fixed point under the current unrolling. The reason behind this situation is that the interpolant $x\%2 = 0$ contains unreachable program states allowing the infeasible assignment $i = 3$. In fact, the safety property of the program is fulfilled as variable i never grows beyond 2, and the problematic statement x++; at line 7 is unreachable.

By contrast, it is easy for data-flow analysis, e.g., one based on the abstract domain of intervals [33], to identify that $0 \leq i \leq 1$ is an invariant at line 5 of the program. If this invariant is injected to strengthen the interpolant $x\%2 = 0$, IMC reaches a fixed point $0 \leq i \leq 1 \wedge x\%2 = 0$ immediately, without further unrolling the program. This reasoning is confirmed by our implementation in CPACHECKER, which injects auxiliary invariants produced by a continuously-refining interval analysis [33] into the plain IMC algorithm [30]. The table below summarizes the first interpolants obtained at line 5 and the numbers of unrollings and interpolation queries needed to prove the program in Fig. 1 for the plain and augmented IMC implementations in CPACHECKER.

Algorithm	First interpolant at line 5	Fixed point	#Unrolling	#Itp-queries
Plain IMC [30]	$x\%2 = 0$	No	3	7
Augmented IMC	$0 \leq i \leq 1 \wedge x\%2 = 0$	Yes	1	2

2 Related Work

Our paper is primarily related to invariant generation and verification approaches aided by auxiliary invariants.

2.1 Invariant Generation

Various approaches exist for invariant generation, ranging from lightweight ones based on data-flow analysis to computationally expensive ones based on SAT/SMT solving, predicate abstraction, or Craig interpolation.

Data-flow analysis has been extensively studied [5, 6, 7, 8, 9, 10]. Classic approaches usually consider program variables over an abstract domain structured as a semi-lattice. A standard fixed-point procedure is used to iteratively explore a program, and abstract states reached by different program traces are merged. When the procedure finishes, the merged abstract state at each program location contains the corresponding program invariant. Common choices of abstract domains include a value domain [36, 37] or an interval domain [38].

Interpolation-based approaches construct invariants by either predicates collected from interpolants [12] or interpolants themselves [11, 13, 14, 15]. Although they are able to build strong invariants, the expensive interpolation queries might hinder their scalability. Iterative SAT solving is used to prove the inductiveness of candidate invariants extracted from simulation data [31]. Invariant generation can also be guided by the safety property [39, 40] or the syntax of the program [41].

2.2 Invariant-Aided Verification

Injecting invariants to enhance model-checking algorithms is a popular technique. There are techniques to restrict the state space in BMC queries by high-level design information [19] or data mining [20]. It is well known that the induction hypothesis of k-induction is often too weak on its own and has to be strengthened by auxiliary invariants from static analysis [21, 23] or property-directed reachability [25, 42]. Alternatively, k-induction can be used to prove candidate invariants instantiated from a template, and the confirmed invariants can be used in an induction proof [43]. Recently, an interval-based analysis that produces continuously-refined invariants is successfully applied to boost the performance of k-induction [22]. Transition relations can be strengthened by invariants from the octagon abstract domain to reduce the number of refinement loops in predicate abstraction [27]. Loop invariants can be used as annotations in symbolic execution [28]. Predicate abstraction is also employed to improve candidate invariants formed by Craig interpolants to reduce the number of refinements [44]. In hardware model checking, IMC is combined with inductive invariants to avoid spurious counterexamples [32]. Different from continuously-refined invariants commonly used in software verification, this approach computes a single invariant by SAT solving and uses it repeatedly throughout the entire verification process. SPACER [45], the backend engine of the verification framework SEA-HORN [29], uses invariants produced by the abstract-interpretation tool IKOS [46] to speed up the analysis.

3   Background

In this section, we provide necessary background knowledge for the rest of the paper. All used logical formulas are quantifier-free and belong to the first-order theory of equality with uninterpreted functions, arrays, bit-vectors, and floats. We consider the satisfiability and validity of a logical formula with respect to this theory. A first-order predicate over state variables is interpreted interchangeably as a set of system states that satisfy the predicate.

3.1   Model Checking

First, we recap the problem formulation of model checking. To simplify the presentation in Sect. 4, we describe model checking with the notation of state-transition systems and view a program as a state-transition system. The implementation of the proposed methods, discussed in Sect. 5, represents a program as a control-flow automaton and verifies C programs.

Describing State-Transition Systems. A state-transition system M is characterized by two predicates $I(s)$ and $T(s, s')$ over state variables: $I(s)$ evaluates to true if state s is an initial state of M; $T(s, s')$ evaluates to true if M can transit from state s to state s'. In the following, a state-transition system is represented by $M = (I(s), T(s, s'))$ and state variables after a transition are denoted with a prime. We write $R(s)$ to represent the set of all reachable states of M.

Verifying Safety Properties. Model checking concerns if a state-transition system fulfills a safety property. A safety property can be described as a predicate $P(s)$ that evaluates to true if state s satisfies the property. Given a state-transition system $M = (I(s), T(s, s'))$ and a safety property $P(s)$, M fulfills $P(s)$

if $R(s) \Rightarrow P(s)$ is valid, namely, the property is satisfied by all reachable states of M. Model-checking algorithms receive $M = (I(s), T(s, s'))$ and $P(s)$ as input and aim at proving or disproving $R(s) \Rightarrow P(s)$.

We discuss two criteria widely used by model-checking algorithms to establish $R(s) \Rightarrow P(s)$. First, some approaches construct an *inductive* set $F(s)$ that implies $P(s)$. That is, $F(s)$ must conform to the constraint below:

$$(I(s) \Rightarrow F(s)) \wedge (F(s) \wedge T(s, s') \Rightarrow F(s')) \wedge (F(s) \Rightarrow P(s)).$$

Since $R(s)$ is the smallest inductive set of M, we know that $R(s) \Rightarrow F(s)$ holds. The implication $R(s) \Rightarrow P(s)$ follows from $R(s) \Rightarrow F(s)$ and $F(s) \Rightarrow P(s)$. Interpolation-based approaches such as IMC [11] and interpolation-sequence-based model checking [14] fall into this category because they try to build an inductive state set $F(s)$ from Craig interpolants [17].

The second criterion takes advantage of *invariants*. An invariant $Inv(s)$ of M is a predicate that holds for all reachable states of M, namely, $R(s) \Rightarrow Inv(s)$ is valid. Given an invariant $Inv(s)$ of M, some methods produce a set $G(s)$ that is *relatively inductive* to $Inv(s)$ and implies $P(s)$. In other words, $G(s)$ must fulfill:

$$(I(s) \Rightarrow G(s)) \wedge (Inv(s) \wedge G(s) \wedge T(s, s') \Rightarrow G(s')) \wedge (G(s) \Rightarrow P(s)). \quad (1)$$

Because $Inv(s)$ holds for every reachable state and $G(s)$ is relatively inductive to $Inv(s)$, we have $R(s) \Rightarrow G(s)$. The implication $R(s) \Rightarrow P(s)$ is thus concluded from $R(s) \Rightarrow G(s)$ and $G(s) \Rightarrow P(s)$. IC3/PDR [40] is a prominent example based on this criterion. It chooses $G(s)$ to be the safety property $P(s)$ and generates clauses to form an invariant to which $P(s)$ is relatively inductive. As will be seen in Sect. 4, instead of producing invariants internally during model checking, we leverage external invariant generators and augment IMC to compute a relatively inductive set to the auxiliary invariants.

3.2   Interpolation-Based Model Checking

IMC is an algorithm proposed by McMillan in 2003 [11]. Originally designed for hardware model checking, it has been adopted to verify software programs recently [30]. IMC extends BMC by constructing inductive invariants (i.e., fixed points) from *Craig interpolants* [17].

Craig Interpolation. Given two formulas A and B, if $A \Rightarrow B$, Craig's interpolation theorem [17] assures the existence of a formula τ such that $A \Rightarrow \tau$ and $\tau \Rightarrow B$ are valid, and τ only involves variables appearing in both A and B. τ is called an *interpolant* of A and B as it is logically between A and B. In the model-checking community, Craig's interpolation theorem is usually stated in the equivalent form below.

Theorem 1. *Given an unsatisfiable formula $A \wedge B$, there exists an interpolant τ of this formula such that (1) $A \Rightarrow \tau$ is valid, (2) $\tau \wedge B$ is unsatisfiable, and (3) τ only refers to the common variables of A and B.*

Computational Stages in IMC. IMC [11] has two nested stages in its computation: The outer stage unrolls a state-transition system and poses BMC queries to satisfiability solvers; the inner stage derives Craig interpolants and constructs fixed points. In the following, we name the outer stage *BMC stage* and the inner stage *interpolation stage*.

In the BMC stage, a state-transition system is unfolded into several copies, which is controlled by an unrolling counter. Suppose the value of the counter is k. A BMC query depicting all possible paths from an initial state to a property-violating state via at most k transitions is posed:

$$\underbrace{I(s_0)T(s_0,s_1)}_{A(s_0,s_1)}\underbrace{T(s_1,s_2)\ldots T(s_{k-1},s_k)(\neg P(s_1)\vee\ldots\vee\neg P(s_k))}_{B(s_1,s_2,\ldots,s_k)}. \tag{2}$$

In the above formula, variable s_i denotes the state variable after the i-th transition. If Eq. (2) is satisfiable, a feasible counterexample to the safety property is found. Otherwise, IMC enters the interpolation stage.

In the interpolation stage, a BMC query such as Eq. (2) is partitioned into two formulas A and B. According to Theorem 1, an interpolant $\tau_1(s_1)$ for Eq. (2) exists such that:

$$I(s_0)T(s_0,s_1)\Rightarrow\tau_1(s_1)\text{ is valid and}$$

$$\tau_1(s_1)\wedge\bigwedge_{i=1}^{k-1}T(s_i,s_{i+1})\wedge\bigvee_{i=1}^{k}\neg P(s_i)\text{ is unsatisfiable.}$$

Conceptually, $\tau_1(s_1)$ summarizes the reason why Eq. (2) is unsatisfiable: It overapproximates the set of states that are (1) reachable via one transition from an initial state and (2) do not violate the safety property within $(k-1)$ transitions.

A fixed point of the state-transition system may be constructed by computing such interpolants iteratively. Substituting variable s_0 into τ_1 and replacing the initial states $I(s_0)$ in Eq. (2) by $\tau_1(s_0)$, the interpolation stage of IMC poses another BMC query from the first interpolant τ_1. If the formula is still unsatisfiable, a second interpolant $\tau_2(s_1)$ can be derived, which overapproximates the set of states reachable from an initial state via two transitions. Replacing τ_1 with τ_2, one can derive the next interpolant τ_3 if the BMC query starting from τ_2 is again unsatisfiable. Suppose the above process is repeated n times, and a list of interpolants $\tau_1,\tau_2,\ldots,\tau_n$ is derived. Upon the derivation of the newest interpolant τ_n, IMC performs a *fixed-point check* to decide whether a fixed point has been reached: If τ_n is contained in the union of the initial states and previous interpolants, namely, $\tau_n\Rightarrow I\vee\bigvee_{j=1}^{n-1}\tau_j$ holds, the set $I\vee\bigvee_{j=1}^{n-1}\tau_j$ is an inductive invariant of the system [11]. The inductive invariant is also safe because every interpolant does not violate the safety property according to the second condition of Theorem 1. As a result, IMC concludes the system preserves the safety property.

If a BMC query starting from an interpolant is satisfiable, the safety property is not definitely violated. The corresponding error path might be spurious because the interpolant could involve unreachable states. In this situation, IMC will return to the BMC stage and increment the unrolling counter to confirm the existence of a feasible error path.

4 Augmenting IMC with Auxiliary Invariants

We propose two methods to augment IMC with auxiliary invariants. The first one strengthens the fixed-point checks, and the second one strengthens the interpolants derived in the interpolation stage. Given a state-transition system $M = (I(s), T(s, s'))$, we further assume that auxiliary invariants are inductive. In other words, an invariant $Inv(s)$ satisfies $I(s) \Rightarrow Inv(s)$ and $Inv(s) \wedge T(s, s') \Rightarrow Inv(s')$. This assumption is reasonable in practice because invariant generators, such as those based on data-flow analysis, usually perform a fixed-point iteration and produce inductive invariants.

4.1 Approach 1: Strengthening Fixed-Point Checks

The first approach strengthens fixed-point checks by restricting the newest interpolant with an auxiliary invariant. Let $\tau_1, \ldots, \tau_n$ be a list of interpolants derived in the interpolation stage. Instead of checking $\tau_n \Rightarrow I \vee \bigvee_{j=1}^{n-1} \tau_j$, we strengthen the check to $Inv \wedge \tau_n \Rightarrow I \vee \bigvee_{j=1}^{n-1} \tau_j$. The intuition is to confine the scope of fixed-point checks mainly within reachable states of the system such that the union of the initial states and previous interpolants is more likely to contain the newest interpolant. The correctness of this approach is stated in Theorem 2.

Theorem 2. *Given a state-transition system $M = (I(s), T(s, s'))$ and a safety property $P(s)$, let $\tau_1, \ldots, \tau_n$ be a list of interpolants derived in the interpolation stage of IMC. If the strengthened fixed-point check $Inv \wedge \tau_n \Rightarrow I \vee \bigvee_{j=1}^{n-1} \tau_j$ holds for some auxiliary invariant Inv, M fulfills the safety property $P(s)$.*

Proof. We rely on the criterion described by Eq. (1) to show that M fulfills P. Defining G to be $I \vee \bigvee_{j=1}^{n-1} \tau_j$, we will prove that G satisfies Eq. (1).

Proving $I \Rightarrow G$ is trivial. Moreover, Theorem 1 guarantees that $\tau_1, \ldots, \tau_n$ do not violate the safety property, which assures that $G \Rightarrow P$ holds.

To show $Inv(s) \wedge G(s) \wedge T(s, s') \Rightarrow G(s')$ is valid, recall $I(s) \wedge T(s, s') \Rightarrow \tau_1(s')$ and $\tau_j(s) \wedge T(s, s') \Rightarrow \tau_{j+1}(s')$ for $j = 1, \ldots, n-1$ both hold according to Theorem 1. Combining these conditions and the inductiveness of the auxiliary invariant Inv, we simplify the implication to

$$Inv(s) \wedge (I(s) \vee \bigvee_{j=1}^{n-1} \tau_j(s)) \wedge T(s, s') \Rightarrow Inv(s') \wedge (\bigvee_{j=1}^{n-1} \tau_j(s') \vee \tau_n(s')).$$

Since the strengthened fixed-point check $Inv \wedge \tau_n \Rightarrow I \vee \bigvee_{j=1}^{n-1} \tau_j$ holds, the right-hand side of the above implication further implies $(Inv(s') \wedge \bigvee_{j=1}^{n-1} \tau_j(s')) \vee (I(s') \vee \bigvee_{j=1}^{n-1} \tau_j(s'))$, which equals $G(s')$.

Therefore, we proved that G satisfies Eq. (1), and hence M fulfills P.

4.2   Approach 2: Strengthening Interpolants

The second approach strengthens interpolants by conjoining them with an auxiliary invariant. Unlike the first approach, which only restricts the newest interpolant with an auxiliary invariant in a fixed-point check, the second approach replaces the original interpolants returned by the interpolation procedure with the strengthened ones. That is, given a BMC query starting from the initial states I or a previously strengthened interpolant $\tau_j \wedge Inv$, the interpolant τ_{j+1} is replaced by $\tau_{j+1} \wedge Inv$. Note that if interpolants are strengthened, fixed-point checks are effectively strengthened as well. Theorem 3 states that $\tau_{j+1} \wedge Inv$ is also an interpolant for the BMC query, and hence the correctness of the approach follows from the plain IMC algorithm.

Theorem 3. *Given a transition system $M = (I(s), T(s, s'))$ and a safety property $P(s)$, consider an unsatisfiable BMC query $\lambda(s_0) \wedge T(s_0, s_1) \wedge \ldots \wedge T(s_{k-1}, s_k) \wedge \bigvee_{i=1}^{k} \neg P(s_i)$ posed in the interpolation stage of IMC, where $\lambda(s_0)$ is either $I(s_0)$ or a previously strengthened interpolant $\tau_j(s_0) \wedge Inv(s_0)$. Suppose $\tau_{j+1}(s_1)$ is an interpolant of the BMC query with $\lambda(s_0) \wedge T(s_0, s_1)$ labeled as formula A and $T(s_1, s_2) \wedge \ldots \wedge T(s_{k-1}, s_k) \wedge \bigvee_{i=1}^{k} \neg P(s_i)$ labeled as formula B. Then $\tau_{j+1}(s_1) \wedge Inv(s_1)$ is also an interpolant of the BMC query.*

Proof. To prove that $\tau_{j+1}(s_1) \wedge Inv(s_1)$ is also an interpolant, we have to show that it satisfies the three conditions in Theorem 1. Namely, (1) $\lambda(s_0) \wedge T(s_0, s_1) \Rightarrow \tau_{j+1}(s_1) \wedge Inv(s_1)$ is valid; (2) $\tau_{j+1}(s_1) \wedge Inv(s_1) \wedge T(s_1, s_2) \wedge \ldots \wedge T(s_{k-1}, s_k) \wedge \bigvee_{i=1}^{k} \neg P(s_i)$ is unsatisfiable; and (3) $\tau_{j+1}(s_1) \wedge Inv(s_1)$ only refers to the common variables of formulas A and B. Condition (2) holds because $\tau_{j+1}(s_1) \wedge T(s_1, s_2) \wedge \ldots \wedge T(s_{k-1}, s_k) \wedge \bigvee_{i=1}^{k} \neg P(s_i)$ is already unsatisfiable according to Theorem 1. Condition (3) holds because $\tau_{j+1}(s_1) \wedge Inv(s_1)$ only uses the common variable s_1. In the following, we prove condition (1) by splitting into two cases.

When $\lambda(s_0) = I(s_0)$, our proof goal is $I(s_0) \wedge T(s_0, s_1) \Rightarrow \tau_{j+1}(s_1) \wedge Inv(s_1)$. Applying Theorem 1 to τ_{j+1}, we have $I(s_0) \wedge T(s_0, s_1) \Rightarrow \tau_{j+1}(s_1)$. Since Inv is an invariant and hence contains all reachable states, we also have $I(s_0) \wedge T(s_0, s_1) \Rightarrow Inv(s_1)$. Therefore, our proof goal is achieved.

When $\lambda(s_0) = \tau_j(s_0) \wedge Inv(s_0)$, our proof goal is $\tau_j(s_0) \wedge Inv(s_0) \wedge T(s_0, s_1) \Rightarrow \tau_{j+1}(s_1) \wedge Inv(s_1)$. Applying Theorem 1 to τ_{j+1}, we have $\tau_j(s_0) \wedge Inv(s_0) \wedge T(s_0, s_1) \Rightarrow \tau_{j+1}(s_1)$. Since Inv is inductive, we also have $Inv(s_0) \wedge T(s_0, s_1) \Rightarrow Inv(s_1)$. Our goal follows from the two implications.

As both cases are proved, we conclude that $\tau_{j+1}(s_1) \wedge Inv(s_1)$ is also an interpolant of the BMC query and can be used in the interpolation stage of IMC.

4.3   Comparison between the Two Approaches

The design spirit behind the two approaches is to make IMC capable of proving the safety of a system with fewer iterations in the BMC and interpolation stages. We attain this objective with the help of auxiliary invariants. In the first approach, instead of an inductive invariant, IMC generates a relatively inductive set of states, which might require fewer interpolation calls to produce. In the second approach, in addition to strengthening fixed-point checks, strengthened interpolants also make BMC queries in the interpolation stage more likely to be unsatisfiable. That is, IMC will remain in the interpolation stage more often, searching for a proof of the safety property. The second approach is more "aggressive" than the first one because strengthened interpolants will change the interpolants obtained later, and IMC will follow a different computational footprint. By contrast, the interpolants encountered in the first approach are identical to the plain IMC algorithm if the interpolation procedure is the same. Moreover, the first approach can in principle be adopted by other verification algorithms with similar fixed-point checks, whereas the second approach is specifically tailored towards IMC. In our evaluation, we also experimented with a basic injection approach that conjoins the safety property with auxiliary invariants. However, this approach performed roughly the same as the plain IMC algorithm.

5   Implementation

We implemented the two approaches discussed in Sect. 4 on top of the plain IMC algorithm in CPACHECKER [34], a software-verification framework for the programming language C. The pseudo code of the implemented procedure is outlined in the extended technical report [1]. The plain IMC implementation [30] in CPACHECKER extracts transition relations from programs with *large-block encoding* [47]. Error labels in a program are used to specify the negation of safety properties. In addition to an IMC implementation, CPACHECKER also has a continuously-refining invariant generator based on intervals [33], which is used to enhance k-induction [22]. Before continuing the discussion, we emphasize that the proposed approaches discussed in Sect. 4 are independent of the implementation framework and not limited to specific invariant generators.

CPACHECKER is based on *configurable program analysis* (CPA) [48], which defines an abstract domain used in a program analysis. The plain IMC algorithm is implemented using several CPAs, with one of them tracking path formulas between program locations in its abstract states [49]. This CPA is configured to perform large-block encoding, which constructs formulas for complete loop unrollings as transition relations [30]. Thanks to its flexibility, it underpins the implementations of other SMT-based algorithms in CPACHECKER [50], including predicate analysis, IMPACT, and k-induction. We will compare the proposed algorithm to these techniques in Sect. 6.

The continuously-refining invariant generator in CPACHECKER uses a CPA with an abstract domain based on expressions over intervals [33]. Compared to methods concerning single intervals only, it can represent complex ranges like the disjunction and conjunction of intervals, e.g., $(x < 5 \lor x > 7) \land (1 \leq y \leq 8)$. Its precision includes a set of important variables and a maximum depth of expressions. The precision can be dynamically adjusted during the analysis [51] to produce more refined invariants.

Given a CPA and an initial abstract state, CPACHECKER constructs a set of reachable abstract states by iteratively processing states and computing their abstract successors. To perform a specific algorithm, required information can be collected from the abstract states. For example, IMC collects the path formulas to assemble BMC queries and derive interpolants [30]. The interval-based invariant generator outputs the union of expressions as invariants [33]. In our implementation, we run IMC and the invariant generator in parallel and inject auxiliary invariants to augment IMC.

6   Evaluation

To understand the effects of the proposed invariant-injection approaches on the proof-finding ability of IMC, we pose the following research questions:

- Part 1: augmented IMC vs. plain IMC
  - **RQ1**: Can auxiliary invariants reduce the numbers of program unrollings and interpolation queries?
  - **RQ2**: Can augmented IMC prove the correctness of additional programs?
  - **RQ3**: Can invariant injection improve the run-time efficiency?
- Part 2: augmented IMC vs. other approaches and tools
  - **RQ4**: Can augmented IMC find more proofs than other state-of-the-art software-verification algorithms and tools?

6.1   Evaluated Approaches and Tools

To answer the above research questions, we evaluated the proposed invariant-injection methods for IMC against (1) its plain version plus three other SMT-based algorithms in the same verification framework CPACHECKER [34] and (2) two other state-of-the-art software verifiers.

The plain IMC algorithm was recently adopted to verify programs [30]. The three other approaches in CPACHECKER are predicate abstraction [12,52], IMPACT [13], and k-induction boosted by continuously-refined invariants [22], whose characteristics were compared in a recent article on SMT-based software verification [50]. All proposed methods and compared algorithms above are implemented in the same framework, so the confounding variables are kept to a minimum (identical frontend, program encodings, SMT solvers, etc.) to facilitate the comparison of algorithmic differences. For invariant generation, we used the continuously-refining data-flow analysis (DF) [33] described in Sect. 5. The invariant generator can prove the safety properties of some programs on its own. In our experiments, we ignored the answers computed by DF because our goal is to study the effects of auxiliary invariants on the main analyses, namely, IMC and k-induction. In the following, we denote the injection of continuously-refined invariants generated by DF into k-induction as KI←⊕-DF, and the augmented IMC algorithms with fixed-point checks and interpolants strengthened by auxiliary invariants as IMC$_f$←⊕-DF and IMC$_i$←⊕-DF, respectively.

To reflect the state of the art on software verification and invariant injection, we further compared our approaches to 2LS [23], which has a mature implementation of k-induction boosted by auxiliary invariants, and SYMBIOTIC [53], which is based on symbolic execution and performs well in the Intl. Competitions on Software Verification [35] (the overall winner in 2022).

6.2   Benchmark Set

Since our goal is to investigate whether auxiliary invariants can improve IMC's capability of delivering correctness proofs, we selected verification tasks without known violation to their reachability-safety properties from the 2022 Intl. Competition on Software Verification (SV-COMP '22) [35]. Furthermore, to concentrate the evaluation on hard verification problems, we excluded the trivial tasks solvable by BMC of CPACHECKER. To keep the program encodings consistent across all evaluated approaches in CPACHECKER, we only considered single-loop programs in our experiments because the IMC implementation in CPACHECKER needs to transform a multi-loop program into a single-loop program as preprocessing [30]. However, this is not a limitation of the proposed augmenting methods, since they can work on multi-loop programs as well if single-loop transformation [54, 55] is applied. In total, we collected 1 623 verification tasks, each of which contains a single-loop program and a challenging safety property.

6.3   Experimental Setup

All experiments were conducted on machines having 33 GB of RAM and a 3.4 GHz CPU (Intel Xeon E3-1230 v5) with 8 processing units. The operating system was Ubuntu 22.04 (64 bit), running Linux 5.15. Each verification task was limited to 4 CPU cores, 15 min of CPU time, and 15 GB of RAM. In our evaluation, we utilized the benchmarking framework BENCHEXEC [56] to ensure reproducibility of our results, and CPACHECKER at revision 42901. All SMT and interpolation queries in CPACHECKER were handled by MATHSAT5 [57]. Since we want to observe IMC's behavior in the presence of auxiliary invariants, we limited the CPU time allocated to the invariant generator to 2.5 min such that IMC had enough time to perform its analysis. For the comparison to other software verifiers, we downloaded 2LS and SYMBIOTIC from the tool archives of SV-COMP '22 [58].

6.4   Results

RQ1: Reduction of program unrollings and interpolation queries. To understand whether the invariant-injection techniques could reduce the numbers of program unrollings and interpolation queries required by plain IMC to find a proof, we conducted a case study on $IMC_i{\leftarrow}{\ominus}$-DF. As discussed in Sect. 4.3, $IMC_i{\leftarrow}{\ominus}$-DF derives different interpolants from plain IMC and is supposed to exhibit distinct computational behavior. We further identified 870 tasks for which DF was able to generate non-trivial and inductive invariants[4] because trivial invariants provide no additional information to help IMC. The scatter plots Fig. 2a and Fig. 2b show the comparison of $IMC_i{\leftarrow}{\ominus}$-DF and IMC on the 870 tasks on the numbers of program unrollings and interpolation queries, respectively. A data point (x, y) in the plots means that there is a task solvable by both IMC and $IMC_i{\leftarrow}{\ominus}$-DF, and IMC (resp. $IMC_i{\leftarrow}{\ominus}$-DF) requires x (resp. y) times of the indicated operation. The color of a data point shows the number of tasks falling into this coordinate. Data points under the diagonal represent the tasks for which $IMC_i{\leftarrow}{\ominus}$-DF needed fewer operations than plain IMC. Observe

---

[4] The trivial invariant $\top$ represents the entire state space.

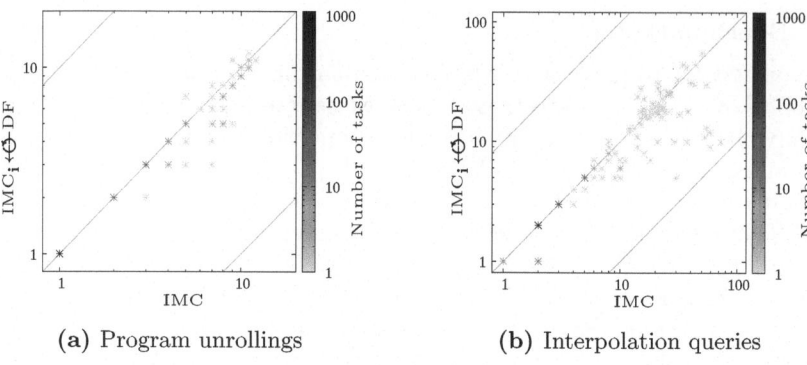

(a) Program unrollings   (b) Interpolation queries

Fig. 2: Comparison of the numbers of program unrollings and interpolation queries

Table 1: Statistics of tasks where $IMC_i \leftrightarrow DF$ has significant improvement over IMC (time unit: s with two significant digits; ✓ for solved, and − for timeout)

Task	IMC					$IMC_i \leftrightarrow DF$				
	result	#unroll	#itp	wall-time	itp-time	result	#unroll	#itp	wall-time	itp-time
Problem03_label03	−	8	50	870	740	✓	5	11	27	13
Problem03_label15	−	7	54	880	750	✓	5	11	35	19
Problem03_label51	−	10	57	880	750	✓	5	11	24	10
benchmark37_conj	−	317	316	890	850	✓	1	2	1.8	0.0080
pals_lcr-var-start	−	11	16	880	630	✓	10	24	760	480
pals_lcr.4.ufo.UNB	✓	11	23	200	160	✓	10	16	120	92
phases_2-2	−	1	1	900	900	✓	1	2	2.2	0.057
s3_srvr_1a.BV.c.cil	−	64	441	890	540	✓	5	13	6.3	1.0
s3_srvr_2a.BV.c.cil	−	39	290	890	790	✓	35	252	640	540
s3_srvr_2a_alt.BV	−	37	278	880	780	✓	19	125	72	52

that $IMC_i \leftrightarrow DF$ often required fewer program unrollings and interpolation calls. For programs correctly proved by both, our augmenting approach reduced the number of program unrollings in 35 tasks, and the number of interpolation calls in 56 tasks.

Table 1 shows the tasks for which $IMC_i \leftrightarrow DF$ exhibited significant performance improvement. (A similar table for $IMC_f \leftrightarrow DF$ can be found in the extended technical report [1].) Upon these tasks, plain IMC either ran into timeouts or required considerably more run-time, whereas the augmented version was able to deliver proofs efficiently. Under each algorithm, the table lists the verification result (solved or timeout), the numbers of program unrollings and interpolation queries, wall-time, and interpolation time (the time spent on all interpolation queries). Note that since plain IMC suffered from timeouts in most of these tasks, the reported numbers of unrollings and interpolation queries are lower bounds of the actual numbers required to compute proofs, which could be much higher.

RQ2: Effectiveness of augmented IMC vs. plain IMC. We compared the number of tasks solved by augmented IMC versus that by plain IMC to evaluate the effectiveness of the proposed approaches. Figure 3b shows the results on the tasks with non-trivial invariants. A data point (x, y) in the plots indicates that x tasks were correctly solved by the respective algorithm within a time bound of y seconds. (The exact numbers can be found in the extended

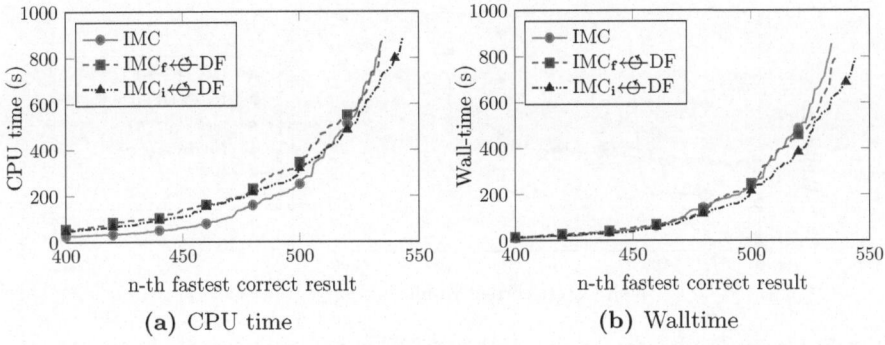

(a) CPU time   (b) Walltime

Fig. 3: Quantile plots comparing the run-time of plain and augmented IMC

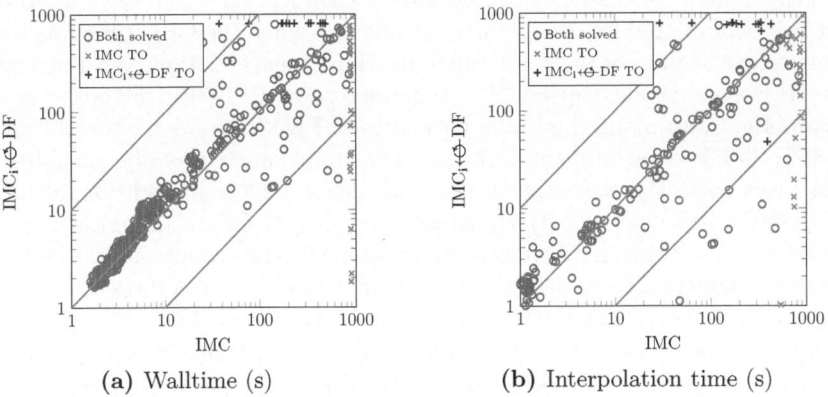

(a) Walltime (s)   (b) Interpolation time (s)

Fig. 4: Comparison of wall-time and interpolation time (TO stands for timeout)

technical report [1].) Observe that both augmented approaches solved more tasks than plain IMC. Specifically, $IMC_i \leftarrow \ominus\text{-}DF$ (resp. $IMC_f \leftarrow \ominus\text{-}DF$) found proofs for 23 (resp. 6) tasks where plain IMC ran into timeouts. However, it was not able to solve 13 (resp. 4) tasks proven by plain IMC, partially due to the extra time spent on invariant generation (see Sect. 6.5 for more discussion). Overall, $IMC_i \leftarrow \ominus\text{-}DF$ (resp. $IMC_f \leftarrow \ominus\text{-}DF$) proved 10 (resp. 2) additional tasks compared to plain IMC, and there were 16 tasks solvable by $IMC_i \leftarrow \ominus\text{-}DF$ but not by plain IMC or DF. Although not extraordinary, the increase shows invariant injection can improve the effectiveness of plain IMC.

To account for the randomness in SMT solving and interpolation, we repeated the experiment with five different random seeds for the underlying SMT solver. In our evaluation, $IMC_i \leftarrow \ominus\text{-}DF$ consistently delivered more proofs than plain IMC with each seed, demonstrating the robustness of the proposed invariant-injection approaches. The results are available in the extended technical report [1].

RQ3: Run-time efficiency of augmented IMC vs. plain IMC. Quantile plots comparing the CPU time and wall-time usage of plain and augmented IMC on the tasks with non-trivial invariants are shown in Fig. 3a and Fig. 3b, respectively. To focus on the performance differences for difficult tasks, we crop the first 400 tasks

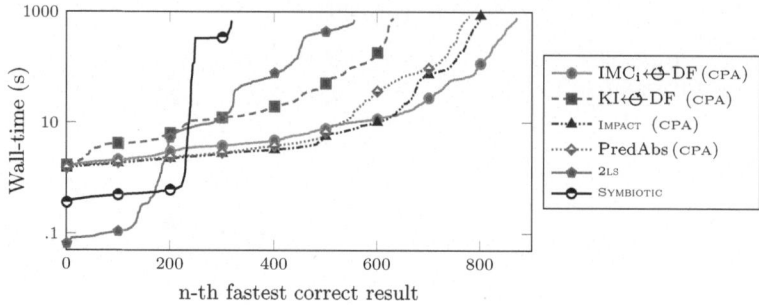

Fig. 5: Quantile plot comparing $IMC_i \leftarrow \hookleftarrow$-DF with other approaches and tools

from the figures because they can be solved within 1 min. From Fig. 3a, observe that IMC dominated its two augmented variants when the CPU time was below 400 s because it did not have a parallel invariant generator that consumed extra time. However, as the elapsed CPU time increased, $IMC_i \leftarrow \hookleftarrow$-DF took over and became the best among the three. By contrast, Fig. 3b shows a clear advantage of $IMC_i \leftarrow \hookleftarrow$-DF regarding the wall-time consumption. It was able to solve the most tasks when the elapsed wall-time was beyond 200 s. In addition, although $IMC_f \leftarrow \hookleftarrow$-DF did not not perform as well as $IMC_i \leftarrow \hookleftarrow$-DF on these tasks, it still showed some wall-time improvement over plain IMC. The results indicate that our proposed augmenting approaches can not only prove the correctness of additional programs but also reduce the wall-time for solving difficult tasks.

The scatter plots in Fig. 4 compare the wall-time and interpolation time consumed by IMC and $IMC_i \leftarrow \hookleftarrow$-DF on the tasks with non-trivial invariants. A data point (x, y) in the plots indicates a task that can be solved by IMC or $IMC_i \leftarrow \hookleftarrow$-DF, for which the former took x seconds of wall-time (resp. interpolation time), and the latter took y seconds of wall-time (resp. interpolation time). Observe that invariant injection helps improve plain IMC's wall-time efficiency for time-consuming tasks because more data points are below the diagonal in the top-right quarter of Fig. 4a. For interpolation time, invariant injection generally lowered the summation of time spent on all queries, as can be seen from Fig. 4b. Moreover, there were 20 tasks showing reduction by an order of magnitude.

RQ4: Comparison with other algorithms and tools. To study whether the proposed invariant-injection approaches for IMC could deliver more proofs than other mature SMT-based methods in CPACHECKER as well as other software verifiers, we evaluated $IMC_i \leftarrow \hookleftarrow$-DF against KI$\leftarrow \hookleftarrow$-DF [22], IMPACT [13], and predicate abstraction (PredAbs) [12, 52] in CPACHECKER, along with 2LS [23] and SYMBIOTIC [53].[5] Figure 5 plots the wall-time usage of the compared algorithms and tools, and shows that $IMC_i \leftarrow \hookleftarrow$-DF solved the most tasks among them. (The exact numbers are summarized in the extended technical report [1].) For the comparison within CPACHECKER, $IMC_i \leftarrow \hookleftarrow$-DF outperformed other approaches in terms of both the number of solved tasks and the run-time efficiency. There were 324, 88, and

---

[5] Plain IMC has been compared with other approaches in a technical report [30].

143 additional tasks solvable by IMC_i⊬⊖-DF, but not by KI⊬⊖-DF, IMPACT, and predicate abstraction, respectively. In total, IMC_i⊬⊖-DF uniquely solved 10 tasks among the four compared algorithms. For the comparison with 2LS and SYMBIOTIC, IMC_i⊬⊖-DF found significantly more proofs than both of them. In particular, 423 and 624 additional tasks were solved by IMC_i⊬⊖-DF, but not by 2LS and SYMBIOTIC, respectively, and a total of 342 tasks were uniquely solved by IMC_i⊬⊖-DF. The results demonstrate the value of our invariant-injection methods for IMC and justify our contribution to the state of the art of software verification.

6.5   Discussion

In our experiments, strengthened interpolants decreased the number of interpolation queries (Fig. 2b) and the total interpolation time (Fig. 4b), but the effect came at a price: SMT and interpolation queries may become more challenging. There were several tasks solvable by plain IMC in a minute but stuck at one difficult SMT query with a strengthened interpolant for hundreds of seconds. Therefore, the effect of invariant injection on IMC is mixed: A similar trade-off between fewer refinement steps and extra computation effort to achieve the reduction was reported for the verifier UFO [44]. In the evaluation, we observed a clear decrease in the numbers of unrollings and interpolation calls, but the number of solved tasks did not increase remarkably. Similar observations that the run-time does not necessarily benefit from auxiliary invariants and that the number of solved tasks might not increase significantly have also been reported in previous publications [23, 32] on invariant-aided verification.

7   Conclusion

We augmented IMC, an interpolation-based model-checking algorithm published by McMillan in 2003 [11], via injecting auxiliary invariants to reduce the numbers of program unrollings and interpolation queries needed to prove the correctness of programs. Invariants are used to strengthen (1) the checks for determining whether a fixed point is reached, or (2) the interpolants derived during the procedure. We rigorously proved the correctness of the proposed approaches and implemented both techniques in the verification tool CPACHECKER. We empirically evaluated our implementations against four SMT-based verification approaches in CPACHECKER and two state-of-the-art software verifiers over C programs whose safety properties are hard to prove. Our experiments show that the proposed techniques effectively reduced the numbers of program unrollings and interpolation calls and therefore reduced the wall-time usage compared to plain IMC. Furthermore, the proposed augmentation helped IMC deliver more proofs than the compared SMT-based algorithms in CPACHECKER and solve 342 tasks unsolvable by the compared tools. For future work, as the strengthened interpolants might lead to extra time spent on SMT solving or interpolation, we plan to devise a strategy to selectively inject invariants into IMC only when they are likely to be helpful, in order to further improve the performance of the proposed methods.

Data-Availability Statement. To enhance the verifiability and transparency of the paper, all relevant materials, including used tools, benchmark tasks, and raw experimental data, are available in a supplemental reproduction package [59]. More information is available at https://www.sosy-lab.org/research/imc-df/.

Funding Statement. This project was funded in part by the Deutsche Forschungs-gemeinschaft (DFG) – 378803395 (ConVeY).

References

1. Beyer, D., Chien, P.C., Lee, N.Z.: Augmenting interpolation-based model checking with auxiliary invariants (Extended version). arXiv/CoRR **2403**(07821) (March 2024). https://doi.org/10.48550/arXiv.2403.07821
2. Myers, G.J., Sandler, C., Badgett, T.: The Art of Software Testing. Wiley, 3rd edn. (2011). https://www.worldcat.org/isbn/978-1-119-20248-6
3. Jhala, R., Majumdar, R.: Software model checking. ACM Computing Surveys **41**(4) (2009). https://doi.org/10.1145/1592434.1592438
4. Kildall, G.A.: A unified approach to global program optimization. In: Proc. POPL. pp. 194–206. ACM (1973). https://doi.org/10.1145/512927.512945
5. Kam, J., Ullman, J.: Global data-flow analysis and iterative algorithms. J. ACM **23**, 158–171 (1976). https://doi.org/10.1145/321921.321938
6. Sharir, M., Pnueli, A.: Two approaches to interprocedural data-flow analysis. In: Program Flow Analysis: Theory and Applications. pp. 189–233. Prentice-Hall (1981). https://www.worldcat.org/isbn/978-0-137-29681-1
7. Kennedy, K.: A survey of data-flow analysis techniques. In: Program Flow Analysis: Theory and Applications, pp. 5–54. Prentice Hall (1981). https://www.worldcat.org/isbn/978-0-137-29681-1
8. Jones, N.D., Muchnick, S.S.: A flexible approach to interprocedural data-flow analysis and programs with recursive data structures. In: Proc. POPL. pp. 66–74. ACM (1982). https://doi.org/10.1145/582153.582161
9. Ryder, B.G.: Incremental data-flow analysis. In: Proc. POPL. pp. 167–176. ACM (1983). https://doi.org/10.1145/567067.567084
10. Reps, T.W., Horwitz, S., Sagiv, M.: Precise interprocedural data-flow analysis via graph reachability. In: Proc. POPL. pp. 49–61. ACM (1995). https://doi.org/10.1145/199448.199462
11. McMillan, K.L.: Interpolation and SAT-based model checking. In: Proc. CAV. pp. 1–13. LNCS 2725, Springer (2003). https://doi.org/10.1007/978-3-540-45069-6_1
12. Henzinger, T.A., Jhala, R., Majumdar, R., McMillan, K.L.: Abstractions from proofs. In: Proc. POPL. pp. 232–244. ACM (2004). https://doi.org/10.1145/964001.964021
13. McMillan, K.L.: Lazy abstraction with interpolants. In: Proc. CAV. pp. 123–136. LNCS 4144, Springer (2006). https://doi.org/10.1007/11817963_14
14. Vizel, Y., Grumberg, O.: Interpolation-sequence based model checking. In: Proc. FMCAD. pp. 1–8. IEEE (2009). https://doi.org/10.1109/FMCAD.2009.5351148
15. McMillan, K.L.: Lazy annotation for program testing and verification. In: Proc. CAV. pp. 104–118. LNCS 6174, Springer (2010). https://doi.org/10.1007/978-3-642-14295-6_10
16. Cimatti, A., Griggio, A.: Software model checking via IC3. In: Proc. CAV. pp. 277–293. LNCS 7358, Springer (2012). https://doi.org/10.1007/978-3-642-31424-7_23

17. Craig, W.: Linear reasoning. A new form of the Herbrand-Gentzen theorem. J. Symb. Log. **22**(3), 250–268 (1957). https://doi.org/10.2307/2963593

18. Awedh, M., Somenzi, F.: Automatic invariant strengthening to prove properties in bounded model checking. In: Proc. DAC. pp. 1073–1076. ACM (2006). https://doi.org/10.1145/1146909.1147180

19. Ganai, M.K., Gupta, A.: Accelerating high-level bounded model checking. In: Proc. ICCAD. pp. 794–801. ACM (2006). https://doi.org/10.1145/1233501.1233664

20. Cheng, X., Hsiao, M.S.: Simulation-directed invariant mining for software verification. In: Proc. DATE. pp. 682–687. ACM (2008). https://doi.org/10.1109/DATE.2008.4484757

21. Donaldson, A.F., Haller, L., Kröning, D.: Strengthening induction-based race checking with lightweight static analysis. In: Proc. VMCAI. pp. 169–183. LNCS 6538, Springer (2011). https://doi.org/10.1007/978-3-642-18275-4_13

22. Beyer, D., Dangl, M., Wendler, P.: Boosting k-induction with continuously-refined invariants. In: Proc. CAV. pp. 622–640. LNCS 9206, Springer (2015). https://doi.org/10.1007/978-3-319-21690-4_42

23. Brain, M., Joshi, S., Kröning, D., Schrammel, P.: Safety verification and refutation by k-invariants and k-induction. In: Proc. SAS. pp. 145–161. LNCS 9291, Springer (2015). https://doi.org/10.1007/978-3-662-48288-9_9

24. Rocha, H., Ismail, H.I., Cordeiro, L.C., Barreto, R.S.: Model checking embedded C software using k-induction and invariants. In: Proc. SBESC. pp. 90–95. IEEE (2015). https://doi.org/10.1109/SBESC.2015.24

25. Jovanovic, D., Dutertre, B.: Property-directed k-induction. In: Proc. FMCAD. pp. 85–92. IEEE (2016). https://doi.org/10.1109/FMCAD.2016.7886665

26. Fischer, J., Jhala, R., Majumdar, R.: Joining data flow with predicates. In: Proc. FSE. pp. 227–236. ACM (2005). https://doi.org/10.1145/1081706.1081742

27. Jain, H., Ivancic, F., Gupta, A., Shlyakhter, I., Wang, C.: Using statically computed invariants inside the predicate abstraction and refinement loop. In: Proc. CAV. pp. 137–151. LNCS 4144, Springer (2006). https://doi.org/10.1007/11817963_15

28. Pasareanu, C.S., Visser, W.: Verification of Java programs using symbolic execution and invariant generation. In: Proc. SPIN. pp. 164–181. LNCS 2989, Springer (2004). https://doi.org/10.1007/978-3-540-24732-6_13

29. Gurfinkel, A., Kahsai, T., Komuravelli, A., Navas, J.A.: The SEAHORN verification framework. In: Proc. CAV. pp. 343–361. LNCS 9206, Springer (2015). https://doi.org/10.1007/978-3-319-21690-4_20

30. Beyer, D., Lee, N.Z., Wendler, P.: Interpolation and SAT-based model checking revisited: Adoption to software verification. J. Autom. Reasoning (2024), accepted, preprint available via https://doi.org/10.48550/arXiv.2208.05046

31. Case, M.L., Mishchenko, A., Brayton, R.K.: Automated extraction of inductive invariants to aid model checking. In: Proc. FMCAD. pp. 165–172 (2007). https://doi.org/10.1109/FAMCAD.2007.12

32. Cabodi, G., Nocco, S., Quer, S.: Strengthening model checking techniques with inductive invariants. IEEE Trans. on CAD of Integrated Circuits and Systems **28**(1), 154–158 (2009). https://doi.org/10.1109/TCAD.2008.2009147

33. Beyer, D., Chien, P.C., Lee, N.Z.: CPA-DF: A tool for configurable interval analysis to boost program verification. In: Proc. ASE. pp. 2050–2053. IEEE (2023). https://doi.org/10.1109/ASE56229.2023.00213

34. Beyer, D., Keremoglu, M.E.: CPACHECKER: A tool for configurable software verification. In: Proc. CAV. pp. 184–190. LNCS 6806, Springer (2011). https://doi.org/10.1007/978-3-642-22110-1_16

35. Beyer, D.: Progress on software verification: SV-COMP 2022. In: Proc. TACAS (2). pp. 375–402. LNCS 13244, Springer (2022). https://doi.org/10.1007/978-3-030-99527-0_20

36. Rosen, B.K., Wegman, M.N., Zadeck, F.K.: Global value numbers and redundant computations. In: Proc. POPL. pp. 12–27. ACM (1988). https://doi.org/10.1145/73560.73562

37. Bodik, R., Anik, S.: Path-sensitive value-flow analysis. In: Proc. POPL. pp. 237–251. ACM (1998). https://doi.org/10.1145/268946.268966

38. Cousot, P., Cousot, R.: Static determination of dynamic properties of programs. In: Proc. Int. Symp. on Programming. pp. 106–130. Dunod (1976). https://www.di.ens.fr/~cousot/COUSOTpapers/publications.www/CousotCousot-ISOP-76-Dunod-p106-130-1976.pdf

39. Bradley, A.R., Manna, Z.: Property-directed incremental invariant generation. Formal Asp. Comput. 20(4-5), 379–405 (2008). https://doi.org/10.1007/s00165-008-0080-9

40. Bradley, A.R.: SAT-based model checking without unrolling. In: Proc. VM-CAI. pp. 70–87. LNCS 6538, Springer (2011). https://doi.org/10.1007/978-3-642-18275-4_7

41. Fedyukovich, G., Bodík, R.: Accelerating syntax-guided invariant synthesis. In: Proc. TACAS. pp. 251–269. LNCS 10805, Springer (2018). https://doi.org/10.1007/978-3-319-89960-2_14

42. Beyer, D., Dangl, M.: Software verification with PDR: An implementation of the state of the art. In: Proc. TACAS (1). pp. 3–21. LNCS 12078, Springer (2020). https://doi.org/10.1007/978-3-030-45190-5_1

43. Kahsai, T., Tinelli, C.: PKIND: A parallel k-induction based model checker. In: Proc. Int. Workshop on Parallel and Distributed Methods in Verification. pp. 55–62. EPTCS 72, EPTCS (2011). https://doi.org/10.4204/EPTCS.72.6

44. Albarghouthi, A., Li, Y., Gurfinkel, A., Chechik, M.: UFO: A framework for abstraction- and interpolation-based software verification. In: Proc. CAV, pp. 672–678. LNCS 7358, Springer (2012). https://doi.org/10.1007/978-3-642-31424-7_48

45. Komuravelli, A., Gurfinkel, A., Chaki, S., Clarke, E.M.: Automatic abstraction in SMT-based unbounded software model checking. In: Proc. CAV. pp. 846–862. LNCS 8044, Springer (2013). https://doi.org/10.1007/978-3-642-39799-8_59

46. Brat, G., Navas, J.A., Shi, N., Venet, A.: IKOS: A framework for static analysis based on abstract interpretation. In: Proc. SEFM. pp. 271–277. LNCS 8702, Springer (2014). https://doi.org/10.1007/978-3-319-10431-7_20

47. Beyer, D., Cimatti, A., Griggio, A., Keremoglu, M.E., Sebastiani, R.: Software model checking via large-block encoding. In: Proc. FMCAD. pp. 25–32. IEEE (2009). https://doi.org/10.1109/FMCAD.2009.5351147

48. Beyer, D., Henzinger, T.A., Théoduloz, G.: Configurable software verification: Concretizing the convergence of model checking and program analysis. In: Proc. CAV. pp. 504–518. LNCS 4590, Springer (2007). https://doi.org/10.1007/978-3-540-73368-3_51

49. Beyer, D., Keremoglu, M.E., Wendler, P.: Predicate abstraction with adjustable-block encoding. In: Proc. FMCAD. pp. 189–197. FMCAD (2010). https://dl.acm.org/doi/10.5555/1998496.1998532

50. Beyer, D., Dangl, M., Wendler, P.: A unifying view on SMT-based software verification. J. Autom. Reasoning 60(3), 299–335 (2018). https://doi.org/10.1007/s10817-017-9432-6

51. Beyer, D., Henzinger, T.A., Théoduloz, G.: Program analysis with dynamic precision adjustment. In: Proc. ASE. pp. 29–38. IEEE (2008). https://doi.org/10.1109/ASE.2008.13

52. Graf, S., Saïdi, H.: Construction of abstract state graphs with Pvs. In: Proc. CAV. pp. 72–83. LNCS 1254, Springer (1997). https://doi.org/10.1007/3-540-63166-6_10

53. Slabý, J., Strejček, J., Trtík, M.: Checking properties described by state machines: On synergy of instrumentation, slicing, and symbolic execution. In: Proc. FMICS. pp. 207–221. LNCS 7437, Springer (2012). https://doi.org/10.1007/978-3-642-32469-7_14

54. Aho, A.V., Sethi, R., Ullman, J.D.: Compilers: Principles, Techniques, and Tools. Addison-Wesley (1986). https://www.worldcat.org/isbn/978-0-201-10088-4

55. Donaldson, A.F., Kröning, D., Rümmer, P.: Automatic analysis of DMA races using model checking and k-induction. FMSD **39**(1), 83–113 (2011). https://doi.org/10.1007/s10703-011-0124-2

56. Beyer, D., Löwe, S., Wendler, P.: Reliable benchmarking: Requirements and solutions. Int. J. Softw. Tools Technol. Transfer **21**(1), 1–29 (2019). https://doi.org/10.1007/s10009-017-0469-y

57. Cimatti, A., Griggio, A., Schaafsma, B.J., Sebastiani, R.: The MathSAT5 SMT solver. In: Proc. TACAS. pp. 93–107. LNCS 7795, Springer (2013). https://doi.org/10.1007/978-3-642-36742-7_7

58. Beyer, D.: Verifiers and validators of the 11th Intl. Competition on Software Verification (SV-COMP 2022). Zenodo (2022). https://doi.org/10.5281/zenodo.5959149

59. Beyer, D., Chien, P.C., Lee, N.Z.: Reproduction package for SPIN 2024 submission 'Augmenting interpolation-based model checking with auxiliary invariants'. Zenodo (2024). https://doi.org/10.5281/zenodo.10548594

Test-Case Generation with Automata-Based Software Model Checking

Max Barth and Marie-Christine Jakobs[(✉)]

LMU Munich, Munich, Germany
{Max.Barth,M.Jakobs}@lmu.de

Abstract. Software quality is often evaluated by testing the software on an adequate test suite, e.g., a test suite achieving certain or high coverage of the software. Manually generating such test suites is tedious. Thus, several automatic test-case generation approaches were developed to support this task. Approaches based on software model checking typically achieve high coverage and have been shown to be sufficiently efficient in the past. Yet, there does not exist a test-case generation approach that builds upon the automata-based approach to software model checking e.g., successfully used by ULTIMATE AUTOMIZER. To close this methodical gap, we present ULTIMATE TESTGEN, a test-case generator built on ULTIMATE AUTOMIZER. An experimental comparison of ULTIMATE TESTGEN against a closely related, up-to-date test-case generation approach reveals that ULTIMATE TESTGEN generates test suites that achieve the same or higher branch coverage for nearly 75% of the evaluated programs.

1 Introduction

Testing is a widely adopted technique to inspect software quality and structural coverage criteria are common metrics to judge the adequacy of generated test suites. Since manually generating test cases or even entire test suites is laborious, several automatic test-case generation techniques have been suggested in the past. Many of them generate test suites for structural coverage criteria, in particular branch coverage, thereby focusing on test input generation. They range from random testing and fuzzing [32,33,37,42] over search-based testing [34] to symbolic execution [16,38] and software model checking [7–9,28,29,41].

By design, approaches based on software model checking achieve high coverage because they check the reachability of each individual test goal. While one might think that checking the reachability of test goals is too expensive, advances in verification technology [5] allowed for the development of rather efficient test-case generation approaches like e.g., COVERITEST [9], FUSEBMC [1], or VERIFUZZ [35], which use (bounded) software model checking.

Inspired by the success of these approaches and the success of the software model checker ULTIMATE AUTOMIZER [5], we propose the test-case generation

T. Neele and A. Wijs (Eds.): SPIN 2024, LNCS 14624, pp. 248–267, 2025.
https://doi.org/10.1007/978-3-031-66149-5_14

technique ULTIMATE TESTGEN. ULTIMATE TESTGEN is based on the automata-based approach to software model checking [27] that is employed by ULTIMATE AUTOMIZER [25] and that has not been used for test-case generation before.

To turn the software model checker ULTIMATE AUTOMIZER into a test-case generator, we employ an approach widely-used when generating test cases with verifiers. Namely, we generate test cases from counterexamples [7,41]. Besides transforming counterexamples into tests, we therefore need to encode the hit of a (new) test goal g as a property violation, i.e., ULTIMATE AUTOMIZER must accept program execution traces that reach the test goal g. To become efficient, we follow e.g., COVERITEST [9] and consider multiple test goals at once. Hence, we need ULTIMATE AUTOMIZER to continue verification after detecting a counterexample. Meanwhile, it should avoid reporting counterexamples that only consider already covered test goals. In contrast to existing approaches [9,21,39], which remove test goals, we achieve this by abstracting such counterexample traces by automata and then restricting ULTIMATE AUTOMIZER's exploration to program execution traces that are not accepted by any of those automata.

We implemented the above extensions into ULTIMATE AUTOMIZER. The resulting tool ULTIMATE TESTGEN allows us to choose between two configurations: ALL and INCR. ULTIMATE TESTGEN-ALL starts test-case generation with all test goals at once. In contrast, ULTIMATE TESTGEN-INCR incrementally increases the set of considered test goals, adding more promising test goals earlier. A test goal will be considered more promising if it may cover a higher number of additional test goals when covered. We experimentally compare both configurations on the benchmark programs used in the International Competition on Software Testing (Test-Comp) [6]. Our evaluation shows that ULTIMATE TESTGEN-ALL achieves equal or higher coverage for 97% of the benchmark programs while ULTIMATE TESTGEN-INCR generates significantly fewer test cases. Furthermore, a comparison with COVERITEST, an up-to-date Test-Comp participant, which like ULTIMATE TESTGEN uses counterexample-guided abstraction refinement, reveals that ULTIMATE TESTGEN-ALL achieves the same coverage for about 60% of the benchmarks and even higher coverage for 5% of the programs. Comparing ULTIMATE TESTGEN-ALL against COVERITEST's component based on predicate abstraction, the most similar approach to ULTIMATE TESTGEN, exhibits that ULTIMATE TESTGEN-ALL even achieves higher coverage for 30% of the benchmark programs.

2   The Basics of Software Model Checking with Automata

Our goal is to use an automata-based approach to software model checking [27] to detect feasible error traces (i.e., counterexamples) in programs and then generate tests from them. This section introduces the types of automata used by this approach and fixes the meaning of a feasible error trace.

2.1   Program Automata and Their Feasible Error Traces

We consider programs that sequentially execute statements from a fixed set Σ. For our presentation, we assume that Σ contains assignments and assume state-

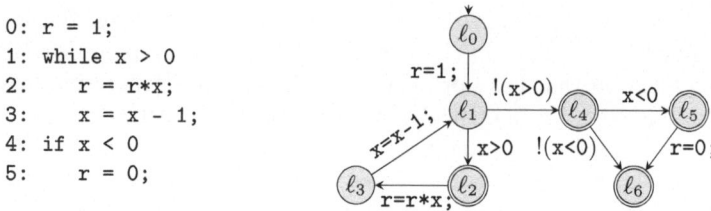

```
0: r = 1;
1: while x > 0
2:    r = r*x;
3:    x = x - 1;
4: if x < 0
5:    r = 0;
```

Fig. 1. Source code (left) and program automaton (right) of our example program `fac`

ments over a set V of (integer) variables.[1] Following literature [27], we then model a program together with its correctness property by a (program) automaton $A_\mathcal{P} = (L, \delta_\mathcal{P}, \ell_0, F_\mathcal{P})$. Thereby, the set L represents the program counter values including the initial program counter $\ell_0 \in L$, while $\delta_\mathcal{P} \subseteq L \times \Sigma \times L$ denotes the control-flow edges, which describe which statements can be executed at a given program location and where to proceed after the statement's execution. In addition, the set $F_\mathcal{P} \subseteq L$ describes the error locations that model the correctness property of a program. More concretely, program executions reaching any of the error locations in $F_\mathcal{P}$ violate the program's correctness property. Thus, $F_\mathcal{P}$ depends on the property to be checked. Since we aim to generate tests from property violations, we use $F_\mathcal{P}$ to characterize the coverage of test goals, in our case program branches (i.e., assume statements in the program automaton).

Figure 1 shows the code and the program automaton of our example program `fac`, which computes the factorial of x if x is non-negative. The automaton contains one edge per assignment and two assume-statement edges per condition of each if or while statement, namely one for each outcome. Moreover, its error locations are the end locations of the program's assume statements (branches).

To generate tests from property violations, our approach inspects error traces, i.e., sequences of statements that are allowed by the program syntax and lead to a property violation. Formally, a sequence of statements $\pi \in \Sigma^*$ is an *error trace* of a program if its program automaton $A_\mathcal{P}$ accepts π. Since error traces only consider the program's syntax, not all error traces can be observed during program execution. An example of a non-observable error trace is the sequence of statements `r=1; ,x>0,r=r*x; ,x=x-1; ,!(x>0) ,x<0,r=0;` of program `fac`. Note that it is unobservable because it visits the loop once (`x>0,r=r*x; ,x=x-1; ,!(x>0)`) and after loop execution, statement `x<0` conflicts with the possible values of variable x, which are not smaller than zero. Due to such potential conflicts between variable values and statements, our approach needs to check that an error trace is feasible. An error trace is feasible if there exists a program execution that executes the exact same sequence of statements as the error trace. Since an error trace is syntactically allowed by the program, to show its feasibility, it remains to be proven that it does not conflict with the variable values during execution, i.e., it respects the statements' semantics. A statement's semantics basically defines for which variable values the statement is executable and what

---

[1] In our implementation, we support C programs.

the variable values resulting from executing the statement are. We represent the variable values using data states that assign to every program variable a value of its domain. On top of the set D of data states, we then define the semantics of a statement stmt by a (partial) function $SP_{\mathrm{stmt}} : D \rightharpoonup D$. For assume statements $\mathrm{stmt}_?$, function $SP_{\mathrm{stmt}_?}$ is a partial identity function that is defined for data state d if the assume (a Boolean expression) evaluates to true in data state d, i.e., $d \models \mathrm{stmt}_?$. For assignments $\mathrm{stmt}_=$, function $SP_{\mathrm{stmt}_=}$ is total and denotes the strongest-post operator of the semantics. Finally, an (error) *trace* $\pi = \mathrm{stmt}_1, \ldots, \mathrm{stmt}_n \in \Sigma^*$ *is feasible* if there exists a sequence of data states $d_0, d_1, \ldots, d_n$ such that $SP_{\mathrm{stmt}_i}(d_{i-1}) = d_i$ holds for all $1 \leq i \leq n$. Technically, we check the feasibility of an error trace π by first converting it into single static assignment (SSA) form [20]. Then, we encode its SSA form into a logic formula φ_π that will be satisfiable if π is feasible. Finally, we use a satisfiable modulo theory (SMT) solver to determine the feasibility of φ_π.

2.2 Abstracting Traces via Interpolant and Error Automata

To generate a test suite that achieves a high coverage, we intend the software model checker to detect one feasible error trace per test goal, i.e., per error location from $F_\mathcal{P}$. To efficiently detect those feasible error traces, we successively exclude irrelevant error traces from the search space of the model checker. More concretely, we exclude error traces for which we know that they (a) are infeasible or (b) do not contribute to the coverage because they end in an already covered test goal. Technically, we build automata $\mathcal{A}_i$ that each accept a subset of irrelevant error traces and subtract them from the program automaton.

Whenever we detect that an error trace π is infeasible, we build an *interpolant automaton* [26] that accepts the analyzed, infeasible trace plus some infeasible traces that have a similar reason for infeasibility. Thereby, each interpolant automaton encodes the reason for infeasibility of the accepted traces. Formally, an interpolant automaton $\mathcal{A}_\mathcal{I} = (Q_\mathcal{I}, \delta_\mathcal{I}, q_0, F_\mathcal{I}, d_\mathcal{I})$ consists of a set of states $Q_\mathcal{I}$ including the initial state q_0 and the accepting states $F_\mathcal{I}$ as well as a transition relation $\delta_\mathcal{I} \subseteq Q_\mathcal{I} \times \Sigma \times Q_\mathcal{I}$. To record the infeasibility argument of accepted traces, it also contains a total function $d_\mathcal{I} : Q_\mathcal{I} \rightarrow 2^D$ that assigns to each state a set of data states that overapproximate the reachable data states while still allowing to prove infeasiblility of accepted traces.[2] To guarantee the infeasibility of accepted traces, each interpolant automaton ensures that (1) the initial state allows any data state $d_\mathcal{I}(q_0) = D$, (2) the transition relation respects the statement's semantics, i.e., $\forall(q, \mathrm{stmt}, q') \in \delta : \{d' \mid \exists d \in d_\mathcal{I}(q) : SP_{\mathrm{stmt}}(d) = d'\} \subseteq d_\mathcal{I}(q')$, and (3) final states do not allow any data state, i.e., $\forall q_f \in F_\mathcal{I} : d_\mathcal{I}(q_f) = \emptyset$. Together properties (1)–(3) ensure that any accepted trace is infeasible.

Figure 2 shows one interpolant automaton for the infeasible error trace $\pi =$r=1;,x>0,r=r*x;,x=x-1;,!(x>0),x<0,r=0; from above, which might be constructed by Ultimate Automizer. The figure uses assertions known from

---

[2] In practice, the sets of data states are represented by assertions.

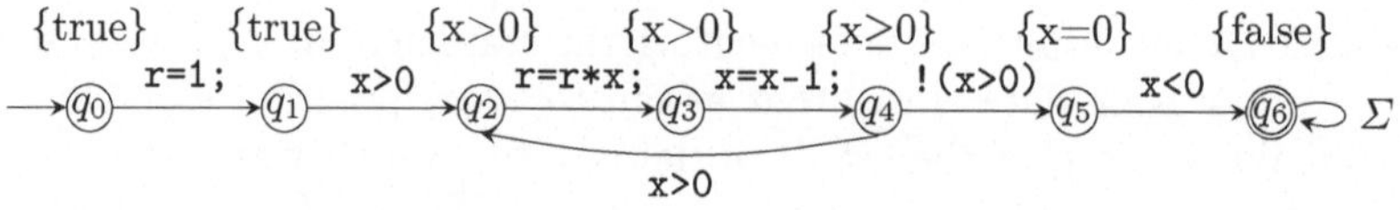

$$\{true\} \quad \{true\} \quad \{x>0\} \quad \{x>0\} \quad \{x\geq 0\} \quad \{x=0\} \quad \{false\}$$

Fig. 2. Example of an interpolant automaton

Hoare logic to describe the set of data states assigned to each state. The automaton encodes the error trace into a corresponding sequence of automaton transitions. In addition, the states are annotated with interpolants (the assertions), which are derived by splitting the encoding φ_π at the position of the state and computing the interpolant. Thereafter, the automaton is enriched by appropriate backward edges resulting from loops that keep the reason of infeasibility. In our case, the automaton accepts any error trace that executes the loop of example program fac at least once and thereafter enters the if branch. We refer to [26] for more details on how to construct an interpolant automaton based on the proof of unsatisfiability of an error trace and interpolation.

Whenever we detect a feasible error trace π, we first generate a test as explained in the next section. The generated test visits all test goals on the error trace. Thus, error traces ending in an error location associated with a visited test goal are irrelevant in future because they only contribute to the coverage when accidentally visiting other uncovered test goals. To exclude such error traces from future analysis, for each of the error locations ℓ_e visited by π we will construct an *error automaton* if no error automaton already exists for ℓ_e. Given an error location, the error automaton accepts any trace that ends in that error location. For the construction of the error automaton, we make use of the following structural properties of the error locations that we consider for test-case generation. First, the error locations will have exactly one incoming transition[3]. Second, we will label statements such that the labeled statement on the incoming transition is unique in the program automaton, i.e., the statement does not occur with the same label on any other transition of the program automaton. Hence, we can use the (labeled) statement to detect whether the error location is reached. Based on these assumption, the error automaton for an error location ℓ_e with corresponding incoming transition $(\ell, \mathtt{stmt}, \ell_e)$ consists of two states, the initial state q_0 and the final state q_e. If the automaton observes the statement stmt, it transitions to the final state q_e and otherwise, it transitions to q_0. Formally, the automaton is defined as follows.

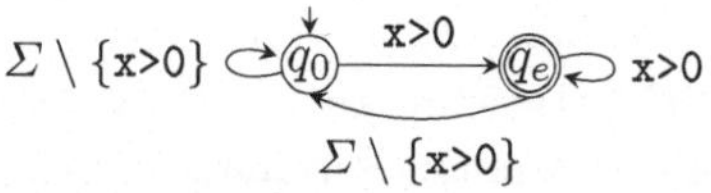

Fig. 3. Example of an error automaton

$$\mathcal{A}_E = \left(\{q_0, q_e\}, \begin{matrix} \{(q_0, \mathtt{stmt}, q_e), (q_e, \mathtt{stmt}, q_e)\} \cup \\ \{(q, \mathtt{stmt}', q_0) \mid (q = q_0 \vee q = q_e) \wedge \mathtt{stmt}' \in \Sigma \setminus \{\mathtt{stmt}\}\} \end{matrix}, \{q_0\}, \{q_e\} \right)$$

---

[3] Note that this property is violated for error location ℓ_6 of our example program. However, our encoding of test goals, which we explain later, ensures this property.

Algorithm 1. ULTIMATE TESTGEN-ALL employing automata-based software model checking [26] for test-case generation

---

Input: Program $A_\mathcal{P} = (L, \delta_\mathcal{P}, \ell_0, \emptyset)$, test goals $G \subseteq \delta_\mathcal{P}$
1: $A_{\mathcal{P}_{\text{test}}} := \text{encode\_test\_goals}(A_\mathcal{P}, G)$;
2: $\texttt{test\_suite} := \emptyset$; $A_{\mathcal{P}_{\text{rel}}} := A_{\mathcal{P}_{\text{test}}}$
3: **while** $\mathcal{L}(A_{\mathcal{P}_{\text{rel}}}) \neq \emptyset$ **do**
4:     determine $\pi \in \mathcal{L}(A_{\mathcal{P}_{\text{rel}}})$
5:     (result, witness) := check_feasibility(π);
6:     **if** result = true **then**
7:         $\texttt{test\_suite} := \texttt{test\_suite} \cup \text{generate\_test\_case}(\pi, \text{witness})$;
8:         $A_\pi := \text{generate\_error\_automata}(\pi)$;
9:     **else**
10:         $A_\pi := \text{generate\_interpolant\_automata}(\pi, \text{witness})$;
11:     $A_{\mathcal{P}_{\text{rel}}} := A_{\mathcal{P}_{\text{rel}}} \setminus A_\pi$;
12: return $\texttt{test\_suite}$;

Figure 3 shows the error automaton for the feasible error trace r=1;,x>0 of program fac. The next section explains how to use the automata introduced in this section for test-case generation.

3   From Model Checking to Test-Case Generation

Our goal is to turn the software model checker ULTIMATE AUTOMIZER into a test-case generator that detects one feasible error trace per test goal and transforms it into a test case. To this end, we adapt its model checking approach.

Before we explain the adaption, let us recapture the original approach [26]. Given a program automaton, ULTIMATE AUTOMIZER iteratively refines an overapproximation of the feasible error traces until the overapproximation is empty, i.e., it proves that the program is correct, or ULTIMATE AUTOMIZER detects a feasible error trace, i.e., it finds a property violation. First, the initial overapproximation becomes the program automaton, which accepts all error traces of the program. Next, each iteration selects an error trace π from the current overapproximation and checks its feasibility. If π is feasible, a violation is found and π is returned as a counterexample. If π is infeasible, an interpolant automaton is constructed with the help of the infeasibility proof and the overapproximation is refined. More concretely, the traces accepted by the interpolant automaton are excluded from the overapproximation and then the subsequent iteration starts.

Next, let us discuss how to adapt the above procedure for test-case generation. Algorithm ULTIMATE TESTGEN-ALL (Algorithm 1) describes the adapted procedure. For now, let us assume that we already encoded the test goals into the program and let us focus on lines 2–12. The algorithm maintains two important data structures: test_suite, which contains the generated test cases, and the overapproximation $A_{\mathcal{P}_{\text{rel}}}$ of feasible and relevant error traces. Like the original approach, the initial overapproximation becomes the program automaton.

In addition, the initial test suite is empty, i.e., no test cases have been generated. Furthermore, the while loop realizes the iterative refinement. At the beginning of each iteration, the algorithm checks whether the overapproximation $A_{\mathcal{P}_{rel}}$ contains any error trace. Since the overapproximation is described by an automaton, the algorithm checks whether the language of the automaton is empty. If the language is empty, no relevant error traces exist, i.e., all error locations are covered[4], and we return the test suite. Otherwise, we perform the next loop iteration. First, we use an A^* algorithm [23,24] to determine a relevant error trace π of the program (line 4) and, then, check π's feasibility as explained in the previous section. The feasibility check returns the result and a witness, a model in case of feasibility and the unsatisfiability proof otherwise. If error trace π is feasible, we deviate from the original approach. Instead of returning π and reporting a property violation, we generate a test case from π and the witness (a model). Afterwards, we generate the error automaton A_π for the error trace π as described in the previous section. Remember the error automaton accepts all traces that end in the same error location as π and, thus, helps us to avoid that traces that end in already covered error locations (test goals) are found in future. If π is infeasible, we proceed as the original approach and generate an interpolant automaton A_π from π and the witness (a proof of unsatisfiability). In both cases, we refine the overapproximation by subtracting the generated automaton A_π and continue with the next iteration.

Generating Test Cases for Feasible Error Traces. After describing the test-case generation procedure, we now explain in detail how to derive a test case from a feasible error trace. Like many other tools [1,7,9,16,35], our test cases only provide test inputs. Hence, they must specify the values for input parameters and external functions like e.g., random, scanf. To compute these input values, we use an approach similar to the one of Blast [7]. For the feasibility check, we already computed a SSA-based formula encoding φ_π of the error trace π. Since we generate test cases from feasible error traces, the witness returned by the feasibility check is a model of φ_π. To generate the test case, we only need to identify the variables in the formula φ_π that correspond to inputs, look up their values in the model, and export their values in the order of the variables' occurrence in φ_π. For the export, we utilize the format[5] used by the International Competition on Software Testing [6], which allows test execution with TESTCOV [13]. For example, consider error trace $\pi =$r=1;,x>0 and corresponding formula $\varphi_\pi = r_1 = 1 \land x_0 > 0$. The first and only input in φ_π is x_0. Given the model $\{r_1 \mapsto 1, x_0 \mapsto 2\}$ of φ_π, we derive the following test case:

```
<testcase><input>2</input></testcase>
```

Encoding Test Goals. So far, we assumed that the test goals are already encoded into the program. Next, we explain how to actually encode them, i.e.,

---

[4] We encode the test goals into the program such that all test goals will be covered if all error locations are covered. Hence, we will abort if all test goals are covered.

[5] https://gitlab.com/sosy-lab/test-comp/test-format/-/blob/main/doc/Format.md

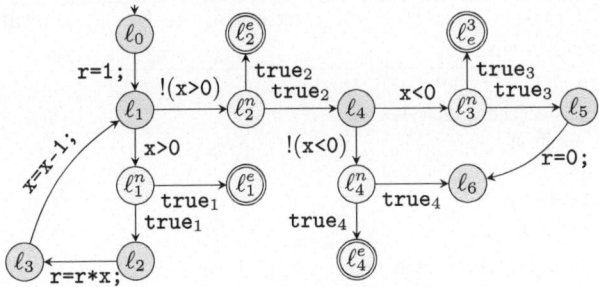

Fig. 4. Program automaton resulting from encoding branch coverage into our example

we explain line 1 of Algorithm 1. Like e.g., CoVeriTest [9], we are interested in structural coverage, in particular branch coverage, that can be expressed as the coverage of a set $G \subseteq \delta_{\mathcal{P}}$ of control-flow edges. For branch coverage, the set G of test goals contains all control-flow edges with an assume statement. However, control-flow edges are not the appropriate format to specify property violations because program automata specify property violations with error locations. Thus, we need to perform a testability transformation [22] and translate the coverage of control-flow edges into the coverage of error locations.

A naive translation would use the predecessors or successors of the test goal edges as error locations. For branch coverage, the predecessors are decision nodes forking into two assume statements (i.e., test goals) and are, thus, inappropriate. Next, we show that successors are inappropriate as well. For this, consider Fig. 1. We observe that the coverage of assume statement !(x<0) is associated with error location ℓ_6. However, location ℓ_6 can be reached via the incoming edge with statement r=0; without executing the assume statement !(x<0). Hence, we require a more intricate transformation.

To ensure that test goal coverage is equivalent to coverage of error locations, our idea is to replace any edge $g = (\ell, \text{stmt}, \ell')$ in the test goals by two successive edges $(\ell, \text{stmt}, \ell_g^n), (\ell_g^n, \text{true}, \ell')$, thereby introducing a new (intermediate) location ℓ_g^n. Since an assume statement does not change the data state and assume true is satisfied by any data state, the two successive edges $(\ell, \text{stmt}, \ell_g^n), (\ell_g^n, \text{true}, \ell')$ are semantically identical to g. While we could use locations ℓ_g^n as error locations, constructing appropriate error automata is rather complex. The error automata from the previous section require unique statements. Thus, we further extend our translation. First, we use labeled assumptions true_g, which are semantically equivalent to assumption true, but make them unique statements and allow us to use the error automata from the previous section. While this is sufficient, it forces an exploration algorithm to either detect shallow goals first and, thus, miss the potential of covering multiple goals at once, or be able to leave accepting states during exploration. To overcome this, we use non-determinism. More concretely, we introduce an additional edge $(\ell_g^n, \text{true}_g, \ell_g^e)$ and use ℓ_g^e instead of ℓ_g^n as error location. This leads us to the following transformation.

Algorithm 2. ULTIMATE TESTGEN-INCR extending test-case generation with support for collateral coverage

---

Input: Program $A_\mathcal{P} = (L, \delta_\mathcal{P}, \ell_0, \emptyset)$, test goals $G \subseteq \delta_\mathcal{P}$
1: $A_{\mathcal{P}_\text{test}} :=$ encode_test_goals$(A_\mathcal{P}, G)$;
2: test_suite $:= \emptyset$; $A_{\mathcal{P}_\text{rel}} := A_{\mathcal{P}_\text{test}}$; goals $:= F_{\mathcal{P}_\text{test}}$;
3: **while** goals $\neq \emptyset \vee \mathcal{L}(A_{\mathcal{P}_\text{rel}}) \neq \emptyset$ **do**
4:    **if** goals $\neq \emptyset$ **then**
5:       pop ℓ_g^e from goals;
6:    select $\pi \in A_{\mathcal{P}_\text{rel}}$ ending with true_g s.t. $\ell_g^e \in F_{\mathcal{P}_\text{test}} \setminus$ goals;
7:    (result, witness) $:=$ check_feasibility(π);
8:    **if** result $=$ true **then**
9:       test_suite $:=$ test_suite $\cup$ generate_test_case$(\pi, \text{witness})$;
10:       $A_\pi := \bigcup_{\pi_p \in \text{error-prefixes}(\pi)}$ generate_error_automata(π_p);
11:       goals $:=$ goals$\setminus \{\ell_g^e \mid \text{true}_g \text{ occurs in } \pi\}$;
12:    **else**
13:       $A_\pi :=$ generate_interpolant_automata$(\pi, \text{witness})$;
14:    $A_{\mathcal{P}_\text{rel}} := A_{\mathcal{P}_\text{rel}} \setminus A_\pi$;
15: **return** test_suite;

Definition 1. *Given program automaton* $A_\mathcal{P} = (L, \delta_\mathcal{P}, \ell_0, \emptyset)$ *and test goals* $G \subseteq \delta_\mathcal{P}$, *we define* encode_test_goals$(A_\mathcal{P}, G) = (L_{\mathcal{P}_\text{test}}, \delta_{\mathcal{P}_\text{test}}, \ell_0, F_{\mathcal{P}_\text{test}})$, *where*

- $F_{\mathcal{P}_\text{test}} = \{\ell_g^e \mid g \in G\}$,
- $L_{\mathcal{P}_\text{test}} = L \cup F_{\mathcal{P}_\text{test}} \cup \{\ell_g^n \mid g \in G\}$, *and*
- $\delta_{\mathcal{P}_\text{test}} = (\delta_\mathcal{P} \setminus G) \cup \{(\ell_g^n, \text{true}_g, \ell_g^e), (\ell, stmt, \ell_g^n), (\ell_g^n, \text{true}_g, \ell') \mid (\ell, stmt, \ell') \in G\}$.

Figure 4 shows the transformed program automaton $A_{\mathcal{P}_\text{test}}$ for our example program `fac` from Fig. 1. The newly added states are highlighted in yellow. Note that like our implementation, Fig. 4 uses unique integer numbers to name the locations instead of the edges g used in the above formalization.

Towards Exploiting Collateral Coverage. Often, shorter error traces are prefixes of longer error traces. For example, `r=1;,!(x>0),true₂` is a prefix of `r=1;,!(x>0),true₂,x<0,true₃`. In addition, a test case that covers the longer trace implicitly covers its shorter prefixes and, thus, may cover additional test goals like `!(x>0)`, which is known as collateral coverage. Hence, covering deeper test goals (error locations) first and considering collateral coverage may reduce the size of the test suite and potentially also the test-case generation time. However, proving unreachability of an error location can be impossible and even if possible it is typically more expensive than detecting a feasible error trace. Therefore, we should not indefinitely focus on a certain subset of test goals. Our suggestion to the problem is to incrementally increase the set of considered goals, thereby adding deeper goals first.

Algorithm ULTIMATE TESTGEN-INCR shown in Algorithm 2 describes test-case generation with an incrementally increasing goal set. To this end, ULTIMATE TESTGEN-INCR uses an additional data structure goals to keep track of the goals that need to be added. Initially, all goals need to be added. Furthermore,

ULTIMATE TESTGEN-INCR extends the test-case generation loop. At the beginning of each iteration, it adds a new goal in case unconsidered goals still exist. Our implementation aims to select deeper goals (error locations) first. Since deeper states have higher IDs, we select the goal (error location) with the highest ID. To ensure that we only consider traces that are accepted by one of the currently considered goal states, we extend the error trace selection with a constraint on the selected trace. Due to the design of the test goal encoding, our constraint achieves the desired purpose. Note that we could have used a similar constraint to exclude traces to already covered error locations. However, we believe that our automata-based exclusion is better because it reduces the state space. Getting back to the discussion of our additions, we also consider the collateral coverage when generating a test case from a feasible error trace. Note that by design of the test goal encoding, we will cover test goal g if we execute statement true_g and every prefix of the error trace π that ends in a statement true_g is an error trace with corresponding error location ℓ_g^e. We use this insight to compute multiple error automata that also exclude covering collaterally covered test goals. To this end, line 10 computes the union of all error automata built for all prefixes π_p of the error trace π that are error traces themselves.[6] The last addition in ULTIMATE TESTGEN-INCR is line 11, which excludes all error locations corresponding to collaterally covered test goals from the set of unconsidered test goals.

Implementation. We realized Algorithms 1 and 2 in the verifier ULTIMATE AUTOMIZER [25], which already supports automata-based software model checking. ULTIMATE AUTOMIZER uses an A^* algorithm [23,24] to detect error traces. We do not provide A^* with a specific heuristic, but ULTIMATE TESTGEN-INCR guides it with the error locations to consider (i.e., $F_{\mathcal{P}_{\text{test}}} \setminus \text{goals}$). Also, ULTIMATE AUTOMIZER already provides the automata operations, the construction of interpolant and error automata, and the feasibility check for error traces.

To encode the test goals, we extend ULTIMATE AUTOMIZER's translation front-end, which translates C programs into Boogie programs and then builds the program automaton. While our approach allows arbitrary test goals, our implementation only supports branch coverage, i.e., the test goals are all edges with assume statements. To encode these test goals, we add an `assert false` statement with a unique label at the beginning of each branch in the Boogie program. When the program automaton is built, every assert statement is translated into two edges. One edge represents the violation of the assert statement (i.e., !false≡true is true), ends in an error location and corresponds to $(\ell_g^n, \text{true}_g, \ell_g^e)$. The other edge passes the assertion and corresponds to $(\ell_g^n, \text{true}_g, \ell')$. Since `assert false` cannot be passed, we configure ULTIMATE AUTOMIZER to ignore the assert condition when passing an assert statement, which is equivalent to condition true. Furthermore, ULTIMATE AUTOMIZER uses block encoding, which assigns loop free code blocks instead of statements to edges. To ensure that we do not loose inputs due to this encoding, we change ULTIMATE TESTGEN's block encoding and interrupt a block if an input occurs in the program.

---

[6] For efficiency, our implementation computes one single automata for all error traces that is equivalent to the union of their error automata.

For the feasibility check, ULTIMATE AUTOMIZER first uses unbounded integers, as known from mathematics, and a combination of the SMT solvers Z3 [36] and SMTINTERPOL [18]. If the feasibility check returns true and the encoding is not precise because the encoded trace considers floats, doubles, bit-wise operations, etc., we will stop test-case generation with unbounded, mathematical integers and start test-case generation using a bit-vector encoding, which uses the two SMT solvers MATHSAT5 [19] and CVC4 [3]. To generate a test case for a feasible error trace, we use the model provided by the feasibility check and proceed as described above. However, due to the use of e.g., unbounded integers, the values might be out of range of the variable's C type. Therefore, we identify the required C type first and if a value is out of range, we use the value modulo the allowed size. If there exists a variable whose value is out of range and the size of the C type depends on the architecture, we write two test cases. One test case shrinks the values to sizes appropriate for 32-bit architectures and the other for 64-bit architectures.

4   Evaluation

In our evaluation, we aim to investigate the following two research questions.

RQ1 How do the two configurations of ULTIMATE TESTGEN compare in terms of achieved coverage and number of generated test cases?

RQ2 How does ULTIMATE TESTGEN compare to similar, up-to-date competitors in terms of achieved coverage and number of generated test cases?

4.1   Evaluation Setup

Tasks. For our evaluation, we consider the coverage criterion branch coverage and perform the evaluation on the corresponding 2 933 tasks considered in the International Competition on Software Testing (Test-Comp) in 2023 [6].

Tools. We consider the two test-case generation techniques presented in this paper, which are implemented in ULTIMATE AUTOMIZER (TestGeneration branch[7] version 94ac8e0). As competitors, we considered the closely related test-case generators from Test-Comp 2023 [6], i.e., test-case generators that also perform counterexample-guided abstraction refinement (CEGAR) and predicate abstraction. From those, we selected the one that achieved the highest score in the category cover-branches, namely CoVERITEST [9]. CoVERITEST is based on the software analysis framework CPACHECKER [10] and uses a portfolio approach that mainly runs a cyclic combination of predicate abstraction [11] and explicit model checking [14]. In addition, it mutates the test cases generated by the cyclic combination and briefly generates random test cases at the beginning. To compare ULTIMATE TESTGEN against an even more closely related approach, we also consider CoVERITEST's predicate analysis component (named PREDICATE)

---

[7] https://github.com/ultimate-pa/ultimate/tree/TestGeneration.

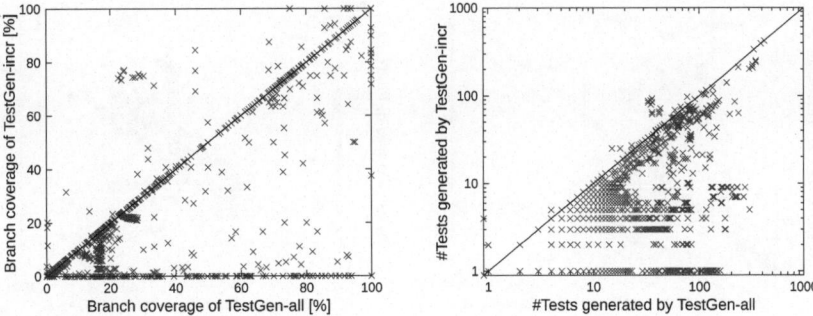

Fig. 5. Comparing achieved branch coverage (left) and number of generated test cases (right) of ULTIMATE TESTGEN-ALL (x-axis) and ULTIMATE TESTGEN-INCR (y-axis)

standalone. For both competitors, we use the CoVeriTest version submitted to Test-Comp 2023. Furthermore, we measure the coverage of the generated test suites with TESTCOV [13] and use the same version as Test-Comp 2023 [6].

Environment. We run all experiments on machines with 33 GB of memory, an Intel Xeon E3-1230 v5 CPU with 8 processing units and a frequency of 3.40 GHz that run Ubuntu 22.04 (Linux kernel 5.15.0). Following Test-Comp, we use BENCHEXEC 3.18 [15] to limit each test-case generation run to 8 cores, 15 min of CPU time and 15 GB of memory and the TESTCOV runs to 2 cores, 10 min of CPU time, and 7 GB of memory.

4.2   RQ 1: Comparison of Ultimate TestGen Configurations

We aim to compare the two configurations of ULTIMATE TESTGEN based on their achieved coverage and the number of generated test cases. In general, we aim for high coverage with a small number of test cases. Figure 5 shows two scatter plots. Its left scatter plot compares for each test task the branch coverage achieved by ULTIMATE TESTGEN-ALL (x-axis) with the branch coverage achieved by ULTIMATE TESTGEN-INCR (y-axis). We observe that a large number of data points is in the lower right half, i.e., ULTIMATE TESTGEN-ALL achieves a higher coverage. A detailed analysis reveals that ULTIMATE TESTGEN-ALL achieves a higher coverage for 1 055 of 2 933 (36%) tasks and the same coverage for 1 778 of 2 933 (61%) tasks. The reason is that ULTIMATE TESTGEN-INCR detects longer error traces at the beginning and if they are infeasible, their interpolant automata are larger, too. Large interpolant automata may prohibit efficient program abstractions and automata operations are more expensive on larger automata. Thus, ULTIMATE TESTGEN-INCR likely detects less feasible error traces in total.

Next, we study the number of generated test cases. The right scatter plot of Fig. 5 compares for each test task the number of test cases generated by ULTI-MATE TESTGEN-ALL (x-axis) with the number of test cases generated by ULTI-MATE TESTGEN-INCR (y-axis). We observe that a large number of data points is in the lower right half, i.e., ULTIMATE TESTGEN-INCR generates fewer test cases. A

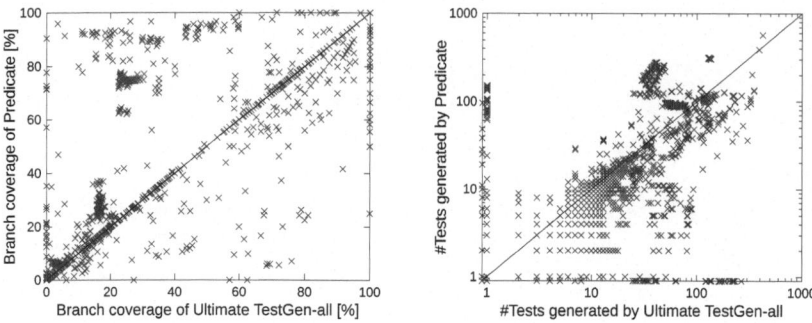

Fig. 6. Comparing achieved branch coverage (left) and number of generated test cases (right) of ULTIMATE TESTGEN-ALL (x-axis) and PREDICATE (y-axis)

detailed analysis reveals that ULTIMATE TESTGEN-INCR generates fewer test cases for 2 258 of 2 933 (77%) tasks and the same number of test cases for 638 of 2 933 (22%) tasks. More importantly for 41% (1 206 of 2 933) of all test tasks, ULTIMATE TESTGEN-INCR covers at least as many branches as ULTIMATE TESTGEN-ALL, but it generates smaller test suites for 64% (1 206 of 1 878) for them.

In sum, ULTIMATE TESTGEN-ALL achieves better coverage at the cost of more tests but about half of the time ULTIMATE TESTGEN-INCR is a valuable alternative.

4.3   RQ 2: Comparison with Competitors

Next, we compare ULTIMATE TESTGEN-ALL, the better of our two approaches, with the closely related up-to-date competitors PREDICATE and COVERITEST. Again, we consider their achieved coverage and the number of generated tests. First, let us compare against PREDICATE, which is the approach most closely related to ours. The left scatter plot of Fig. 6 compares for each test task the branch coverage achieved by ULTIMATE TESTGEN-ALL (x-axis) with the branch coverage achieved by PREDICATE (y-axis). We observe that many data points are in the upper left as well as the lower right half, i.e., there exist many tasks for which ULTIMATE TESTGEN-ALL achieves higher coverage and vice versa. Indeed, both approaches achieve the same coverage for 44% (1 298 of 2 933) of the tasks. PREDICATE achieves a higher coverage for 30% (867 of 2 933) of the tasks, while ULTIMATE TESTGEN-ALL achieves higher coverage for 26% (768 of 2 933) of the tasks. To better understand their strengths and weaknesses, let us study Table 1. For each category considered in Test-Comp, Table 1 contains a row that shows the number of tasks in the category and more importantly for each considered test-case generator the accumulated coverage, i.e., the sum of the coverage achieved for each task in that category. The last row shows the sum of the previous rows. For each row, Table 1 highlights the entry with the highest accumulated coverage in light gray. Studying Table 1, we notice that there exist some categories like e.g., BitVectors, ECA, ProductLines, Sequentialized, or Combinations, in which ULTIMATE TESTGEN-ALL achieves a significantly lower

Table 1. Sum of achieved coverage per task category and test-case generator

Category	# Tasks	ULTIMATE TESTGEN all	incr	PREDICATE	CoVeriTest
Arrays	292	20600	18100	19500	20800
BitVectors	61	3750	3350	4760	4820
ControlFlow	11	32.4	27	26.6	46.1
ECA	29	296	267	488	562
Floats	197	8800	8150	8820	9230
Heap	110	7360	4940	6850	7580
Loops	661	52000	47600	49300	52600
ProductLines	263	5120	3010	7050	7670
Recursive	51	4100	3620	3730	4000
Sequentialized	91	2790	45.8	7380	7350
XCSP	114	11100	10700	6380	11400
Combinations	671	17000	14800	22500	23800
BusyBox	62	982	690	607	1030
DriversLinux64	287	5830	5790	5840	5920
SQLite	1	–	–	–	0.02
Termination	32	2990	2910	2800	3060
Total	2933	142750.4	123999.8	146031.6	159868.12

accumulated coverage than PREDICATE. Looking at the SV-COMP results[8], we can recognize that ULTIMATE AUTOMIZER also struggles with the verification tasks of these categories. It seems these categories are particular difficult for ULTIMATE AUTOMIZER. However, there also exist categories like e.g., Arrays, Heap, Loops, Recursive, XCSP, and BusyBox for which ULTIMATE TESTGEN-ALL performs significantly better than PREDICATE. This is one reason why CPACHECKER and CoVeriTest usually use a combination of analyses.

Next, let us compare their numbers of generated test cases. The right scatter plot of Fig. 6 compares this. Again, we observe that many data points are in the upper left as well as the lower right half, i.e., there exist many tasks for which ULTIMATE TESTGEN-ALL generates fewer test cases and vice versa. Still, PREDICATE generates fewer test cases more often, namely for 68% (2 004 of 2 933 of the tasks). In addition, in 58% (1 263 of 2 165) of the tasks for which PREDICATE achieves the same or more coverage, PREDICATE also generates smaller test suites. Nevertheless, ULTIMATE TESTGEN-ALL is complementary to PREDICATE.

---

[8] https://sv-comp.sosy-lab.org/2023/results/results-verified/.

Finally, let us compare ULTIMATE TESTGEN-ALL against COVERITEST. The scatter plot of Fig. 7 compares their achieved branch coverage. Once more, we observe that there exist data points in the upper left as well as the lower right half, i.e., there exist tasks for which ULTIMATE TESTGEN-ALL achieves higher coverage and vice versa. While COVERITEST achieves the same coverage in 59% of the cases (1 725 of 2 933 tasks) and even better coverage in 36% of the cases (1 044 of 2 933) of the tasks, ULTIMATE TESTGEN-ALL sometimes achieves better coverage, too. Again, looking at Table 1, we notice that there exist

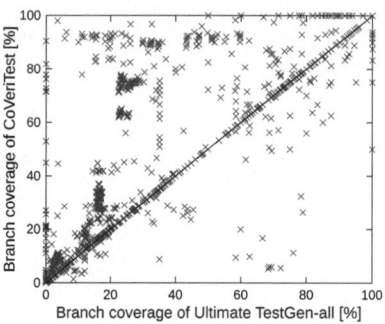

Fig. 7. Comparing achieved branch coverage of ULTIMATE TESTGEN-ALL (x-axis) against achieved branch coverage of COVERITEST

some categories like e.g., BitVectors, ECA, ProductLines, Sequentialized, or Combinations, in which ULTIMATE TESTGEN-ALL achieves a significantly lower accumulated coverage as COVERITEST. These are mainly the same categories for which PREDICATE already outperforms ULTIMATE TESTGEN-ALL. In many other categories, ULTIMATE TESTGEN-ALL achieves a similar accumulated coverage and in category Recursive it even achieves a higher accumulated coverage. When studying the number of generated test cases, our raw data reveals that COVERITEST generates more test cases for 76% (2 232 of 2 933) of the tasks because it mutates every test case generated by its model checkers several times. To sum up, ULTIMATE TESTGEN complements related state-of-the-art approaches.

4.4   Threats to Validity

Our results may not generalize for several reasons. First, we use benchmark programs in our evaluation. Although the benchmark is well-established and contains diverse programs, it may not reflect the characteristics of (all) real-world applications. Second, we used fixed resource limits for the execution of the tasks. Different resource limits may change the results. Third, we compared against two closely related competitors. We expect different results when comparing ULTIMATE TESTGEN to more diverse test-case generators.

Also, the coverage results might be imprecise. On the one hand, TESTCOV may contain bugs that result in wrong coverage numbers. However, TESTCOV will make the same mistakes for all approaches. In addition, TESTCOV is an established validator of Test-Comp. Therefore, we expect that significant bugs would have been detected already. On the other hand, TESTCOV sometimes runs out of resources and may not execute all tests. This happens rarely for ULTIMATE TESTGEN and sometimes for PREDICATE and COVERITEST. We believe the comparison of the two ULTIMATE TESTGEN approaches is hardly affected by this. Moreover, PREDICATE and COVERITEST still regularly outperform ULTIMATE TESTGEN.

Hence, these observations are not affected by the exhaustion of resource limits. Also, the comparison of the number of generated test cases is not affected by any imprecision of the coverage results.

5 Related Work

Symbolic execution is a well-known test-case generation technique [16,17,30,38] based on verification. Moreover, several approaches have been proposed that generate test cases from counterexamples produced by software model checkers. Some of the early approaches are PATHFINDER [41] and BLAST [7]. PATHFINDER applies explicit-state model checking or symbolic execution. In contrast, BLAST uses predicate abstraction to generate test cases. Also, CPA/TIGER [8,39], abstraction-driven concolic testing [21], and CoVERITEST [9] use predicate abstraction. Other test-case generation techniques like FSHELL [28], CBMC [31], FuSeBMC [1], and VERIFUZZ [35] employ bounded model checking. Furthermore, the conditional tester $\text{tester}_{\text{veri}}^{\text{cyc}}$ (Fig. 14 in [12]) describes a template construction to turn an arbitrary verifier into a test-case generator, but we are the first that generate test-cases using an automata-based approach to software model checking [26,27].

Next to their difference in the applied software model checking technique, the approaches also differ in their encoding of the test goals. Test goals in BLAST [7] are pairs of program location and target predicate. Abstraction-concolic testing [21] and CoVERITEST [9] describe test goals with a set of control-flow edges and monitor the reachability of those edges. FSHELL [28] and CPA/TIGER [8,39] express test goals in FQL [29], in particular its representation as test goal automata. FSHELL encodes the automata into the program, while CPA/TIGER runs the automata in parallel to the analysis. FuSeBMC [1,2] represents test goals by labels in the program. The conditional tester $\text{tester}_{\text{veri}}^{\text{cyc}}$ [12] adds calls to function __VERIFIER_error() into the program such that the function is called whenever a test goal is reached. Similarly, our approach adds additional (assert) statements to the program and then characterizes test goals by error locations reachable when executing the statement, i.e., violating the assertion.

Like ULTIMATE TESTGEN-ALL (Algorithm 1), FSHELL [28], abstraction concolic testing [21], CoVERITEST [9], and the conditional tester $\text{tester}_{\text{veri}}^{\text{cyc}}$ [12] consider all test goals at once and exclude already covered goals from being found again. To this end, CoVERITEST and conditional tester $\text{tester}_{\text{veri}}^{\text{cyc}}$ remove covered goals from their specification, while FSHELL adds SAT constraints on the paths to be detected. In contrast, ULTIMATE TESTGEN uses error automata. Similarly, symbolic execution tools like e.g., KLEE [16], which aim to cover every program path, may use coverage-optimized search strategies that prefer states that likely lead to the coverage of new code. At the other extreme, BLAST [7] and the BMC component of FuSeBMC [2] aim to cover one test goal at a time. Thereby, BLAST considers deeper goals first and FuSeBMC prefers deeper goals but may additionally consider the type of the goal, e.g., whether it is a branch of an if statement or a loop. CPA/TIGER [39] considers a compromise between the two extrema and partitions the set of test goals into subsets either randomly or aiming to cluster test goals with similar prefixes. Goals of one subset are considered at once and covered goals are removed. If CPA/TIGER cyclically runs multiple analyses [40], the uncovered test goals are repartitioned before each analysis run. ULTIMATE TESTGEN-INCR (Algorithm 2) incrementally increases the considered test goals, thereby prioritizing deeper test goals.

6 Conclusion

Testing is a well-established process for quality assurance, which can be supported by automatic test-case generation approaches. We propose ULTIMATE TESTGEN, the first test-case generator based on the automata-based approach to software model checking used by ULTIMATE AUTOMIZER. ULTIMATE TESTGEN first extends the program automaton, the program representation, to encode the reachability of test goals as property violations. Then, it runs the automata-based verification and transforms feasible counterexamples into test cases. While verification typically stops after detecting a feasible counterexample, ULTIMATE TESTGEN aims to cover more than one test goal and, thus, continues verification to detect further counterexamples. To avoid detecting a counterexample triggered by an already covered test goal, ULTIMATE TESTGEN extends the verification approach with error automata that allow us to exclude counterexamples triggered by an already covered test goal. Moreover, ULTIMATE TESTGEN can be configured to either consider all test goals at once or to incrementally add test goals in the decreasing order of their distance to the initial program location. Our experiments reveal that configuration ULTIMATE TESTGEN-ALL regularly achieves higher coverage while configuration ULTIMATE TESTGEN-INCR generates smaller test suites. Also, we show that in 70% of the evaluated tasks ULTIMATE TESTGEN-ALL achieves equal or higher coverage than the most similar competitor PREDICATE. However, it is rarely better than competitor COVERITEST, which combines different approaches.

Data-Availability Statement. All experimental data, all used software as well as the test tasks are publicly available in our supplementary artifact [4].

References

1. Alshmrany, K.M., Aldughaim, M., Bhayat, A., Cordeiro, L.C.: FuSeBMC: an energy-efficient test generator for finding security vulnerabilities in C programs. In: Loulergue, F., Wotawa, F. (eds.) TAP 2021, pp. 85–105. LNCS, vol. 12740. Springer, Cham (2021). https://doi.org/10.1007/978-3-030-79379-1_6
2. Alshmrany, K.M., Aldughaim, M., Bhayat, A., Shmarov, F., Aljaafari, F., Cordeiro, L.C.: FuSeBMC v4: Improving code coverage with smart seeds via fuzzing and static analysis. CoRR abs/2206.14068 (2022). https://doi.org/10.48550/arXiv.2206.14068
3. Barrett, C.W., et al.: CVC4. In: Gopalakrishnan, G., Qadeer, S. (eds.) CAV 2011. LNCS, vol. 6806, pp. 171–177. Springer, Cham (2011). https://doi.org/10.1007/978-3-642-22110-1_14

4. Barth, M., Jakobs, M.: Replication package for paper "Test-case generation with automata-based software model checking" SPIN 24 (2024). https://doi.org/10.5281/zenodo.10574234

5. Beyer, D.: Competition on software verification and witness validation: SV-COMP 2023. In: Sankaranarayanan, S., Sharygina, N. (eds.) TACAS 2023. LNCS, vol. 13994, pp. 495–522. Springer, Cham (2023). https://doi.org/10.1007/978-3-031-30820-8_29

6. Beyer, D.: Software testing: 5th comparative evaluation: test-Comp 2023. In: Lambers, L., Uchitel, S. (eds.) FASE 2023. LNCS, vol. 13991, pp. 309–323. Springer, Cham (2023). https://doi.org/10.1007/978-3-031-30826-0_17

7. Beyer, D., Chlipala, A., Henzinger, T.A., Jhala, R., Majumdar, R.: Generating tests from counterexamples. In: Proc. ICSE, pp. 326–335. IEEE (2004). https://doi.org/10.1109/ICSE.2004.1317455

8. Beyer, D., Holzer, A., Tautschnig, M., Veith, H.: Information reuse for multi-goal reachability analyses. In: Felleisen, M., Gardner, P. (eds.) ESOP 2013, LNCS, vol. 7792, pp. 472–491. Springer, Cham (2013). https://doi.org/10.1007/978-3-642-37036-6_26

9. Beyer, D., Jakobs, M.: CoVeriTest: cooperative verifier-based testing. In: Hähnle, R., van der Aalst, W. (eds.) FASE 2019. LNCS, vol. 11424, pp. 389–408. Springer, Cham (2019). https://doi.org/10.1007/978-3-030-16722-6_23

10. Beyer, D., Keremoglu, M.E.: CPAchecker: a tool for configurable software verification. In: Gopalakrishnan, G., Qadeer, S. (eds.) CAV 2011. LNCS, vol. 6806, pp. 184–190. Springer, Cham (2011). https://doi.org/10.1007/978-3-642-22110-1_16

11. Beyer, D., Keremoglu, M.E., Wendler, P.: Predicate abstraction with adjustable-block encoding. In: Proc. FMCAD, pp. 189–197. IEEE (2010). https://ieeexplore.ieee.org/document/5770949/

12. Beyer, D., Lemberger, T.: Conditional testing - off-the-shelf combination of test-case generators. In: Chen, Y.F., Cheng, C.H., Esparza, J. (eds.) ATVA 2019, pp. 189–208. LNCS, vol. 11781. Springer, Cham (2019). https://doi.org/10.1007/978-3-030-31784-3_11

13. Beyer, D., Lemberger, T.: TestCov: robust test-suite execution and coverage measurement. In: Proc. ASE, pp. 1074–1077. IEEE (2019). https://doi.org/10.1109/ASE.2019.00105

14. Beyer, D., Löwe, S.: Explicit-state software model checking based on CEGAR and interpolation. In: Cortellessa, V., Varró, D. (eds) FASE 2013. LNCS, vol. 7793, pp. 146–162. Springer, Cham (2013). https://doi.org/10.1007/978-3-642-37057-1_11

15. Beyer, D., Löwe, S., Wendler, P.: Reliable benchmarking: Requirements and solutions. STTT **21**(1), 1–29 (2019). https://doi.org/10.1007/s10009-017-0469-y

16. Cadar, C., Dunbar, D., Engler, D.R.: KLEE: unassisted and automatic generation of high-coverage tests for complex systems programs. In: Proc. OSDI, pp. 209–224. USENIX Association (2008). http://www.usenix.org/events/osdi08/tech/full_papers/cadar/cadar.pdf

17. Chalupa, M., Vitovská, M., Strejcek, J.: SYMBIOTIC 5: boosted instrumentation - (competition contribution). In: Beyer, D., Huisman, M. (eds.) TACAS 2018, pp. 442–446. LNCS, vol. 10806. Springer, Cham (2018). https://doi.org/10.1007/978-3-319-89963-3_29

18. Christ, J., Hoenicke, J., Nutz, A.: SMTInterpol: an interpolating SMT solver. In: Donaldson, A., Parker, D. (eds.) SPIN. LNCS, vol. 7385, pp. 248–254. Springer, Cham (2012). https://doi.org/10.1007/978-3-642-31759-0_19

19. Cimatti, A., Griggio, A., Schaafsma, B.J., Sebastiani, R.: The MathSAT5 SMT solver. In: Piterman, N., Smolka, S.A. (eds.) TACAS 2013, pp. 93–107. LNCS, vol. 7795. Springer, Cham (2013). https://doi.org/10.1007/978-3-642-36742-7_7

20. Cytron, R., Ferrante, J., Rosen, B.K., Wegman, M.N., Zadeck, F.K.: Efficiently computing static single assignment form and the control dependence graph. TOPLAS **13**(4), 451–490 (1991). https://doi.org/10.1145/115372.115320

21. Daca, P., Gupta, A., Henzinger, T.A.: Abstraction-driven concolic testing. In: Jobstmann, B., Leino, K. (eds.) VMCAI 2016. LNCS, vol. 9583, pp. 328–347. Springer, Cham (2016). https://doi.org/10.1007/978-3-662-49122-5_16

22. Harman, M., et al.: Testability transformation. IEEE TSE **30**(1), 3–16 (2004). https://doi.org/10.1109/TSE.2004.1265732

23. Hart, P.E., Nilsson, N.J., Raphael, B.: A formal basis for the heuristic determination of minimum cost paths. TSSC **4**(2), 100–107 (1968). https://doi.org/10.1109/TSSC.1968.300136

24. Hart, P.E., Nilsson, N.J., Raphael, B.: Correction to "a formal basis for the heuristic determination of minimum cost paths." SIGART Newsl. **37**, 28–29 (1972). https://doi.org/10.1145/1056777.1056779

25. Heizmann, M., et al.: Ultimate automizer and the CommuHash normal form - (competition contribution). In: Sankaranarayanan, S., Sharygina, N. (eds.) TACAS 2023. LNCS, vol. 13994, pp. 577–581. Springer, Cham (2023). https://doi.org/10.1007/978-3-031-30820-8_39

26. Heizmann, M., Hoenicke, J., Podelski, A.: Refinement of trace abstraction. In: Palsberg, J., Su, Z. (eds.) SAS 2009. LNCS, vol. 5673, pp. 69–85. Springer, Cham (2009). https://doi.org/10.1007/978-3-642-03237-0_7

27. Heizmann, M., Hoenicke, J., Podelski, A.: Software model checking for people who love automata. In: Sharygina, N., Veith, H. (eds.) CAV 2013. LNCS, vol. 8044, pp. 36–52. Springer, Cham (2013). https://doi.org/10.1007/978-3-642-39799-8_2

28. Holzer, A., Schallhart, C., Tautschnig, M., Veith, H.: Query-driven program testing. In: Jones, N.D., Müller-Olm, M. (eds.) VMCAI 2009. LNCS, vol. 5403, pp. 151–166. Springer, Cham (2009). https://doi.org/10.1007/978-3-540-93900-9_15

29. Holzer, A., Schallhart, C., Tautschnig, M., Veith, H.: How did you specify your test suite. In: Proc. ASE, pp. 407–416. ACM (2010). https://doi.org/10.1145/1858996.1859084

30. King, J.C.: Symbolic execution and program testing. Commun. ACM **19**(7), 385–394 (1976). https://doi.org/10.1145/360248.360252

31. Kroening, D., Schrammel, P., Tautschnig, M.: CBMC: the C bounded model checker. CoRR abs/2302.02384 (2023). https://doi.org/10.48550/arXiv.2302.02384

32. Lemberger, T.: Plain random test generation with PRTest. STTT **23**(6), 871–873 (2021). https://doi.org/10.1007/s10009-020-00568-x

33. Li, J., Zhao, B., Zhang, C.: Fuzzing: a survey. Cybersecurity **1**(1), 6 (2018). https://doi.org/10.1186/s42400-018-0002-y

34. McMinn, P.: Search-based software test data generation: a survey. STVR **14**(2), 105–156 (2004). https://doi.org/10.1002/stvr.294

35. Metta, R., Medicherla, R.K., Karmarkar, H.: VeriFuzz: Good seeds for fuzzing (competition contribution). In: Johnsen, E.B., Wimmer, M. (eds.) FASE 2022. LNCS, vol. 13241, pp. 341–346. Springer, Cham (2022). https://doi.org/10.1007/978-3-030-99429-7_20

36. de Moura, L.M., Bjørner, N.S.: Z3: an efficient SMT solver. In: Ramakrishnan, C.R., Rehof, J. (eds.) TACAS 2008. LNCS, vol. 4963, pp. 337–340. Springer, Cham (2008). https://doi.org/10.1007/978-3-540-78800-3_24

37. Pacheco, C., Lahiri, S.K., Ernst, M.D., Ball, T.: Feedback-directed random test generation. In: Proc. ICSE, pp. 75–84. IEEE (2007). https://doi.org/10.1109/ICSE.2007.37
38. Pasareanu, C.S., Visser, W.: A survey of new trends in symbolic execution for software testing and analysis. STTT **11**(4), 339–353 (2009). https://doi.org/10.1007/s10009-009-0118-1
39. Ruland, S., Lochau, M., Fehse, O., Schürr, A.: CPA/Tiger-MGP: test-goal set partitioning for efficient multi-goal test-suite generation. STTT **23**(6), 853–856 (2021). https://doi.org/10.1007/s10009-020-00574-z
40. Ruland, S., Lochau, M., Jakobs, M.: HybridTiger: hybrid model checking and domination-based partitioning for efficient multi-goal test-suite generation (competition contribution). In: Wehrheim, H., Cabot, J. (eds.) Proc. FASE. LNCS, vol. 12076, pp. 520–524. Springer, Cham (2020). https://doi.org/10.1007/978-3-030-45234-6_26
41. Visser, W., Păsăreanu, C.S., Khurshid, S.: Test input generation with Java PathFinder. In: Proc. ISSTA, pp. 97–107. ACM (2004). https://doi.org/10.1145/1007512.1007526
42. Zeller, A., Gopinath, R., Böhme, M., Fraser, G., Holler, C.: The Fuzzing Book. CISPA Helmholtz Center for Information Security (2023). https://www.fuzzingbook.org/, retrieved 2023-01-07 14:37:57+01:00

Author Index

T. Neele and A. Wijs (Eds.): SPIN 2024, LNCS 14624, pp. 269–270, 2025.
https://doi.org/10.1007/978-3-031-66149-5